Stress Relieving Heat Treatments of Welded Steel Constructions

Les Traitements Thermiques de Relaxation des Constructions Soudees en Acier

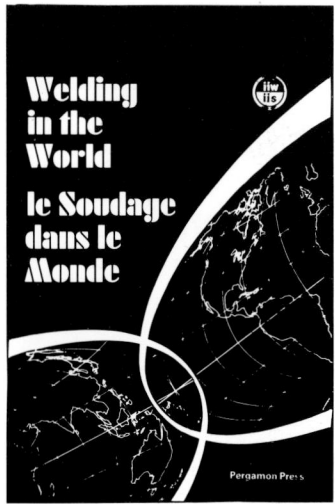

WELDING IN THE WORLD / LE SOUDAGE DANS LE MONDE

Journal of the International Institute of Welding
Revue de l'Institut International de la Soudure

Editors/Rédacteurs en chef: P D BOYD, *Secretary-General of the IIW, 54 Princes Gate, Exhibition Road, London SW7 2PL, UK* and M BRAMAT, *Secrétaire Scientifique et Technique de l'IIS, 32 Boulevard de la Chapelle, Paris 18e, France*

The journal publishes technical reports on all aspects of welding technology. Topics covered by the journal include authoritative reports on:
* Processes, welding equipment and consumables
* Joining, cutting and surfacing by thermal processes
* Arc and resistance welding
* Flux and gas shielded electrical welding processes
* Inspection and behaviour of materials and welded assemblies
* Testing, measurement and control of welds

Commencing with the 1987 volume, *Welding in the World* will publish one additional special number devoted to the publication of extended abstracts of around 200 documents of IIW Working Units and other important commissioned reports and surveys. Topics covered will include: welding processes, welding metallurgy, testing and inspection of welds, particular welding applications, health and safety, terminology and welding instruction. It is intended that this supplement, which will be sent automatically to current subscribers, will form a very useful information tool in maintaining awareness of IIW activities and important R & D programmes around the world.

A selection of papers
X-ray real-time imaging for weld inspection (report by a working group of sub-commission VA), Doc. IIS/IIW-833-85.
Automation and robotization in welding and allied processes: general survey and reflections on the present and the future, J PANISSET.
Re-heat cracking. A review of recent studies, Doc. IIS/IIW-837-85.
Guidelines for the classification of ferritic steel weld metal microstural constituents using the light microscope, Doc. IIS/IIW-835-85.
A review of stress corrosion cracking of welded duplex ferritic/austenitic stainless steels, Doc. IIS/IIW-847-86.
A survey review of weld metal hydrogen cracking - Doc. IIS/IIW-IIS/IIW-851-86.
Wide gap brazing of stainless steel with nickel-base brazing alloys, Doc. I-783-85.
Effect of carbon on weld metal microstructure and properties - Doc. IX-1375-85.
Guidelines for selecting quality assurance levels in welding technology - Doc. V-790-85.
Guide for the welding and weldability of reinforcing steels for concrete structures, Doc. IX-1357-85.

Subscription Information
1987: Volume 25 (7 issues)
Annual subscription (1987)   DM205.00
Two-year rate (1987/88)   DM389.50

FREE SAMPLE COPIES AVAILABLE ON REQUEST
Advertising rate card available on request. Back issues and current subscriptions are also available in microform.
The Deutsche Mark prices shown include postage and insurance. For subscription rates in Africa, Asia, Australasia, UK and Eire and the Americas apply to your nearest Pergamon office. Prices are subject to change without notice.

Pergamon

Headington Hill Hall, Oxford OX3 0BW, UK
Fairview Park, Elmsford, New York 10523, USA

3D 43 11 86

Stress Relieving Heat Treatments of Welded Steel Constructions

Proceedings of the International Conference held in Sofia, Bulgaria,
6–7 July 1987 under the auspices of the International Institute of Welding

Les Traitements Thermiques de Relaxation des Constructions Soudees en Acier

Communications présentées à la Conférence Internationale tenue à Sofia, Bulgarie,
les 6–7 juillet 1987 sous les auspices de l'Institut International de la Soudure

Published on behalf of the INTERNATIONAL INSTITUTE OF WELDING

Edité au nom de L'INSTITUT INTERNATIONAL DE LA SOUDURE

by/par

PERGAMON PRESS

OXFORD · NEW YORK · BEIJING · FRANKFURT
SAO PAULO · SYDNEY · TOKYO · TORONTO

U.K.	Pergamon Press, Headington Hill Hall, Oxford OX3 0BW, England
U.S.A.	Pergamon Press, Maxwell House, Fairview Park, Elmsford, New York 10523, U.S.A.
PEOPLE'S REPUBLIC OF CHINA	Pergamon Press, Room 4037, Qianmen Hotel, Beijing, People's Republic of China
FEDERAL REPUBLIC OF GERMANY	Pergamon Press, Hammerweg 6, D-6242 Kronberg, Federal Republic of Germany
BRAZIL	Pergamon Editora, Rua Eça de Queiros, 346, CEP 04011, Paraiso, São Paulo, Brazil
AUSTRALIA	Pergamon Press Australia, P.O. Box 544, Potts Point, N.S.W. 2011, Australia
JAPAN	Pergamon Press, 8th Floor, Matsuoka Central Building, 1-7-1 Nishishinjuku, Shinjuku-ku, Tokyo 160, Japan
CANADA	Pergamon Press Canada, Suite No. 271, 253 College Street, Toronto, Ontario, Canada M5T 1R5

Copyright © 1987 International Institute of Welding

All Rights Reserved. No part of this publication may be reproduced, stored in a retrieval system or transmitted in any form or by any means: electronic, electrostatic, magnetic tape, mechanical, photocopying, recording or otherwise, without permission in writing from the publishers.

First edition 1987

British Library Cataloguing in Publication Data
Stress relieving heat treatment of welded steel constructions: proceedings of the international conference held in Sofia, Bulgaria, 6–7 July 1987 under the auspices of the International Institute of Welding = Les traitements thermiques de relaxation des constructions soudees en acier: communications preséntées à la conférence internationale tenue à Sofia, Bulgarie, les 6–7 juillet 1987 sous les auspices de l'Institut internationale de la soudure.
1. Welded steel structures 2. Steel-Heat treatment
I. International Institute of Welding
624.1'821   TA684
ISBN 0-08-035900-0

In order to make this volume available as economically and as rapidly as possible the author's typescript has been reproduced in its original form. This method unfortunately has its typographical limitations but it is hoped that they in no way distract the reader.

Printed in Great Britain by A. Wheaton & Co. Ltd., Exeter

Other Pergamon Titles of Interest

ASHBY & BROWN	Perspectives in Creep Fracture
ASHBY & HIRTH	Perspectives in Hydrogen in Metals
COUDURIER et al.	Fundamentals of Metallurgical Processes, 2nd Edition
FROST & ASHBY	Deformation-Mechanism Maps
HAASEN et al.	Decomposition of Alloys—the Early Stages
HEARN	Mechanics of Materials, 2 Vols., 2nd Edition
HULL & BACON	Introduction to Dislocations, 3rd Edition
IIW	Automation and Robotisation in Welding and Allied Processes
IIW	The Physics of Welding, 2nd Edition
IIW	Electron and Laser Beam Welding
IIW	Welding of Tubular Structures
KAZAKOV	Diffusion Bonding of Materials
MASUBUCHI	Analysis of Welded Structures
McQUEEN et al.	Strength of Metals & Alloys, 3 Vols. (ICSMA 7)
MORGAN	Tinplate & Modern Canmaking Technology
NIKU-LARI	Advances in Surface Treatments, Volumes 2–6
OSGOOD	Fatigue Design, 2nd Edition
PARKIN & FLOOD	Welding Craft Practice
REID	Metal Forming and Impact Mechanics
SHEWARD	High Temperature Brazing in Controlled Atmospheres
TAIT & GARRETT	Fracture & Fracture Mechanics: Case Studies
VALLURI et al.	Advances in Fractures Research, 6 Vols. (ICF6)
YAN	Mechanical Behaviour of Materials, 3 Vols (ICM5)

Pergamon Journals of Related Interest
(Free specimen copies gladly sent on request)

Acta Metallurgica

Canadian Metallurgical Quarterly

Computers and Structures

Engineering Fracture Mechanics

Fatigue & Fracture of Engineering Materials & Structures

International Journal of Impact Engineering

Materials Research Bulletin

Robotics & Computer-integrated Manufacturing

Scripta Metallurgica

Vacuum

Welding in the World

PREFACE

La Conférence Internationale de 1987 représente une somme considérable d'informations sur les traitements de relaxation des contraintes résiduelles. En premier lieu le nombre des contributions (conférences, communications, affiches) est très important et il y a lieu de souligner que la sélection en a été faite très soigneusement sur la base de l'intérêt technique qu'elles présentent.

Par ailleurs, il a été jugé indispensable d'aborder les méthodes de relaxation ne faisant pas appel à un traitement thermique (vibration, explosion, surcharge mécanique). Souvent mentionnées, ces techniques sont en définitive mal connues, ce qui en rend l'application industrielle peu fréquente.

Enfin, l'efficacité des traitements thermiques - classiques - de relaxation est maintenant très bien connue grâce à des efforts entrepris récemment pour la caractériser ce qui permet d'ajuster les températures aux plus justes valeurs nécessaires et de réduire à un minimum les altérations métallurgiques.

M BRAMAT
Secrétaire Scientifique et Technique

PREFACE

The 1987 International Conference provides a considerable amount of information on residual stress relieving heat treatments. The number of contributions (lectures, papers, posters) is large and it should be underlined that these contributions were carefully selected on the basis of their technical interest.

In addition, it was deemed absolutely necessary to include stress relieving treatments involving no heat treatment (i.e., treatments using vibration, explosion, mechanical overload). Although often mentioned, these techniques are in fact little known and therefore seldom applied in industry.

Finally, the efficacy of traditional stress relieving heat treatments is now well known, thanks to the recent efforts made to characterise it; this enables the proper heat treatment temperatures to be selected with great accuracy while minimising metallurgical deterioration.

M BRAMAT
Scientific and Technical Secretary

INTERNATIONAL CONFERENCE ON STRESS RELIEVING HEAT TREATMENTS OF WELDED STEEL CONSTRUCTIONS

CONFERENCE INTERNATIONALE SUR LES TRAITEMENTS THERMIQUES DE RELAXATION DES CONSTRUCTIONS SOUDEES EN ACIER

Portevin Lecture	Conference Portevin
Stress relieving heat treatment: a critical appraisal of specification and practice	Les traitements thermiques de relaxation des contraintes: examen critique des spécifications et des pratiques

R. V. SALKIN
(Belgium/Belgique)

Invited Paper	Conference Invitee
Mechanical aspect of stress relieving by heat and non-heat treatment	Aspect mécanique de la relaxation des contraintes par les traitements thermiques et non-thermiques

V. A. VINOKUROV
(USSR/URSS)

Invited Paper	Conference Invitee
Principles of mechanical stress relief treatment	Bases de traitement mécanique de relaxation de contraintes

I. HRIVNAK
(Czechoslovakia/Tchécoslovaquie)

K. YUSHCHENKO
(USSR/URSS)

Sessions	Sessions
I – Metallurgy of stress-relieving heat treatments and mechanical consequences	I – Métallurgie des traitements thermiques de relaxation et conséquences mécaniques
II – Residual stresses generation – Local relieving heat treatments	II – Création des contraintes résiduelles – Traitements locaux de relaxation
III – Mechanical stress relieving treatments	III – Traitements mécaniques de relaxation de contraintes
IV – Advisibility and optimization of stress relieving treatment	IV – Opportunité et optimisation des traitements de relaxation de contraintes

Posters	Affiches
Author Index	Index de l'auteur
Subject Index	Index de sujet

LIST OF PAPERS
LISTE DES COMMUNICATIONS

SESSION I

METALLURGY OF STRESS-RELIEVING HEAT TREATMENTS AND MECHANICAL CONSEQUENCES/METALLURGIE DES TRAITEMENTS THERMIQUES DE RELAXATION ET CONSEQUENCES MECANIQUES

I.1 The effect of stress relief heat treatment of welded joints on changes in structural and mechanical characteristics
(J. Bosansky, L. Mraz — Czechoslovakia/Tchécoslovaquie)   29

I.2 Properties of dissimilar Cr-Mo steel joints as a basis for the selection of the PWHT-temperature
(J.-J. Chene, M. E. J. Shakeshaft — Switzerland/Suisse)   37

I.3 Prevision de l'evolution des caractéristiques mécaniques de la zone affectée et du métal fondu lors du détensionnement
(Ph. Bourges, Ph. Maynier, R. Blondeau — France)   53

I.4 Prediction of HAZ Hardness after PWHT
(M. Okumura, N. Yurioka, T. Kasuya and J.-J. Cotton — Japan/Japon)   61

I.5 Effets secondaires des traitements de relaxation sur la ténacité de la zone fondue de soudures multipasses d'aciers non alliés ou faiblement alliés
(S. Debiez — France)   69

I.6 Influence of post weld treatment on the fracture toughness in welding procedure qualification and component welds
(J. G. Blauel, W. Burget — FRG/RFA)   79

I.7 Effects of heat input and stress-relief annealing on the toughness of the HAZ in the welding of the boiler steels type HI, HII, 17 Mn4 and 19 Mn5
(S. Anik, K. Tulbentçi, A. Dikicioglu — Turkey/Turquie)   91

I.8 Heat treatment and welded structures dimensions stability
(V. M. Sagalevich, V. F. Savelyev — USSR/URSS)   99

I.9 The improvement of fatigue resistance of structure welded joints by heat treatment
(V. I. Trufiakov, P. P. Mikheev, Yu. F. Kudryavtsev — USSR/URSS)   101

SESSION II

RESIDUAL STRESSES GENERATION-LOCAL RELIEVING HEAT TREATMENTS/ CREATION DES CONTRAINTES RESIDUELLES-TRAITEMENTS LOCAUX DE RELAXATION

II.1 Role of preliminary heat treatment on heat-affected zone behaviour as regards relieving of internal stresses
(K. Velkov, L. Vasileva — Bulgaria/Bulgarie)   109

II.2 Minimization of residual welding stresses by the superposition of thermal and transformation stresses
(P. Seyffarth, H. G. Gross, K. M. Gatovskij, S. P. Markov — GDR/RDA; USSR/URSS)   117

II.3 Local heat treatment by induction
(H. H. Müller — FRG/RFA)   125

II.4 Dans certain cas le traitement thermique par induction est non réussi. Quand et pourpuoi?
(S. Cundev — Yugoslavia/Yougoslavie)   131

SESSION III

MECHANICAL STRESS RELIEVING TREATMENTS/TRAITEMENTS MECANIQUES DE RELAXATION DE CONTRAINTES

III.1 Application of gyrotron for heat treatment of materials
(B. E. Paton, V. E. Sklyarevich — USSR/URSS)   141

III.2 Vibratory stabilization of welded constructions — experiments and conclusions
(P. Sedek — Poland/Pologne)   145

III.3 Vibratory lowering of residual stresses in weldments
(M. Jesensky — Czechoslovakia/Tchécoslovaquie)   153

III.4 Reduction and redistribution of the residual welding stresses through local explosive treatment
(S. Hristov, P. Petrov, S. Zvetanov — Bulgaria/Bulgarie)   161

III.5 Investigation on the mechanism of stress relief by explosion treatment — The rule of plastic deformation of metal in the process of explosion
(L. Chen, Z. Si, Y. Wang, H. Chen and X. Dong — China/Chine)   169

III.6 Mechanical stress-relief treatment of welded pressure vessels by warm pressure test
(K. Kalna — Czechoslovakia/Tchécoslovaquie)   177

III.7 Relaxation overstressing of huge spherical storage vessels repaired by welding
(I. Hrivnak, J. Lancos, S. Vejvoda — Czechoslovakia/Tchécoslovaquie)   189

SESSION IV

ADVISABILITY AND OPTIMIZATION OF STRESS RELIEVING TREATMENT/OPPORTUNITE ET OPTIMISATION DES TRAITEMENTS DE RELAXATION DE CONTRAINTES

IV.1 Post weld heat treatment of a high integrity component of complex geometry
(C. M. White, W. P. Carter — United Kingdom/Royaume-Uni)   199

IV.2 Post heat treatment of welds on high strength steel nioval 47[1]
(M. Velikonja — Yugoslavia/Yougoslavie)   207

IV.3 Welding structures heat treatment application experience for stress relieving in heavy machine building industry
(L. P. Eregin et al. — USSR/URSS)   219

IV.4 Dependence of heat treatment parameters VS. Thermal conditions in welding and post-welded heterogeneity of properties
(F. A. Khromchenko — USSR/URSS)   221

IV.5 Comparison of the residual stress distibutions after stress-relief annealing of welded sheets of the high strength structural steel ST E 690 at different temperatures
(J. Heeschen, H. Wohlfahrt — FRG/RFA)   231

IV.6 Influence of the welding technology and the stress-relieving heat treatment on the corrosion cracking resistance of welded nitrogen-Alloyed stainless steel
(L. Kalev, V. Mihailov, A. Krustev — Bulgaria/Bulgarie)   239

IV.7 Relaxation of residual stresses during postweld heat treatment of submerged-arc welds in a C-Mn-Nb-Al steel
(R. H. Leggatt — United Kingdom/Royaume-Uni)   247

IV.8 Estimation methods for studying the degree of relaxation of residual welding stresses at appropriate heat treatment, as well as for evaluation of effect of non-relaxed stresses on the load-carrying capacity of structure members
(V. I. Makhnenko, E. A. Velikoivanenko, V. E. Pochinok — USSR/URRS)   257

IV.9 Qualification d'un traitement de relaxation
(A. Leclou — France)   269

IV.10 Electromagnetic monitoring of residual stress relaxation during heat treatment
(I. M. Zhdanov, V. V. Batyuk, A. A. Khriplivy, R. K. Gachik, K. B. Pastukhov, G. F. Kolot, A. V. Pulyayev — USSR/URSS)   277

IV.11 Postweld heat treatment of C-Mn and microalloyed steels: an evaluation on the basis of C.T.O.D. and wide plate tests
(W. Provost, A. Dhooge, A. Vinckier — Belgium/Belgique)   289

IV.12 Is high tempering always needed for low-carbon and low-alloyed steel structures?
(A. E. Asnis, G. A. Ivashchenko — USSR/URSS)   307

IV.13 Considerations of the post weld heat treatment of pressure parts
(I. G. Hamilton, A. R. G. Abbott — United Kingdom/Royaume-Uni)   313

POSTERS/AFFICHES

Argon-arc treatment application in welded structure fabrication
(A. E. Asnis, G. A. Ivashchenko — USSR/URSS)   321

Recent studies on reheat cracking of Cr-Mo Steels
(K. Tamaki, J. Suzuki — Japan/Japon)   325

Traitement thermique de joints soudés de la conduite principale de circulation d'une centrale nucléaire
(I. I. Varbenov, P. I. Raykov, D. D. Dimitrov — Bulgaria/Bulgarie)   327

Development of alloying systems of structural steels having a high resistance against overheating, eliminating the post electroslag welding normalization
(S. V. Egorova, Yu. A. Sterenbogen, A. V. Yurchishin, E. N. Solina, N. G. Zotava — USSR/URSS)   329

Producing a cast bimetallic die billets by using the electroslag heating with accompanied annealing and selection of heat treatment conditions
(V. A. Nosatov, Yu. A. Sterenbogen, T. Kh. Ovchinnikova, O. G. Kuz'menko, A. V. Denisenko — USSR/URSS)   331

The influence of the thermal treatment on the welding stress relieving and the improving of properties of welding joints of large-sized pressure vessels
(A. G. Lamzin, N. M. Kabanov, P. M. Korol'kov — USSR/URSS)   333

Problems in welding and heat treatment of thick-walled martensitic steel steam lines
(B. Kovacevic — Yugoslavia/Yougoslavie)   339

Amélioration des règles de détensionnement des joints soudés
(Programme de recherche coopératif français — France)   341

Efficiency of application of heat and vibrotreatment to reduce residual stresses in weldments
(V. V. Batyuk, A. A. Khriplivy, S. K. Fomichev, S. N. Minakov, P. N. Gansky — USSR/URSS)   343

Crankshaft overlaying without post heat treatment
(A. M. Slivinsky, V. T. Kotyk, P. F. Petrov, V. I. Prokhorov — USSR/URSS)   347

Thermal deformation process control in welding of thin sheet pieces with application of heat sink
(I. M. Zhdanov, V. V. Lysak, B. V. Medko, V. V. Kononchuk, V. N. Nifantov — USSR/URSS)   349

Evaluation of prefracture state during welding of large-sized thick-wall constructions
(E. A. Kirillov, V. I. Panov, I. J. Ievlev, V. P. Pokatilov, V. V. Volkov)   351

Heat treatment effect on HAZ metal crack-resistance of anti-corrosion cladded low alloy steels
(E. G. Starchenko, S. N. Navolokin, D. M. Shur, A. E. Runov, A. V. Zalinov)   353

PORTEVIN LECTURE   CONFERENCE PORTEVIN

Les Traitements Thermiques de Relaxation des Contraintes après Soudage

The Stress Relaxation Heat Treatments after Welding

R. V. Salkin

Cockerill Mechanical Industries
Belgique/Belgium

1. INTRODUCTION

   Il m'échoit le grand honneur et la tâche périlleuse de prononcer la Conférence dédiée à la mémoire du Professeur PORTEVIN. L'Institut International de la Soudure ouvre de la sorte un colloque international et une Assemblée Annuelle, réunissant des experts réputés venus du monde entier.

   Un honneur de pouvoir me réclamer du nom de ce prestigieux métallurgiste et de ce savant éminent pour aborder un problème sur lequel la Commission de l'I.I.S, que je préside cette année pour la dernière fois, a eu à maintes reprises l'occasion de se pencher. Permettez-moi donc d'associer à cet honneur tous ceux de mes collègues qui ont contribué à l'étude de ses divers aspects.

   Une tâche périlleuse parce qu'un sujet d'une telle ampleur a été traité de manière exhaustive dans de multiples publications, et que le colloque qui y est consacré cette année apportera aux experts qui y prendront part bien plus d'enseignements, d'applications et d'idées nouvelles que je ne pourrais le faire.

   C'est donc avec une prudente humilité que je m'avancerai dans le sujet; mais c'est aussi avec la profonde conviction de son importance et de son actualité que je m'efforcerai de souligner les lignes directrices d'une réflexion qui se prolongera au cours de ces journées.

   C'est une pratique bien établie de soumettre chaque fois que cela est possible - techniquement et économiquement - les constructions soudées à un traitement dans le but d'éliminer ou au moins de réduire les contraintes résiduelles pouvant résulter de la mise en oeuvre et notamment du soudage.

   On peut admettre que de tels traitements, quand ils sont réalisés dans les meilleures conditions, sont généralement utiles et parfois indispensables.

   Mais, depuis un certain temps, et principalement du fait d'études réalisées au cours de ces quinze dernières années, il est apparu de plus en plus clairement que de tels traitements exercent, en plus de leur effet de relaxation proprement dit, des effets métallurgiques qui sont généralement favorables mais qui, dans certains cas, peuvent être complexes et parfois néfastes, et que ces effets métallurgiques peuvent être aussi importants, parfois beaucoup plus importants que l'élimination des contraintes résiduelles.

   De telles constatations ont amené les métallurgistes, les constructeurs et les prescripteurs à se repencher sur les codes et règlementations en vigueur. L'internationalisation du problème a tout naturellement conduit l'Institut International de la Soudure à y consacrer une partie de son activité au sein d'un groupe de travail spécialement constitué à cet effet, et présidé par l'Académicien I.HRIVNAK (Tchécoslovaquie).

Au cours de la dernière décade, sous la pression conjuguée de la demande du marché et de la concurrence entre sidérurgistes et entre fabricants, de nouvelles générations d'aciers ont été développées, des outils performants ont été mis en place pour leur transformation et leur mise en oeuvre, des moyens de plus en plus sophistiqués ont été adoptés par les bureaux d'études pour l'utilisation optimale de leurs propriétés mécaniques.

Dès les années '60, les exigences du marché poussaient au développement des aciers de qualité dans le domaine des tubes et des récipients sous pression. Les limites imposées par la soudabilité et la résistance à la rupture fragile à un accroissement des teneurs en C et en Mn donnaient le jour aux générations d'aciers à grains fins réalisés par traitement thermique et/ou par adjonction d'éléments dispersoïdes. En agissant sur le durcissement par précipitations et sur l'affinage du grain, des additions d'Al, de V, de Nb, de Ti permirent d'obtenir une gamme de résistances plus élevées pour des aciers laminés et traités par voie conventionnelle, sans en affecter la ténacité et la soudabilité.

Les progrès accomplis dans la technique des traitements thermo-mécaniques ont conduit à tirer parti du laminage contrôlé pour l'affinage du grain et l'élévation concommittente des propriétés mécaniques et de la ténacité. Ces traitements thermomécaniques allaient être essentiellement appliqués aux tôles et bandes d'épaisseur moyenne ou faible, tandis que la normalisation ou la trempe et revenu était réservée aux tôles fortes et grosses pièces de forge.

Dans la Conférence HOUDREMONT qu'il a prononcée en Septembre 1982 à l'Assemblée Annuelle de Ljubljana, le Dr. SUZUKI a brillamment décrit les progrès récents accomplis dans le développement des aciers soudables. Il rappelait à ce propos que les progrès accomplis dans le domaine de la soudabilité des aciers ont été avant tout le résultat d'une collaboration entre les sidérurgistes et les utilisateurs.

Il indiquait les astreintes principales imposées par les besoins du marché, et les efforts d'amélioration de la qualité métallurgique des produits en vue de les satisfaire.

Les nuances et qualités résultant de ces améliorations se trouvent à présent à un niveau remarquablement plus élevé qu'il y a quinze ans.

Quant aux méthodes de fabrication, elles ont dû, elles aussi, répondre à la demande pressante du marché et faire face aux épaisseurs toujours plus fortes, aux conceptions toujours plus délicates et aux exigences de qualité toujours plus sévères.

Les progrès accomplis en soudage à haute densité d'énergie, notamment, ont été exposés de manière magistrale dans la Conférence HOUDREMONT prononcée en Juillet 1986 à Tokyo par le Dr. SAYEGH. Il est apparu, à de nombreuses reprises dans ce mémoire et dans les présentations qui ont été faites lors de la Conférence Internationale sur le même thème, combien sont devenues importantes les propriétés d'emplois des aciers du point de vue du soudage, et à quel point la mise en oeuvre d'un acier chez le constructeur constitue une sorte de prolongement des traitements qui lui sont appliqués par le sidérurgiste.

Il ne faut dès lors pas s'étonner si les questions que se posent
les sidérurgistes et les constructeurs à propos des traitements
dits de relaxation des contraintes résiduelles de soudage demeurent
d'une brûlante actualité.

Eu égard à l'évolution des besoins du marché et aux progrès
réalisés pour y faire face, comme se positionnent les
règlementations actuellement en vigueur ?
Quelles méthodes alternatives sont susceptibles d'être appliquées
pour éviter le recours à des traitements thermiques, lorsque
ceux-ci se révèlent inadéquats ou difficiles à appliquer ?
Quels procédés modernes de traitement thermiques après soudage sont
les plus adaptés à répondre aux prescriptions inamovibles des codes
et aux modulations nécessaires pour assurer économiquement une
bonification aussi complète que possible des propriétés globales
des joints soudés ?

Telles sont nos interrogations et la substance de nos réflexions.
Pour leur donner consistance, il convient d'abord de repositionner
les traitements de bonification après soudage à la mesure des buts
poursuivis.

***
*

2. <u>LES EFFETS DES CONTRAINTES RESIDUELLES DE SOUDAGE ET DE LEUR
   RELAXATION</u>

Il demeure acquis, dans l'esprit de nombreux praticiens, que la
valeur principale d'un traitement thermique après soudage est de
détendre les contraintes résiduelles.

Ces contraintes sont présentes dans et au voisinage du joint soudé
et ont le plus souvent un niveau voisin de la limite d'élasticité
du métal, qu'il y ait eu ou non un bridage extérieur lors du
soudage. Sous des conditions de sollicitations multi-axiales, de
chocs et de basses températures, en particulier, elles peuvent
entraîner la rupture fragile d'un joint soudé.

Réduire ou supprimer les contraintes résiduelles de soudage
constitue une protection contre le risque d'amorçage et de
propagation d'une rupture brutale ou par fatigue, contre la
propension à la corrosion du métal, et en particulier la corrosion
sous tension, et contre les instabilités dimensionnelles se
traduisant par des déformations en cours de mise en oeuvre
ultérieure.

Il est cependant notoire que les traitements de relaxation des contraintes exercent des effets métallurgiques sur les propriétés du
métal de base, du métal fondu et de la zone affectée par la
chaleur. Ces effets sont souvent bénéfiques, mais sous certaines
conditions, le détensionnement des contraintes s'accompagne de
dégradations de la ductilité ou de l'intégrité du métal, dont il
importe de cerner toutes les composantes.

2.1. Création des contraintes résiduelles

Sous l'effet du cycle thermique brutal, et localisé auquel est soumis le métal de base lors du soudage, il se produit un régime complexe de dilatations et de contractions thermiques d'origines métallurgique, associée aux transformations allotropiques, et mécanique, liée au retrait et au bridage.

L'étendue, la distribution et le niveau maximum des contraintes résiduelles qui résultent de ces dilatations et de ces contractions peuvent être modélisés. On conçoit toutefois la grande complexité d'une telle analyse, où interviennent les caractéristiques thermiques et spacio-temporelles de la source de chaleur, les caractéristiques physiques et géométriques de la pièce, et l'ensemble des circonstances d'environnement telles que contrôle de la vitesse de refroidissement, complexité de l'assemblage, température de la pièce avant soudage, erreurs d'accostage, etc...

Le principe de la modélisation consiste à établir les distributions dans le temps des températures non uniformes au cours du soudage d'un joint entre deux plaques, en supposant que ces distributions sont bidimensionneles, stationnaires par rapport à un système d'axe accompagnant la source de chaleur, et que les plaques sont minces et infiniment larges. On établit ces distributions de températures en adoptant la solution connue de Rosenthal pour une source de chaleur linéique. Le calcul des contraintes transitoires et résiduelles en cours ou en fin de soudage repose sur la méthode des solutions élastiques successives, par laquelle on établit les contraintes et dilatations à toute distance du cordon en faisant varier par incréments de temps successifs une distribution supposée uniforme sur toute la longueur du cordon (1).

2.2. Evaluation des contraintes résiduelles par voie non destructive

L'évaluation du niveau réel des contraintes résiduelles dans un assemblage soudé a constitué depuis toujours un sujet de fascination pour les spécialistes du contrôle non destructif. Car, si des méthodes de type semi-destructif, telles que la méthode des rosettes ou la méthode de Gunnert présentent l'avantage d'une certaine fiabilité, du moins en ce qui concerne les contraintes résiduelles de surface, elle ne sont pas d'application commode sur produit.

Plusieurs méthodes non destructives ont été étudiées et parfois appliquées avec succès, dans le double but d'identifier les contraintes résiduelles après soudage et l'efficacité des traitements de relaxation (2-3). Certaines méthodes font appel au magnétisme rémanant (4) règnant après soudage du fait du passage de l'arc électrique; son intensité dans les matériaux ferro-magnétiques est liée au champ de contrainte du fait de l'orientation des domaines.

Ailleurs, c'est la simple magnéto-élasticité qui est mesurée.
D'autres méthodes utilisent les modifications des
caractéristiques de propagation des ultrasons en présence d'un
champ de contraintes élastiques. Les caractéristiques visées sont
la vitesse, l'atténuation et l'angle critique de génération
d'ondes longitudinales, transversales et de Rayleigh.

Il faut également signaler l'observation de franges de Moiré
transcrites par holographie, et les procédés d'évaluation des
paramètres géométriques du niveau cristallin par diffraction de
rayons X.

On doit cependant admettre que l'application industrielle de ces
méthodes demeure encore lointaine, spécialement pour l'évaluation
des contraintes subsuperficielles, en raison des risques
d'interférence avec d'autres causes physiques influençant les
phénomènes observés, ainsi que des difficultés d'interprétation.

2.3. Effets sur la ténacité vis-à-vis de la rupture fraile

2.3.1. Effets des contraintes

Les effets des contraintes résiduelles sur la résistance à la
rupture fragile sont les mêmes que ceux résultant d'autres
états de contrainte de même nature mais d'origine différente.
Toutefois, la distribution des contraintes résiduelles, qui
doit respecter les lois de l'équilibre statique, est très
difficile à calculer ou à prévoir de façon précise et peut en
outre être modifiée par les traitements thermiques. Pour cette
raison, les effets qui en dépendent sont délicats à prédire
quantitativement.

En ce qui concerne l'amorçage des fissures, les facteurs
métallurgique sont généralement prédominants. Les contraintes
résiduelles peuvent cependant exercer leur influence de deux
manières différentes.

---

(1) J.E.Agapakis, K.Masubushi, Weld.Jnl USA (1984) <u>63</u> n° 6, pp. 1870-1963
(2) A.J.A.Pavlane, in "Residual Stresses in Welded Construction and their Effect" - The Welding Institute, London, 15-17 Nov. 1977
(3) K.Stahlkopf et al - Ibid
(4) Non-Destructive Estimation of Residual Stress in Welded Pressure Vessel Steels by means of Remanant Magnetization Measurement (Doc. Nr. X.801.76)

D'une part, les déformations plastiques engendrées lors de la
mise en charge en service peuvent être suffisantes pour
entraîner une fragilisation locale du métal par écrouissage ou
être de l'ordre de grandeur des déformations maximales
admissibles sans rupture à cette température. La formation de
fissures peut donc en résulter ou l'amorçage ultérieur de la
rupture en être facilité. Ceci pourra se produire par exemple
dans des pièces fragilisées à basse température ou en présence
d'entailles ou de défauts au voisinage des joints soudés.
D'autre part, la triaxialité des contraintes se traduit par des
effets analogues à ceux d'une diminution de la température sur
la fragilité des matériaux; l'état triaxial de traction qui est
associé aux constructions de fortes épaisseurs, est à cet égard
le plus dangereux. Les contraintes résiduelles peuvent
renforcer cet effet au plan local, en fonction de leur
distribution, de la présence d'entailles ou de concentrations
des contraintes.
La propagation des fissures, quant à elle, peut être stimulée
et orientée par les contraintes résiduelles. Le phénomène est
en effet contrôlé principalement par la quantité d'énergie
élastique libérée par modification de la répartition des
déformations élastiques due à ces contraintes. Quant aux
températures d'arrêt de propagation, on considère bien souvent
qu'elles sont peu influencées par les contraintes résiduelles
seules, étant donné la prépondérance qu'excercent, ici encore,
le facteurs métallurgiques locaux.

2.3.2. Effets d'un traitement après soudage

Parallèlement à la relaxation des contraintes résiduelles de
soudage, un traitement thermique en-dessous de la température
Ac_1 provoque généralement un adoucissement du métal de base et
peut également avoir une influence plus ou moins prononcée sur
sa résistance à la rupture fragile. Les effets sont cependant
très variables selon les nuances et qualités, d'une part, et
selon les conditions de soudage d'autre part.

Dans les aciers au carbone-manganèse, on peut assister à une
globulisation des carbures et au grossissement du grain
ferritique si le traitement thermique est de longue durée à des
températures élevées, dépassant 625°C. Ces traitements peuvent
entraîner une baisse de la limite d'élasticité et de la
résistance à la traction, suffisamment sensible pour qu'il soit
nécessaire d'en tenir compte, surtout si les températures
dépassent 600°C lors de la mise en oeuvre. A noter qu'un
traitement thermique d'une heure à 600°C provoque une baisse de
la limite d'élasticité de l'ordre de 2 %. En ce qui concerne la
résistance à la rupture fragile, l'effet du traitement
thermique sur la modification de la température de transition
du métal de base est peu important. Il n'en est pas de même
pour les zones écrouies et pour les zones affectées par la
chaleur (ZAC).

Les aciers semi-calmés sont très susceptibles au phénomène de fragilisation tenso-thermique inhérent à leur mise en oeuvre par soudage. Le vieillissement provoqué par un écrouissage à des températures entre 150 et 350°C entraîne une forte fragilisation de l'acier et, en présence de défauts et des contraintes résiduelles de traction élevées, donne naissance à des ruptures fragiles à basses contraintes. Un traitement thermique à des températures au-delà de 450°C peut, à cet égard, donner lieu à une restauration très nette.

Par ailleurs, si on applique un traitement thermique à une température en-dessous de Ac_1 à des zones soumises à écrouissage critique (4 à 5 %) qui, d'ordinaire, ne recquièrent pas de traitement de normalisation, il peut en résulter un grossissement de grain. Il y a dès lors lieu de recuire à plus basse température les nuances d'aciers sensibles à cette dégradation.

En ce qui concerne les structures présentes dans la Z.A.C, le traitement thermique conduit toujours à un adoucissement plus ou moins prononcé. Le cycle thermique de soudage est tel que dans le plupart des cas, la Z.A.C conserve après le traitement thermique une limite d'élasticité et une résistance à la traction supérieure à celle du métal de base après le revenu. Le traitement thermique a presque toujours un effet bénéfique sur le comportement à la rupture fragile, parce que dans les aciers C-Mn, la fragilisation dans la Z.A.C après soudage est souvent liée à la présence de constituants de trempe. Si la structure est à gros grains, l'effet du traitement thermique est cependant négligeable.

Dans le cas des aciers à dispersoïdes, l'évolution des propriétés mécaniques au cours du traitement thermique est conditionnée par la mise en solution ou la précipitation des carbures et des nitrures. En fait, il a été constaté que l'entrée en solution de l'azote et du carbone augmente lorsque s'accroît la température du traitement. Pour les aciers au Niobium, il est impossible de discriminer les pourcentages respectifs de nitrures et de carbures ainsi concernés. Toutefois, on sait que les carbo-nitrures de niobium se dissolvent peu ou pas du tout en-dessous de températures de l'ordre de 950°C. Pour les aciers au vanadium, on assiste à une remise en solution des nitrures à partir de températures relativement basses, dans certains cas dès 550°C. Dans les deux cas (aciers au vanadium ou au niobium), une telle évolution peut être contrecarrée par la présence des éléments de calmage : selon que l'on a affaire à un acier calmé à l'Al, calmé au Si ou semi-calmé, le processus d'évolution de l'azote dépend largement de son affinité pour les éléments en présence. Il est par ailleurs connu que les temps de maintien suffisamment longs provoquent également des remises en solution de ces précipités.

Ces diverses évolutions se traduisent en général par une diminution discrète de la limite élastique et par une augmentation parfois sensible de la température de transition Charpy V. Néanmoins, ce dernier effet est largement compensé par une température de transition presque toujours très basse pour ce type d'acier.

Lors du soudage, l'auténitisation qui se produit dans la Z.A.C s'accompagne d'une mise en solution massive des précipités tels que les carbonitrures de Nb ou de V. Ultérieurement, lors d'un traitement thermique de relaxation, il se produit une précipitation de carbures ou de carbonitrures qui peut conduire à la fragilisation d'une zone mince proche de la ligne de fusion. Ce phénomène est facilement constaté en utilisant des éprouvettes de simulation. Du fait que cette zone est très petite, le traitement thermique provoque en général une amélioration du comportement global du joint réel en y diminuant la taille de la zone fragilisée. Cependant, si les joints sont exécutés par des procédés de soudage avec fort apport de chaleur, cette amélioration n'est pas évidente.

Les aciers faiblement alliés sont le siège de phénomènes complexes attribuables à leur analyse chimique et aux t raitements subis probablement au traitement thermique après soudage.

Les aciers au Cr, Ni, Mo étant en général livrés à l'état normalisé ou à l'état normalisé - revenu, leurs caractéristiques après relaxation dépendront essentiellement de l'état initial (analyse, traitement thermique, microstructure) et de la température et la durée du traitement de revenu.

Déjà, sur le métal de base, se manifestent divers effets.

- On peut s'attendre à ce qu'un traitement thermique effectué à une température inférieure de 30 à 50°C à la température de revenu du métal n'aura que peu d'effet sur le structure métallographique, les propriétés mécaniques et la résilience. Une température plus proche de celle du revenu, maintenue pendant une longue durée, conduira à une structure moins aciculaire par une polygonisation plus prononcée des structures de trempe; cette pratique qui produit un abaissement graduel de la résistance à la traction avec, en général, une amélioration de la résistance à la rupture fragile, est cependant d'application délicate en raison des risques de surrevenu. L'intégration des paramètres du recuit de détente dans la notion de paramètre de revenu se révèle, à cet égard, intéressante pour autant que l'on soit sûr des niveaux de température et des durées d'exposition réellement atteints par l'acier lors de son élaboration, et par la construction au cours de ses traitements successifs.

- Une fragilisation irréversible des aciers faiblement alliés au Cr-Ni-Mo peut survenir après un temps de maintien suffisamment long à la température du traitement. Elle se traduit par une perte prononcée de résilience. Cette perte répond à la loi de Larson-Miller et est susceptible d'intervenir dans le cas où de nombreux recuits successifs de détentionnement sont imposés lors de la soudure des réservoirs de forte épaisseur.

- A une température de traitement plus basse - généralement de l'ordre de 450°C - on observe parfois un autre type de fragilisation caractérisé par une rupture intergranulaire. Cette fragilisation décroît toutefois rapidement lorsque la température augmente. Il s'agit d'un phénomène essentiellement réversible qui peut être estompé par un maintien à température plus élevée.

Les effets du traitement thermique après soudage sur le métal fondu peuvent se traduire par des dégradations de la résilience. Dans le cas d'addition de Ni (2 %) prévues pour accroître la résilience à basse température (5), il se produit après traitement de relaxation des précipitations de carbures associées à une présence plus prononcée d'une phase martensito-austenitique.
De tels effets sont atténués par la durée du palier du traitement, aux températures de traitement les plus élevées et par un refroidissement plus rapide de la pièce (6). En d'autres termes, il y a lieu de se préoccuper des faibles vitesses de refroidissement mais aussi on peut douter du principe d'équivalence appliqué lors de la qualification des modes opératoires, où l'effet de plusieurs traitements successifs est simulé par un seul traitement de plus longue durée.

Mais, s'est surtout la zone affectée par la chaleur qui fait l'objet des préoccupations les plus manifestes.
La fissuration au réchauffage est un sujet de préoccupation sur lequel l'I.I.S a produit un important travail de synthèse au cours des récentes années (7-8-9).

On définit le phénomène comme une fissuration intergrannulaire dans la zone affectée par la chaleur ou dans le métal fondu d'un joint soudé, prenant naissance au cours d'un traitement thermique ou d'un service à une température suffisamment élevée. Ce sont les aciers inoxydables qui, les premiers, ont attiré l'attention sur son existence, dès le milieu des années '50.

- Une fragilisation irréversible des aciers faiblement alliés au Cr-Ni-Mo peut survenir après un temps de maintien suffisamment long à la température du traitement. Elle se traduit par une perte prononcée de résilience. Cette perte répond à la loi de Larson-Miller et est susceptible d'intervenir dans le cas où de nombreux recuits successifs de détentionnement sont imposés lors de la soudure des réservoirs de forte épaisseur.

- A une température de traitement plus basse - généralement de l'ordre de 450°C - on observe parfois un autre type de fragilisation caractérisé par une rupture intergranulaire. Cette fragilisation décroît toutefois rapidement lorsque la température augmente. Il s'agit d'un phénomène essentiellement réversible qui peut être estompé par un maintien à température plus élevée.

Les effets du traitement thermique après soudage sur le métal fondu peuvent se traduire par des dégradations de la résilience. Dans le cas d'addition de Ni (2 %) prévues pour accroître la résilience à basse température (5), il se produit après traitement de relaxation des précipitations de carbures associées à une présence plus prononcée d'une phase martensito-austenitique.

De tels effets sont atténués par la durée du palier du traitement, aux températures de traitement les plus élevées et par un refroidissement plus rapide de la pièce (6). En d'autres termes, il y a lieu de se préoccuper des faibles vitesses de refroidissement mais aussi on peut douter du principe d'équivalence appliqué lors de la qualification des modes opératoires, où l'effet de plusieurs traitements successifs est simulé par un seul traitement de plus longue durée.

Mais, s'est surtout la zone affectée par la chaleur qui fait l'objet des préoccupations les plus manifestes.

La fissuration au réchauffage est un sujet de préoccupation sur lequel l'I.I.S a produit un important travail de synthèse au cours des récentes années (7-8-9).

On définit le phénomène comme une fissuration intergrannulaire dans la zone affectée par la chaleur ou dans le métal fondu d'un joint soudé, prenant naissance au cours d'un traitement thermique ou d'un service à une température suffisamment élevée. Ce sont les aciers inoxydables qui, les premiers, ont attiré l'attention sur son existence, dès le milieu des années '50.

Les travaux ont porté, à l'époque, sur des cas de fissuration
rencontrés dans des conduites de vapeur en acier type AISI
347 stabilisé au Nb. Puis, ce fut le tour des aciers
ferritiques au CrMo dans les assemblages soudés de tubes
surchauffeurs au cours des années '60. Plus tard, on s'est
rendu compte que des fissures dites de réchauffage pouvaient
survenir dans la zone affectée par le soudage d'aciers
faiblement alliés utilisés dans la construction de récipients
sous pression ainsi que dans la zone sous-jacente au
revêtement de tels aciers par dépôt d'acier inoxydable.
Enfin, il est apparu que le métal fondu n'était pas exempt de
tout danger, en particulier dans le cas de dépôts à 2 1/4 Cr
1 Mo.

En général, on attribue le risque de fissuration au
réchauffage d'un métal ayant subi un cycle de soudage au
mécanisme suivant. Lors du soudage, le cycle thermique auquel
est soumise la zone affectée par la chaleur du métal de base
ou du métal fondu réputé sensible provoque une ségrégation
aux joints de grains d'une solution sursaturée en carbone et
en impuretés. Cette même zone est sujette, lors du dépôt du
cordon suivant, à un cycle thermique de moindre amplitude,
qui dans une gamme comprise entre 450° et Ac_1, induit une
fragilisation intergrannulaire par écrouissage. Lors du
traitement de relaxation des contraintes, la capacité de
résistance de la zone incriminée, éventuellement affaiblie
encore par une fragilisation au revenu, est de beaucoup
inférieure à celle des grains, en particulier dans les aciers
comportant du Cr. Les contraintes résiduelles présentes y
induisent des dilatations incompatibles avec la capacité de
déformation qui y règne et provoquent une rupture avant même
qu'elles ne soient résolues.

(5) J.E.M.Braid, J.A.Gianetto - The Effects of PWHT on the Toughness
of Shielded Metal Arc Weld Metals for Use in Canadian Offshore
Structure Fabrication - Doc.IX.1385.86/X.1101.86.

(6) D.Cartoud, S.Debiez, Institut de Soudure - Novembre 1984.

(7) A.Dhooge, A.Vinckier - Stress Relief Cracking - Status Report -
Doc.IX.1250.82.

(8) I.Hrivnak et al - Mathematical Evaluation of Steel Susceptibility
for Reheat Cracking - Doc.IX.1346.85.

(9) A.Dhooge, A.Vinckier - Reheat Cracking - A Review of Recent
Studies - Doc.IX.1373.85

2.4. Effets sur la sécurité à la fatigue

La présence de fissures de fatigue dans une construction soudée constitue un danger vis-à-vis duquel il faut se prémunir au mieux. Les contraintes résiduelles influencent le comportement en fatigue des assemblages soudés car les contraintes locales résultent de la superposition des contraintes résiduelles de soudage et des contraintes variables de service.
Dans le domaine d'endurance, pour les assemblages soudés et non relaxés, les contraintes variables réellement présentes sont caractérisées par des valeurs maximales S max et minimale S min données par les expressions suivantes :

$$S\ max = Re$$
$$et$$
$$S\ min = Re - dS$$

où Re est la limite d'élasticité et dS, l'étendue des contraintes de service. Si la ductilité du matériau au droit de l'apparition éventuelle de la fissure de fatigue est suffisante pour que la plastification, due au premier cycle de contrainte, se développe sans fissuration, les contraintes variables de service peuvent être définies par la seule valeur de l'étendue de contrainte ds. Dans le domaine d'endurance, il a été montré expérimentalement que la relaxation des contraintes résiduelles de soudage permet une augmentation de la limite d'endurance qui est au plus de 20 % et très généralement située entre 10 et 15 %. Une diminution sensible, voire une suppression de l'effet d'entaille est, à cet égard, de loin plus efficace qu'un traitement de relaxation.

Dans le cas de la fatigue oligocyclique, les contraintes résiduelles non relaxées jouent un rôle effacé car elles sont redistribuées lors des premiers cycles de déformations plastiques. L'un des facteurs dominants de la résistance à la fissuration devient la capacité du matériau à supporter des déformations plastiques cycliques dont chacune consomme une partie de sa réserve de ductilité. Cette réserve de ductilité peut, dans certains cas, être sensiblement améliorée par un traitement thermique de restauration métallurgique judicieusement appliqué.

2.5. Effets sur la résistance à la corrosion

La mise en oeuvre de constructions soudées à montré qu'il existait des types de corrosion spécifiques à la soudure.

C'est ainsi que lorsque les circonstances sont défavorables, on peut parfois observer une attaque sélective importante du matériau de base au voisinage des cordons de soudure.

Parmi les facteurs principaux qui régissent ce phénomène, on peut
citer :

- la nature du matériau de base,
- la nature du métal déposé,
- la nature de l'agent corrosif, sa concentration, sa
  température, son agitation,
- l'hétérogénéité tant macroscopique que microscopique des
  matériaux en présence (défauts, joints des grains, lacunes,
  dislocations...),
- les conditions de soudage, depuis la préparation des pièces,
  séquence de soudage, etc... jusqu'au préchauffage et au
  traitement de recuit final,
- l'état de contrainte,
- l'état de surface.

A côté des facteurs principaux qui régissent le phénomène, les
contraintes macroscopiques et microscopiques (les premières sont
engendrées par la soudure et les sollicitations extérieures, les
secondes proviennent de la structure même du métal) peuvent
influencer grandement l'importance de la corrosion sélective du
métal de base au voisinage des cordons de soudure.
La concentration de contraintes au droit de défauts de surface,
même de faible importance, accélère le processus.
Combinant leurs effets à la corrosion proprement dite, les
contraintes résiduelles peuvent donner naissance au phénomène de
corrosion fissurante si redoutable quant à ses effets, en
particulier dans le cas des aciers à haute limite élastique.
Ici, encore, le traitement de relaxation des contraintes
résiduelles, lorsqu'il est praticable, permet d'éviter le
développement d'une corrosion sous tension dangereuse.
Un cas particulier où le traitement de relaxation se révèle
particulièrement efficace est celui de la sensibilité à la
fissuration par l'hydrogène dite fissuration différée. L'effet de
la post chauffe immédiatement après le soudage est d'accélérer la
diffusion de l'hydrogène hors des structures dangereuses. L'effet
de la température élevée est d'opérer un revenu des structures
martensitiques réputées telles.

2.6. Effets sur la stabilité globale

Les constructions chaudronnées ou tubulaires sont rarement le
siège de phénomènes d'instabilité, en raison du type d'efforts
appliqués et de la raideur des pièces. De sorte que l'existence
de contraintes résiduelles n'a généralement pas d'importance pour
le calcul des éléments qui les constituent.
Il peut cependant exister certaines circonstances où la stabilité
globale d'une construction soudée pose un problème.

Dans les constructions à élancement moyen ou élevé, lorsque les
contraintes de service s'ajoutent localement à des contraintes
résiduelles de même signe, la limite d'élasticité du matériau
peut être atteinte prématurément et les déformations plastiques
locales correspondantes tendent à diminuer la rigidité de
l'élément. Il s'agit là d'une cause de réduction de la charge
d'instabilité de l'élément dont il importe de tenir compte par le
recours aux courbes de flambement pour les phénomènes de
flambement et de déversement, ou à une méthode de calcul
spécifique, pour le calcul des semelles comprimées des caissons.

Le recuit après soudage est susceptible de supprimer les
contraintes résiduelles qui sont localement de même signe que
celles de service et donc de fournir un remède efficace à la
réduction des charges d'instabilité associées aux déformations
plastiques. L'efficacité de ce remède est, dans certains cas,
explicitée dans les normes de calcul.

Cependant, il faut que le recuit pratiqué soit tel qu'il
n'engendre pas de dilatations différentielles dues à un gradient
thermique trop important lors de la montée en température, car
les distorsions résultantes pourraient être plus néfastes que les
contraintes résiduelles qu'on cherchait à éliminer. Ce problème
se pose en particulier en cas de traitement de pièces comportant
des parties d'épaisseurs fortement différentes.

De façon générale, on peut dire que l'effet du recuit sur la
stabilité globale n'est cependant pas à sens unique et que chaque
cas particulier demande une étude nuancée.

***
*

3. ANALYSE DES PRESCRIPTIONS DES CODES ET DES REGLEMENTATIONS

L'Institut International de la Soudure s'est, à plusieurs reprises,
penché sur les réglementations et codes en vigueur en matière de
traitement thermique de relaxation (10-11-13). Le rôle que joue
l'I.I.S dans la normalisation internationale est, il convient de le
rappeler, déterminant. Ce rôle a été réaffirmé récemment par l'ISO,
dans une prise de position instituant l'I.I.S comme organe officiel
habilité à préparer les projets de normalisation internationale.

Etant donné la grande diversité des points de vue exprimés dans les
différents codes et réglementations en vigueur, il ne faut pas
s'étonner que les nombreux auteurs se soient attachés à comparer
leurs prescriptions selon les applications, les catégories d'aciers
et les buts recherchés (12).

Il ne faut pas s'étonner non plus que les pays membres de l'I.I.S
ont tenté d'accorder leurs points de vue, en prenant appui sur les
considérations métallurgiques qui, de plus en plus, ont été rendues
obligatoires par l'usage de nouvelles familles d'aciers.

Une remarque préliminaire s'impose. Un code constitue un ensemble homogène, à utiliser globalement. Il est difficile d'en extraire une partie, ou d'envisager des parallélismes, en omettant le contexte général du document ainsi que les dispositions d'autres articles.

Bien que l'examen des champs d'application prévus par les codes révèle une grande diversité des points de vue, il apparait de manière assez nette que les traitements thermiques de relaxation sont obligatoires dans le cas des appareils soumis à risque de corrosion sous tension ou à risque de rupture fragile ou par fatigue dans les conditions de l'épreuve ou du service.

L'analyse conduit à retenir quatre critères essentiels de comparaison :

1) Les épaisseurs maximales admissibles sans traitement thermique,
2) La température de traitement,
3) Les temps de maintien minimum à température maximum,
4) Les vitesses maximum de chauffage ou de refroidissement et les températures maximum d'enfournement et minimum de défournement.

L'analyse conduit également à distinguer les aciers au carbone ou au C-Mn, les aciers faiblement alliés au CrMo et les aciers alliés au Ni et prévus pour les applications à basse température.

3.1. Epaisseur critique

L'épaisseur de paroi est un facteur déterminant, dans les prescriptions des codes pour l'application des traitements thermiques après soudage aux récipients sous pression. En effet, sous une certaine épaisseur, le traitement peut-être omis, alors qu'il se révèle obligatoire au-dessus d'une certaine limite.

L'examen comparé de divers code révèle une large dispersion dans la fixation de cette épaisseur limite, quel que soit le type d'acier. On ne peut certes s'empêcher de s'interroger sur la signification d'une valeur passe-passe pas. Mais on peut certes également s'interroger sur les base scientifiques ou techniques qui ont présidé au choix de valeurs aussi arbitraires. On a rassemblé dans le tableau I les impositions de quelques codes applicables aux aciers de qualité pour récipients sous pression.

Les aciers au C et au CMn peuvent ne pas être soumis à un traitement thermique après soudage lorsque l'épaisseur de paroi est inférieure à une valeur qui selon les codes, varie de 20 à 50 mm. Des valeurs plus élevées peuvent être néanmoins être acceptées sans traitement s'il est établi qu'il n'y a pas de risque de rupture fragile, et que les aciers sont à grains fins. Une tendance à aligner les valeurs limites autour de 35 mm, épaisseur sous laquelle le traitement serait facultatif, se fait jour.

TABLEAU I

ACIERS	ASME VIII/2	BS 5500	AD HP 7 Re < 370	AD HP 7 430 > Re > 370	ISO DIS 2694	NBN F11-001 NL	Rules NF 32-105 Rm < 560 Mpa	NF 32-105 Rm < 700 Mpa	DIN 17155	IIS Rec
C et CMn	32 / 38	35 / 40	30 / 38 / 50	30	30 / 38 / 50	20 / 30	32	40	30	35
0,35 Mo	16	20	30-38-50	30	30-38-50	30	32		30	20
0,5 Mo	16	20	30-38-50	30	20	30	0			20
1 Cr - 0,5 Mo	16	0	0		15	15	0			0
2,25 Cr - 1Mo	0	0	0		0	0	0			0
Cr Mo V	–	0	0		0	0	0			15

En ce qui concerne les aciers microalliés au Nb et au V, un groupe
de travail au sein de l'IIS s'active à définir les pistes de
réflexion (13). Il est certain que des facteurs de décision
nouveaux, notamment l'apport calorifique de soudage, devront
intervenir dans la définition de la limite d'épaisseur jusqu'à
laquelle un traitement n'est pas requis.

Les aciers faiblement alliés au Mo, CrMo, CrMoV sont également
l'objet de prescriptions très différentes selon les codes. Si on
constate, en général, que l'épaisseur limite maximum est réduite au
fur et à mesure qu'augmente le degré d'alliage de l'acier, il faut
reconnaître que règne le plus grande confusion pour les aciers
C-Mo. Ici aussi, une tendance se fait jour à aligner les valeurs
sur des épaisseurs unifiées, au-delà desquelles le traitement
thermique est obligatoire et en-deçà desquelles il est facultatif.

3.2. Temperature du traitement

Le cycle thermique du traitement après soudage est sanctionné
dans les codes par :

- la température maximum d'enfournement
- la vitesse maximum de chauffage
- le temps de maintien
- la vitesse maximum de refroidissement
- la température maximum de défournement.

C'est la température de maintien qu'est considérée comme le point
de mire pour l'obtention d'une bonne relaxation des contraintes.
La limite supérieure de température est régie par la résistance à
la déformation des portions non supportées. Compte tenu de la
durée du maintien à cette température, et de la contribution des
phases de montée en température et de refroidissement, il faut
admettre que c'est aussi la température de maintien qui est le
point de mire pour ce qui concerne les effets métalliques du
traitement de détente.

Par température, on entend toujours celle que doit attendre le
matériau traité, et non pas la température de l'atmosphère du
four. En fonction de la géométrie des pièces à traiter et de la
géométrie de ces fours, le savoir faire du praticien doit
permettre de se placer dans les conditions optimales.

---

(10) Opportunités du recours à des traitements thermiques de
relaxation - Doc.IX.1059.77/X.913.78.
(11) Traitements après soudage des appareils à pression en aciers -
Etat de la question - Doc.IIS.798.84, ex Doc.XI.420.84.
(12) Traitement thermique des constructions chaudronnées - colloque
Fabrication-Soudage-Contrôle - Paris, Octobre 1983.
(13) WG Influence of Stress Relief Heat Treatment.
Statut report Doc IX.1359-85

Le régime thermique d'une construction ou d'un assemblage soumis
à un traitement thermique dans un four n'est pas uniforme en
chacun de ses points, quand bien même le four aurait une
répartition homogène des températures dans le temps et dans
l'espace. Les portions les plus minces de l'assemblage tendent à
suivre le régime de température imposé. Les portions les plus
massives, par contre, sont caractérisées par une inertie
thermique importante, que ce soit durant la période de chauffage
ou durant la période de refroidissement. Ces mêmes portions
peuvent en outre être sujettes à des gradients thermiques entre
leur surface et leur coeur.

Considérons le cas simplifié de deux tôles séparées de grandes
dimensions et d'épaisseurs, e_1 et e_2 soumises à un cycle
thermique de chauffage, de maintien à une température T et de
refroidissement. Au cours du chauffage, l'écart des températures
instantanées entre le coeur des tôles augmente avec la différence
d'épaisseur et avec la vitesse de chauffage du four.

Ces écarts peuvent être importants et conduire à des contraintes
thermiques de compression dans les membrures minces d'un
assemblage réel (dans lequel toutefois le phénomène est atténué
par la conduction thermique qui s'établit via la liaison entre
membrures minces et membrures épaisses).

Lorsque le membrure ou la tôle la plus mince a atteint la
température maximale du traitement, cette température doit être
maintenue un temps suffisamment long pour permettre aux membrures
les plus épaisses d'atteindre à leur tour le niveau de
température souhaité. Cette durée de maintien indispensable est
fonction des conditions d'enfournement et de montée en
température du four, et des épaisseurs relatives des pièces à
traiter. Si l'on prévoit que le traitement thermique s'effectue à
une température élevée, à laquelle une relaxation presque
complète des contraintes résiduelles se produit en un temps très
bref, il convient d'évaluer la durée de maintien qui correspond à
une homogénéisation complète des températures en vue d'évaluer les
dangers de fragilisation irréversible qui sont susceptibles
d'intervenir, et les risques de voir correspondre cette durée
avec celle d'un minimum de restauration des zones fragilisées. Si
l'on prévoit que le traitement thermique s'effectue à plus basse
température, mais avec un temps de maintien prolongé, la durée
d'homogénéisation des températures ne sera pas grandement
réduite, mais son influence relative sur la dégradation
métallurgique à long terme sera moindre. Au cours du
refroidissement en four, les gradients thermiques entre la
portion la plus massive et les membrures les plus minces sont
faibles. Lorsque l'on procède à un défournement à, par exemple,
300°C, les membrures les plus minces se refroidissent plus
brutalement et sont de ce fait soumises à traction. Si les
gradients thermiques sont importants, ces membrures peuvent être
amenées en certains de leurs point faibles à un dépassement de la
limite d'élasticité à la température où surviennent les maxima
d'écarts thermiques.

Il s'ensuit une déformation plastique et, à l'issue du refroidissement l'apparition de tensions résiduelles. En outre, les états de tension en cours de refroidissement sont susceptibles de causer des fissurations au départ de défauts ou de microfissures au voisinage desquels a pu se produire une fragilisation irréversible ou une recristallisation. Des cas de fissurations de pièces épaisses en cours de traitement de recuit peuvent certainement être attribués (en dehors de la fissuration au réchauffage) à des différences de températures exagérées soit lors de l'échauffement, soit plus probablement lors du refroidissement. Il est donc opportun de procéder en continu à des mesures de températures par thermocouples sur le produit lui-même et spécialement en les zones entre lesquelles s'établiraient les plus grands écarts de températures au cours du chauffage et du refroidissement; les thermocouples doivent être efficacement protégés.

La mesure en cours de chauffage permet d'apprécier l'écart maximal des températures atteintes en phase transitoire, et la durée qu'il faut respecter à température maximale pour atteindre le degré de relaxation souhaité, ou, lorsque le cas se produit, le pic de restauration le plus approprié.

La mesure en cours de refroidissement fournit une indication sur les gradients thermiques maxima atteints et sur les contraintes qui correspondent, et, partant, sur la localisation des zones où l'on peut suspecter des fissurations. Toutefois, ces mesures ne fournissent que des indications sur les températures en surface des produits, et il faut encore tenir compte des gradients thermiques entre la peau et le coeur de ceux-ci. De telles mesures sont utilement complétées par une estimation préalable des conditions les plus favorables du traitement. Une telle estimation est possible par voie mathématique connaissant les épaisseurs relatives en cause, et le programme général du traitement. En particulier, on peut parfaitement envisager de disposer d'abaques fournissant pour le cas le plus défavorable, les régimes de chauffage, de maintien d'homogénéisation et de refroidissement en four, et les températures adéquates d'enfournement qui correspondent à des valeurs acceptables des contraintes thermiques transitoires.

Revenant à l'examen comparatif des codes, on constate que ceux-ci sont caractérisés par la plus grande dispersion dans les températures à atteindre.

<u>Les acier au C et au C-Mn</u> font en général état d'un détensionnement de 80 à 90 % pour des traitements de 550 à 575 °C. Il ne faut dès lors pas s'étonner si des codes récents recommandent des températures de traitement assez basses comformes à ces résultats (ISO, CODAP). C'est dans ce sens, également, que l'IIS a pris position, dès 1977. Il n'empêche que, dans l'ensemble, les températures de maintien s'échelonnent entre 530° et 650°C, comme en témoigne le tableau II.

Le code yougoslave recommande que la température soit inférieure à 560°C. Une gamme de température cimprise entre 530° et 580°C est recommandée :

- par les AD Merkblatt selon SEW089 pour les aciers dont la limite élastique est inférieure à 370 Mpa, et pour les aciers à grains fins en épaisseur supérieure à 30 mm.
- par le CODAP 80-F2 pour les aciers CMn et microalliés au V ou au Nb, jusqu'aux tensions de ruptures de 530 MPa, en épaisseurs supérieures à 40 mm, et jusqu'aux tensions de rupture de 590 MPa en épaisseurs supérieures à 30 mm.
- par la DIN 17155 pour les aciers STE460

Dans une gamme légèrement supérieure, on identifie :

- 550°-625°C dans les règles de conception francaises PWR.S1000 applicables aux aciers de toutes catégories dont la tension de rupture est inférieure à 500 M.pa et l'épaisseur supérieure à 35 mm.

- 560°-600°C dans l'ISO DS 2694.

- 550°-620°C dans la DIN 17155 pour les aciers ST48-52 en épaisseurs supérieures à 30 mm, la limite supérieure étant ramenée à 600°C pour les aciers STE690.

- 580° - 620°C dans la norme belge NBN 629, dans la BS5500 pour les aciers de catégories M0 et M1 en épaisseurs supérieures à 35 mm.

Par ailleurs, le code américain ASME VIII recommande pour les aciers P1 en épaisseur de 30 mm et au delà (40 mm s'il y a préchauffe supérieure à 95°C) une température de 593°C.

Le code néérlandais impose un minimum de 600°C, de même que le code japonais pour les aciers P1, P2 et P3 selon la norme JIS B 8243.

Dans la gamme des températures mini-maxi les plus élevées, on rencontre :

- AD Merkblatt selon DIN pour les aciers ST37 à 45 : 600-650°C. le standard tcheque CSN 420284 qui recommande la même gamme, mais, fait intéressant, qui prend en compte les types de métaux déposés, et fait relever à 620°C la limite inférieure lorsqu'on est en présence de métal fondu allié (p.exemple au Mo).

- Les codes soviétiques OP 02CS-66 et RTM-1S-73 qui imposent une gamme de 630°C - 660°C et de 650°C-680°C respectivement.

TABLEAU II

CODE	PAYS	T° C
CODAP 80F2	F	530° - 580°C
DIS 2694	ISO	560° - 600°C
BS 5500	UK	580° - 620°C
ASME VIII	USA	593°C
AD M DIN	RFA	600° - 650°C
+ SEW 089	RFA	530° - 580°C
JIS B 8243	J	600°C
RULES	NL	600°C
NBN 629	B	580° - 620°C
CODE	Yu	< 560°C
CS 420284	Cs	600° - 650°C
OP-02CS-66	URSS	630° - 660°C
Doc X-861-77	IIS	540° - 580°C

Au vu des considérations développées au début de l'exposé, on comprend que la tendance à mettre sur le marché des aciers a grains fins et plus délicats sur le plan métallurgique incite à abaisser les températures mini-maxi du traitement, quitte à concéder la permanence d'un certain niveau de contraintes résiduelles. Il est cependant bien rare que l'on se préoccupe de l'apport calorifique du soudage, ou de l'incidence du métal fondu.

Les aciers faiblement alliés au CrMo et au CrMoV sont en général traités thermiquement après soudage, essentiellement en vue d'améliorer la tenacité du métal fondu et des zones affectés par la chaleur. L'examen de quelques codes montre, ici encore, des discordances assez grandes, mais qui s'estompent lorsque que s'accroit la teneur en Cr (3 % Cr ou 5 % Cr).

La tendance est à l'augmentation de la température minimum de traitement par rapport à l'ASME VIII, qui préconise 593°C pour les 1/2 Cr - 1/2 Mo, 1 Cr-1/2 Mo et 1 1/4 Cr - 1/2 Mo-Si et 675°C pour les 2 1/4 Cr - 1 Mo et 3 Cr - 1 Mo. Le CODAP préconise respectivement 630°-680°C et 670°-710°C, et l'ISO 620°-660°C pour les 1/2 Cr 1/2 Mo et 1 Cr 1/2 Mo et 625°-750°C pour les 2 1/4 Cr 1 Mo, ce qui est une fourchette énorme, à comparer avec la fourchette étroite de 600°-650°C de l'AD Merkblatt, pour le même acier.

Les aciers au CMo se distinguent également par une remarquable dispersion dans les fourchettes de températures maxi-mini :

- au moins 593°C pour l'ASME
- 580°-620°C pour l'ISO/DIS 2694 et le CODAP
- 600°-650°C pour la NBN F 11-001 et l'AD Merkblatt
- 650°-680°C pour la BS 5500

Quant aux aciers au Ni, s'il y a unité de vue pour préconiser pour les aciers où le Ni < 3,5 % le même traitement que pour les aciers non alliés, les recommandations relatives aux aciers à 3,5 % Ni ne sont pas homogènes :

550°-610°C pour l'ANNCC (Sd 5/1) si plus de 15 mm
580°-620°C pour la BS 5500   épaisseur en option
594°-635°C pour l'ASME VIII-2   (P9b) si plus de 16 mm.

Les aciers à 5 % est à 9 % Ni présentent un risque de fragilité réversible de revenu, de sorte que les codes ne prévoient un traitement thermique qu'au delà de 50 mm d'épaisseur (Italie 40 mm pour le 9 Ni et 30 mm pour le 5 Ni, mais des dérogations sont possibles).

L'IIS propose de ne pas procéder à un traitement thermique après soudage pour les aciers à 5 % et à 9 % Ni, et d'assimiler les aciers à < 3,5 % Ni aux aciers CMn pour basses températures.

3.3. Temps minimum de maintien a température maxi

Il existe une relation inverse entre la température maximum nécessaire à la relaxation des containtes et la durée d'exposition à cette température. On utilise à cet effet le paramètre d'Hollonmon-Jaffé et le paramètre dérivé des lois de diffusion (14), qui permettent par ailleurs d'identifier un cycle thermique réel à un cycle rectangulaire équivalent, et d'utiliser à bon escient la période de montée en température et de refroidissement de manière à réduire la durée d'exposition à température maximum, lorsque le code le permet.

Les prescriptions des codes et de normes associent généralement la durée d'exposition à l'épaisseur et la nuance de l'acier.

Pour les aciers au C ou au CMn, on constate que de nombreux codes (CODAP, AD Merkblatt DIN et SEW, AFNOR S 1000, NBN F 11-001, ISO DIS 2694, etc) s'accordent sur une durée de 2 min/mm dans une gamme d'épaisseur comprise entre l'épaisseur minimale requérant un traitement et environ 60 mm. Aux épaisseurs les plus faibles, les codes imposent souvent un temps minimum de maintien (20 à 60 min), lorsque le traitement y est nécessaire. Au delà de 60 mm, certains codes atténuent proportionnellement la durée de maintien. Le code BN 5500 impose des données de maintien plus longues (2,5 min/mm), mais admet la possibilité d'y intégrer les temps de montée et de descente en température. Ces données les plus faibles correspondent aux gammes de températures les plus élevées requises par certains codes (ADM - DIN, CS 420284).

Les aciers microalliés ne donnent pas lieu à des impositions spécifiques.

Quant aux aciers faiblement alliés, ils sont soumis à des prescriptions du même ordre, mais les durées minimales de traitement aux températures requises sont pour certains codes, et selon la nuance de l'acier, allongées. On observera, à titre anecdotique, la disparité entre les codes BS 5500 et CODAP, qui pour les aciers 2 1/4 Cr 1 Mo et, à même durée de maintien au delà de 72 mm, imposent des durées minimum respectives de 30 min et de 180 min pour les épaisseurs inférieures à 15 mm.

3.4. Montée et descente en température

Au cours de la montée en température, la première phase d'échauffement vers 200-400°C présente une importance particulière. En effet, c'est la gamme de température au cours de laquelle les tensions internes dues à des dilatations différentielles se superposent aux contraintes résiduelles dans des microstructures de transformation qui n'ont pas encore atteint un état de revenu de défragilisation (martensite, bainite).

Il est donc capital de veiller à ce que la vitesse de chauffage soit lente, au moins jusque 300-400°C, en particulier lorsque l'épaisseur est forte et que croit le danger de microstructures sensibles et de dilatations différentielles élevées. Les vitesses maximum de chauffage imposées par les codes sont assez homogènes entre 25 et 100 mm; elles sont exprimées en °/h par v = 5550/emm ou 200 à 220/t".

---

(14) T. Maynier, M. Toitot, J. Dollet, P. Bastien Mem. Sc. Revue de Metallurgie n°7/8 - 1972

La discordance se manifeste en dessous de 25 mm :

- max 200°/h pour ISO DIS 2649
- max 220°/h pour DODAP 80-F2, BS 5500 et JIS Z 3700-80
- max non préciser pour NBN, ASME...

et dans une moindre mesure pour vitesses minimum autorisées.

- min 55°/h pour ISO DIS 2649, BS 5500, NBN F 11-001,
- min 50°/h pour JIS Z 3700-80,
- min 38°/h pour ASME VIII.

Les vitesses autorisées pour le refroidissement suivent sensiblement les mêmes tendances, tout en admettant des vitesses maximales plus rapides, comprises entre 200°/h (ASME III) à 275°/h (NBN).

Il faut remarquer que les codes ne contiennent pas des prescriptions sur la vitesse de chauffage depuis la température ambiante jusqu'à la température maximum du four à l'enfournement, laquelle est le plus souvent limitée à 300°-400°C.

Il faut aussi remarquer que les températues maximales de traitement étant comprises entre 540°C et 650°C, les durées nécessaires à atteindre la température prescrite peuvent être sensiblement différentes, avec les conséquences économiques que l'on devine. Il est donc raisonnable de tenir compte autant que possible, comme le BS 5500, de la phase de mise à température de la pièce dans le bilan global du traitement.

***
*

4. APERCU DE LA PRATIQUE INDUSTRIELLE DES TRAITEMENTS DE RELAXATION APRES SOUDAGE

Le traitement de relaxation après soudage est réalisé par des procédés thermiques et par des procédés mécaniques.

Les procédés thermiques comportent en général trois familles de traitements :

- les traitements globaux, par lesquels l'ensemble de la structure est informément chauffé, chacun de ses points atteignant théoriquement la même température au même instant.

- les traitements locaux uniformes, où seules les soudures ou les zones adjacentes sont portées à une température réputée uniforme en tout point.

- les traitements locaux progressifs, où les soudures et zones adjacentes sont échauffée par une source en mouvement, ce qui conduit à une température variable en tout point.

Quant aux procédés mécaniques de relaxation des contraintes ils comportent essentiellement :
- les traitements de sursollicitation, correspondant à une déformation plastique brutale ou lente du métal, obtenue par explosion ou par sollicitation mécanique progressive.
- les traitements par vibration conduisant à une relaxation progressive des contraintes par cycles d'hystérésis mécanique successifs.

Ces procédés peuvent être appliqués globalement au localement.
- les traitements locaux superficiels résultant du grenaillage ou du martelage, qui ont pour effet de redistribuer les contraintes superficielles.

4.1. PROCEDES DE DETENSIONNEMENT THERMIQUE

Les traitements globaux ou par tronçons sont appliqués dans les fours à recuire qui sont conçus en tenant compte des points suivants :
- Précision et régularité suffisante de la température en tous les points.
- Facilité de réglage
- Facilité d'enfournement et de défournement
- Possibilité d'enregistrement automatique de la température.

En général, le chauffage des fours modernes se fait au moyen de flammes (fuel ou gaz) ou de l'électricité.

- Fours à flamme : chauffage par gaz enflammé, en général au moyen de brûleurs à combustibles liquides ou gazeux.
- Fours électriques : utilisation de résistances électriques ou chauffage par induction. Le chauffage par induction à basse fréquence a l'avantage de chauffer la pièce dans la masse, ce qui élimine certains gradients thermiques. L'emploi de résistances rayonnantes permet de concevoir des fours de type modulaire, particulièrement adaptés aux économies d'énergie.
- Fours à bain : pour traitements spéciaux.

Le chauffage de la pièce peut être réalisé dans l'atmosphère directe du four ou dans une enceinte éventuellement remplie d'une atmosphère déterminée (traitement sous atmosphère).

Selon la conception, on distingue :

- Le four roulant en forme de tunnel, avec une porte à chaque extrémité, qui permet un traitement continu.
- Le four tournant (à sole circulaire) : type généralement conçu pour le traitement de petites pièces.
- Le four à sole mobile : comme le four roulant, il peut être pourvu de deux portes qui permettent d'effectuer des traitements sur pièces dont la longueur dépasse celle du four (traitement partiel répété). La sole mobile ne sert que pour faciliter l'enfournement et le défournement de pièces lourdes.
- Les fours de conception spéciale sous forme de cloche ou de moufle.

En ce qui concerne la mesure de la température, on dispose
d'appareils pyrométriques, d'une précision et d'une sensibilité
telles qu'ils dépassent largment les exigences et tolérances de
contrôle de température d'un four de recuit.

La question du réglage de la température dans le four est
beaucoup plus critique. Ce réglage peut se faire globalement ou
par zones distinctes. Le cycle de traitement est réalisable par
contrôle manuel ou sans intervention de l'opérateur (four
complètement automatique). Un bon four automatique permet à vide
une stabilité de température dont la précision est 5 à 10 °C
autour de la température de consigne en n'importe quelle zone du
four. Dans certains cas, l'inertie calorifique du four joue un
rôle important en ce qui concerne la stabilité de la température
et les vitesses de chauffage et de refroidissement. Notons
néanmoins que le contrôle et le réglage adéquat de la température
du four n'assurent pas l'égalité de température dans tous les
points de la pièce, surtout quand celle-ci est composée
d'éléments d'épaisseurs très différentes. On a montré
précédemment que des différences peuvent se produire entre
température du four et température de la pièce. Ce décalage est
fonction de la vitesse de montée et de descente de la
température, de la proportion entre volume ou masse à recuire et
volume ou capacité du four et de son calorifugeage. La mesure de
la température au moyen de thermocouples dûment étalonnés fixés
sur la construction même et convenablement protégés est utile
voire indispensable, selon la classe de l'appareil.

La conception et la qualité du four, ainsi que l'expérience que
l'on peut en avoir, sont des facteurs non négligeables en ce qui
concerne la régulation et la stabilité de la température.
L'enregistrement automatique du déroulement du traitement a pour
but de s'assurer que l'appareil a bien subi le traitement prévu.

D'autres causes de préoccupations importantes sont :

- l'emplacement de la pièce dans les fours de manière à obtenir
  une circulation régulière du flux calorifiques autour de la
  pièce,
- la protection de la construction contre le contact direct de la
  source de chaleur ou contre une oxydation excessive,
- l'étanchéité du four ou la protection des parties sortant
  partiellement du four (limitation du gradient de températures
  et déformations éventuelles).

Finalement, on doit tenir compte de la possibilité de
déformations éventuelles de la construction lors d'un traitement;
il importe donc de prévoir des supports adéquats.

Une variante des traitements globaux applicable aux récipients sous pression consiste en une disposition où l'appareil est calorifugé extérieurement et chauffé intérieurement par tout moyen permettant de respecter les conditions prévues aux codes utilisés et applicables au cas du traitement global par chauffage externe.

Lorsque les traitements globaux sont appliqués par tronçons successifs en raison de l'incompatibilité des dimensions du four avec celles de la pièce, on préconise une longueur de recouvrement de l'ordre de 1,5 m et un calorifugeage des parties situées à l'extérieur du four pour éviter un gradient thermique trop brutal. Ces parties ne peuvent en outre comporter de cloison intérieure, de fond ou de plaque tubulaire.

Dans certains cas, on préfère procéder à des traitements globaux de tronçons séparés suivi de leur assemblage final et de leur traitement thermique local. On impose une largeur de traitement d'au moins trois fois l'épaisseur de paroi et un calorifugage d'une largeur d'au moins six fois l'épaisseur de la paroi plus épaisse.

Les traitements thermiques locaux s'appliquent essentiellement aux corps de révolution. Ils sont réalisés par des dispositifs de chauffage à la flamme ou électrique et, dans leur version moderne, agissent par chauffage d'air, par rayonnement ou par induction. La largeur minimum sur laquelle doit s'étendre la zone chauffée de part et d'autre de la zone à traiter est régie par certains codes, notamment le code français qui fixe cette largeur à quatre fois l'épaisseur de paroi. Le calorifugeage s'étend, quant à lui, de part et d'autre de la zône à traiter sur une largeur de dix fois cette épaisseur. D'autres codes, comme le code hollandais, préconisent les implantations des thermocouples de contrôle de température dans le plan médian et à certaines distances de ce plan, ainsi que les écarts de température admissibles entre les différents points de mesure.

Les réalisations industrielles de traitement thermique locaux sont nombreuses et très diversifiées : elles dépassent de loin le souci de relaxation des contraintes résiduelles de soudage et couvrent toute la gamme des températures, de la préchauffe ou de la postchauffe aux recuits de normalisation.

Les performances atteintes par les traitements de relaxation des contraintes de soudage par chauffage local ne doivent pas être seulement évaluées du seul point de vue de la relaxation. Des améliorations sensibles de la tenacité du joint soudé, comparables à celles que procurent des traitements globaux, peuvent être enregistrées, même si, au demeurant, les contraintes résiduelles de soudage n'ont pas été totalement résolues.

***
*

4.2. PROCEDES DE DETENSIONNEMENT MECANIQUE

Dans le cas où il se révèle inutile , peu pratique ou inadéquat de procéder à un détentionnement de contraintes par voie thermiques, il arrive que l'on fasse appel aux propriétés de plasticité à température ambiante du matériau pour y effacer les pics de contraintes résiduelles.

Nous relevons trois catégories de procédés :

- la sursollication globale ou locale, appliquée de manière lente ou brutale, destinée à plastifier le métal en état de tension et créer, le cas échéant, après l'opération, un état de compression résiduelle.

- la mise sous vibrations mécaniques, au cours de laquelle les contraintes résiduelles sont chapeautées par des contraintes alternées qui en modifient peu à peu le niveau.

- les traitements locaux de surface résultant du martelage, du grenaillage et du sablage, grâce auxquels il se crée une zone de déformation plastique superficielle, siège de contraintes résiduelles de compression, qui compensent les contraintes de traction préexistantes.

Les traitements mécaniques par sursollicitation sont surtout intéressants pour les appareils soumis à pression - leur justification repose sur la constation que, lors de la montée en pression, les contraintes externes se superposent aux contraintes internes, entraînant des plastifications locales. Lors de la descente en pression, le retour élastique provoque dans les zones plastifiées des contraintes résiduelles jouant un rôle de précontrainte favorable. La correcte application de ce traitement est aisément contrôlable puisque c'est la pression appliquée qui la détermine. Par ailleurs, certaines règlementations, telles que la réglementation française, ont prévu d'autoriser l'application de traitements mécaniques à contraintes imposées en substitution à un traitement thermique de relaxation, pour autant que la contrainte atteinte en tout point des joints soudés soit au moins égale à 85 % de la limite d'élasticité à la température d'essai. Si elle est la plus souvent réalisée par mise en pression lente, la sursollicitation peut être également le résultat local d'une extension mécanique provoquée par l'énergie d'explosion d'une charge, posée par exemple au voisinage de cordons de soudure. D'un type assez particulier, ce traitement ne présente de l'intérêt que dans un nombre limité d'applications.

Le détentionement par vibrations mécaniques est appliqué depuis de nombreuses années principalement en vue de stabiliser la forme des assemblages mécanosoudés. Il se pratique généralement en fixant un vibreur à fréquence variable sur un élément de la structure à détentionner et en ajustant la fréquence sur la fréquence naturelle de cet élément jusqu'à résonnance.

Cette dernière condition ne semble cependant pas être indispensable (15). C'est la nature cyclique de la sollicitation appliquée, et l'effet Bauschinger y associé, qui est à l'origine du détentionnement. Cette explication est cependant simpliste et masque un ensemble de phénomènes complexes sur les plans macroscopiques et microscopiques. A l'échelle des dislocation, il est probable que les vibrations mécaniques produisent des désancrages et des mouvements d'atomes intersticiels, conduisant à accroître leur mobilité. On a pu observer (16) un relèvement de la striction lors de l'essai de traction sur éprouvettes soumises au préalable à vibration, ce qui conduit à penser que le détentionnement par vibration agit également sur la capacité de déformation plastique d'un métal, au delà d'une simple relaxation des contraintes. A l'échelle macroscopique, le métal subit une série de cycles de chargement et déchargement, qui se distingue de la mise en charge monotone par une entrée en plasticité beaucoup plus précoce. Ceci a pour effet de relaxer les contraintes, et s'accompagne dans le cas des aciers ferritiques d'un adoucissement.

Les applications du détentionement par vibrations sont nombreuses. On peut signaler à titre non limitatif (17) des échangeurs de chaleur à plaque, des carters de machines, des cadres supports d'éléments de précision, pour lesquels des traitements thermiques provoqueraient des distorsions inadmissibles. Moyennant des précautions d'utilisation, les objections que l'on peut formuler sur ce procédé à propos de fissuration par fatigue ou d'amorçage de ruptures fragiles peuvent être aisément levées. Il faut cependant reconnaître que les modes opératoire sont mal définis et que la matérialité de la relaxation des contraintes, qui peut être constatée par une modification de l'intensité de courant nécessaire pour imposer une vibration à la pièce, n'a qu'un caractère qualitatif.

Quant aux traitements par impact, qui induisent des contraintes résiduelles de compression sous les zones où l'on applique un martelage ou un grenaillage, ils ont un caractère essentiellement local, et sont le plus souvent utilisés pour améliorer la résistance à la fatigue et à la corrosion sous tension des joints soudés. Les contraintes réellement atteintes sont difficiles à préciser.

---

(15) P. Barbarin et al CETIM information n°84 - Février 1984, pp 43-48
(16) I. Rak et Al, doc IX-1249-82
(17) G.G. Saunders, R.A. Clayton in "Residual Stresses in Welded construction and their effects". Publ. the Welding Institute London 15-17 Nov. 1977

CONCLUSION

A la lumière des considérations précédentes, il apparaît que les effets métallurgiques occasionnées par un recuit dit de relaxation dépassent très souvent la relaxation proprement dite des contraintes résiduelles, soit en bien, soit en mal.

Il apparaît aussi que si l'objectif poursuivi de détendre les contraintes, ou de n'en point créer de nouvelles, impose au traitement thermique un certain nombre de règles portant sur la température d'enfournement, la vitesse de chauffage, la température maximum du traitement, la durée de maintien à cette température et la vitesse de refroidissement, ces mêmes règles agissent sur l'intégrité du métal de base, du métal fondu et de la ZAC.

Il apparaît enfin que la sauvegarde ou le renforcement de la tenacité de la réserve de ductilité des joints soudés d'une construction prime dans la majorité des cas sur une suppresion complète des contraintes résiduelles qui s'accompagnerait d'une dégradation de ces propriétés.

L'Institut International de la Soudure se doit de poursuivre son rôle d'unification des vues sur le bien fondé et la bonne application des traitements de relaxation.

Une uniformisation des codes est nécessaire, mais elle ne peut se régler en un temps bref. Les positions prises par l'IIS sont, à ce point de vue, claires : abaisser la température maximum de traitement, réduire la durée d'exposition à température maximale au risque de ne pas résoudre entièrement la relaxation des contraintes, intégrer le cycle de montée et de descente de la température à l'effet du maintien à température maximale, se pencher sur les effers pervers de montée non contrôlée en température lors de l'enfournement, élargir au métal fondu et à la zone affectée par la chaleur les appréciations de l'impact des valeurs recommandées par les codes sur le métal de base, nuancer ces mêmes valeurs en fonction des grades d'aciers comportant des microéléments d'alliage et des aciers traités thermomécaniquement, et des circonstances de leur soudage, poursuivre l'analyse de la relaxation des contraintes.

Le champ est vaste et la bonne volonté est générale.

Aux commissions spécialisées de jouer.

THE STRESS RELAXATION HEAT TREATMENTS AFTER WELDING

1. INTRODUCTION

   I am gratified with the honour and the delicate task of presenting the Conference dedicated to Professor PORTEVIN. By such a presentation, the International Institute of Welding is opening its International Conference and Annual Assembly, with the attendance of worldwide prominent experts.

   It is an honour to rely to the name of this prestigious metallurgist and utmost scientist, to approach a subject upon which a Commission of IIW I chair for the last time this year has been working several times over the past years. I would like, therefore, to associate to this honour any of my colleagues who contributed to the study of its various aspects.

   It is also a delicate task, as such a wide subject has been tackled and extensively studied in several publications, and as the conference devoted to it this year will carry to the attending experts more informations, applications and new ideas than I would ever do.

   Then, I will enter the subject with a cautious humility. Though, with a deep thrust of its importance and actuality, I will try to underline the driving factors of its evaluation, to be eventually covered during the two following days.

   It is a well established practice to submit - whenever technically and economically possible - a welded construction to a treatment aimed at eliminating, or at least at reducing, the residual stresses resulting from manufacturing, and namely for welding. It is accepted that, when achieved in the best conditions, such treatments are generally useful, and sometimes compulsory.

   However, since a few years, and namely, as resulting from the investigations of the past fifteen years, it appeared more and more that such treatments are exerting metallurgical effects beside pure relaxation effects. These metallurgical effects are generaly favourable, though in some cases very complex and even deleterious, and their importance may be as or even more important than the relaxation of residual stresses.

   Such an assessment stimulated the metallurgist, the manufacturers and the prescribers to reevaluate the codes and standards. Because of its international implications, the problem was quite naturally taken in consideration by the IIW within a group specially dedicated to this task, and chaired by Academician I. HRIVNAK from Czecoslovakia.

During the past decade, the joint pressure of market requirements
and of competition between the steelmakers and between the
manufacturers, has stimulated the developpment of new steels
grades, of transformation and manufacturing original processes,
and of more and more sophisticated engineering approaches to
optimize the properties of use of these materials.

By the sixties, already, the market pull was driving the
development of quality steels for tubes and pressure vessels.
The limitation to the increase of C and Mn contents resulting from
weldability and brittle fracture resistance requirements gave rise
to new generations of fine grain steels obtained through thermal
treatment and/or microalloying of some elements.

Higher strengths in as rolled steels were acheved through
addition of Al, V, Nb, Ti which were acting on the precipitation
hardening and grain refinement of the matrix, without deleterious
effect on toughness and weldability.

The progress in thermo-mechanical treatments led to an efficient
usage of controlled rolling for grain refinement and associated
improvement of strength and toughness. These treatments were to be
applied essentially to plates and strips in the range of medium
and small thicknesses, while heavy plates and forgings were to be
more often normalized or quenched and tempered.

In the HOUDREMONT Lecture given at the Annual Assembly of
Ljubljana in September 1982, Dr. SUZUKI (Japan) masterfully
described the recent progresses completed in the development of
weldable steels. He reminded that most of the achievements
in the area of weldability of steels are resulting from a
cooperation between the steel producer and the steel user.
He indicated the main contraints of the market requirements, and
the related efforts to improve the metallurgical quality of
products. The inferring grades and qualities are at the present
time at a remarkably higher level than fifteen years ago.

The manufacturing process, on the other hand, had also to react to
the pressure of the market requirements, and, accordingly, to
face with heavier thicknesses, tricky designs and more and more
stringent quality regulations.

Typically, the progresses in high density welding were presented
at the HOUDREMONT Lecture given in July 1986 at the Assembly of
Tokyo by Dr. SAYEGH (France). This presentation, and many of the
papers introduced at the International Conference following on the
same subject, enlightened the importance acquired by the
properties of use of steels subjected to welding. They indicated
also the extent to which the operations at manufacturing stage are
a follow-up of the thermal treatments induced at steelmaking
stage.

Hence, it is not surprising that the question raised by both the
steelmakers and the fabricators on the so called relaxation
treatment of welding residual stresses are still remaining under
the spotlights.

Considering the evolution of market needs and associated
technical progresses, where are lying the present regulations ?
Which alternate methods to thermal treatments have to be applied,
if they are not adequate, or difficult to operate ?

Which modern postweld heat treatments are the most appropriate to
respond to the fixed requirements of the codes, and to necessary
adjustments to provide economically to the best possible extend an
improvement of the overall properties of the weldments ?

These are our interrogations, and the bases of our reflexion.
To give them further consideration, it is appropriate to
reevaluate the influence of postweld heat treatments with respect
to the improvements expected.

***
*

2. THE EFFECTS OF THE WELDING RESIDUAL STRESSES AND OF THEIR RELIEF

In the mind of numerous praticians, the principal value of a heat
treatment after welding lies in the relieving of the residual
stresses.

These stresses are located in the welded joint and in its direct
vicinity. Generally they reach a level close to the yield
strength of the metal, an external constraint having or not
occured during welding.

Under conditions of multi-axial loadings, they can result in the
brittle fracture of a welded joint.

To reduce or to suppress the welding residual stresses constitues
a protection against the risk of initiation and of propagation of
a brittle facture or a fatigue failure. In addition, it
represents a protection against the tendency to the metal
corrosion, and particularly the stress corrosion, also against
the dimensional instabilities producing deformation during
further working of the pieces.

However, it is to be noted that the stress relief treatments
produce metallurgical effects on the properties of the base metal,
of the fused metal and of the heat affected zone.

The effects are often beneficial; but under certain conditions,
the relieving of stresses is accompanied by a decay of the
ductility or of the metal integrity, decay of which it is
essential to identify all the components.

2.1. Creation of residual stresses

Under the effect of a sharp and localized thermal transient the base metal undergoes during welding, a complex pattern of thermal expansions and shrinkages of metallurgical (associated to allotropic transformations) and of mechanical origins (related to shrinkage and clamping) arises.

The concerned area, the distribution and the highest level of the residual stresses which originate in these expansions and shrinkages can be modelized. However, it is easy to conceive the high complexity of such an analysis, where are taking place the thermal as well as the space and time dependent characteristics of the heat source; the physical and geometrical characteristics of the piece, and all the environmetal circumstances such as the control of the cooling rate, the complexity of the assembly, temperature of the piece before welding, mounting alignment erros, and so on...

The principle of modelizing consists in determining, versus the time, the non-uniform temperature distributions during the welding of a joint between two plates, supposing that these distributions are bidimensional, unvarying with regard to a system of axes linked to the heat source; and that the plates are thin and infinitely wide.

These temperature distributions are established by adopting the well known Rosenthal solution for a linear heat source. The calculation of transient residual stresses during or a the end of welding uses the method of successive elastic solutions, by means of which the stresses and strains at any distance from the joint are determined, by varying (by successive increments of time) a distribution, supposed to be uniform, along the whole length of the seam (1).

2.2. Evaluation of the residual stresses by non destructive methods

The evaluation of the actual level of residual stresses in a welded assembly has always been a fascinating subject for the experts of the non destructive examination. Methods of semi-destructive type such as the strain-gage method (using rosettes) or the Gunnert method present the advantage of a certain reliability, at least with regard to the surface residual stresses; however they are not of easy application on the product.
Several non destructive methods have been studied and sometimes successfully applied, in the aim to identify both the residual stresses after welding and the efficiency of the stress relief treatments (2-3). Some methods are using the remanent induced magnetism (4), after welding, due to the displacement of the welding arc; its intensity in the ferromagnetic materials is depending on the stress field because of the orientation of the domains.

Elsewhere, the simple magneto-elasticity is measured. Other
methods use the modification in the propagation
characteristics of the ultrasonic waves in the presence of
a field of elastic stresses. The concerned characteristics
are : the velocity, the damping and the critical angle of
generation of longitudinal, shear and Rayleigh waves. The
observation of the Moiré fringes transcribed by holography
must also be mentioned, as well as the processes to evaluate
the geometrical parameters of the crystal lattice by X-rays
diffraction. However, it must be admitted that these methods
still remain far from the industrial applications, especially
for the evaluation of subsurface stresses, due to the risks
of interference with other physical causes influencing the
observed phenomena and also due to the difficulties of
interpretation.

2.3. Effects on the toughness with regard to brittle failure

2.3.1. Effects of the stresses

The effects of residual stresses on the resistance to
brittle fracture are the same as those resulting from other
stress states of the same nature but of different origin.
However, the distribution of residual stresses, which must
comply with the laws of static equilibrium, is very
difficult to calculate or to foresee with accuracy.
Moreover, this distribution can be modified by the heat
treatments. For this reason, the effects which depend upon
it, are uneasy to predict quantitatively. With regard to
the crack initiation; the metallurgical factors generally
prevail. The residual stresses may however exert their
influence in two different ways. On the one hand, the
plastic deformation generated during the "in service"
loadings can be sufficient to entail a local embrittlement
of the metal by strain hardening or to be of the order of
magnitude of the maximum allowable strains without fracture
at this temperature. This can result in the generation of
cracks or facilitate the further initiation of the fracture.

---

(1) J.E. Agapakis, K. Masubushi, Weld.Jnl USA (1984) $\underline{63}$ n°6,
 pp. 1870-1963.
(2) A.J.A. Pavlane, in "Residual Stresses in Welded Construction
 and their Effect" - The Welding Institute, London 15-17
 Nov. 1977
(3) K. Stahlkopf et Al - Ibid
(4) . Non-Destructive Estimation of Residual Stress in Welded
 Pressure Vessel Steels by means of Remanant Magnetization
 Measurement (Doc. Nr. X.801.76)

This could occur, for instance, in pieces embrittled at low temperature or in presence of notches or defects in the vicinity of welded joint. On the other hand, the triaxial state of stresses results in effects similar to those due to a diminution of the temperature on the brittleness of the materials; the state of tension triaxiality which is associated to the thick member constructions, is in this respect the most dangerous. The residual stresses can locally reinforce this effect according to their distribution, the presence of notches or stress concentrations. The crack propagation, as for it, can be stimulated and oriented by the residual stresses. Indeed, the phenomenon is mainly governed by the amount of elastic energy released by modification in the distribution of the elastic deformations due to these stresses. As for the crack arrest temperatures, it is very often considered that they are slightly influenced by the residual stresses only, given the preponderance which, here again, is exerted by the local metallurgical factors.

2.3.2. Effects of a postweld heat treatment

In addition to the relieving of the welding residual stresses a heat treatment below the AC_1 temperature generally produces a softening of the base metal and can also have an influence more or less pronounced on the resistance to the britte failure. The effects are however very different according to the grades and qualities, on the one hand, and according to the welding conditions on the other hand.

In the Carbon-Manganese steels, it can be observed a carbide spheroidizing and a growing of the ferritic grain size if the holding time of the heat treatment is long, at high temperatures, exceeding 625°C. These treatments can result in a reduction of the yield strength and of the tensile strength; reduction so significant that it is necessary to take it into account, mainly if the temperatures exceed 600°C during working of the piece. To be mentioned that a heat treatment of one hour at 600°C results in a diminution of the yield strength of about 2 %. With regard to the resistance to the brittle failure, the effect of the heat treatment on the change in the transition temperature of the base metal is of minor importance. This is not the case for the strain-hardened and the heat affected zones.
The semi-killed steels are very sensitive to the phenomenon of "stress-temperature" embrittlement due welding operations. The aging produced by a strain-hardening at temperatures ranging from 150 to 350°C results in a high embrittlement of the steel and, in presence of defects and of high tension residual stresses, gives rise to brittle fracture under low stress level. A heat treatment at temperature above 450°C can, in that respect, produce a very evident restoring.

Moreover, if a heat treatment is applied, at a temperature below AC_1, to zones having undergone a critical strain-hardening (4 to 5 %) which, usually, do not require normalizing treatment, this can result in grain growth. Therefore, the annealing of steel grades sensitive to this degradation must be performed at a lower temperature.
With regard to the structures existing in the HAZ, the heat treatment always leads to a more or less pronounced softening. The thermal cycle of welding is such that in most of the cases, the HAZ keeps, after the heat treatment, a yield strength and a tensile strength higher than those of the base metal after tempering. The heat treatment has nearly always a beneficial effect on the behaviour with regard to the brittle fracture; because, in Carbon-Manganese steels, the embrittlement in the HAZ after welding is often linked to the presence of quenching constituents. However, in case of a coarse grain s tructure, the effect of the heat treatment is negligible.

In the case of micro-alloyed steels, the evolution of the mechanical properties during the heat treatment is conditioned by the solubilization or precipitation of the carbides and of the nitrides. Indeed, it has been observed that the solubilization of the nitrogen and of the carbon increases when the treatment temperature increases. For the Niobium steels, it is impossible to discriminate the respective percentages of the so concerned nitrides and carbides.

Nevertheless, it is well known that the niobium carbonitrides dissolve a little or not at all below temperatures of about 950°C.
For vanadium steels, it can be observed a resolubilizing of the nitrides from fairly low temperatures, in some cases, from 550°C. In both cases (Vanadium or Niobium Steels), such an evolution can be hindered by the presence of deoxidizing elements : depending on whether an aluminium-killed, or a silicium-killed or semi-killed steel is concerned, the process of nitrogen evolution largely depends upon its affinity for the elements in presence.
In other respects, it is known that sufficiently long holding times also produce resolubilizing of these precipitates.
Those various evolutions generally result in a slight diminution of the yield strength and an increase, sometimes significant, of the Charpy V transition temperature. Nevertheless, this last effect is largely counterbalanced by a transition temperature which, originally, is nearly always very low for this type of steel.

During welding, the austenitizing occuring in the HAZ is accompanied by a massive solubilizing of the precipitates such as niobium or vanadium carbonitrides.

Further, during a stress relief heat treatment, a precipitation of carbides or of carbonitrides occurs, which can lead to an embrittlement of a thin zone close to the fusion line. This phenomenon is easily oberserved by using simulation samples. Due to the fact that this zone is very small, the heat treatment generally produces an improvement of the global behaviour of the actual joint by reducing the size of the embrittled zone. However, if the joints are welded with a process involving a high energy input, this improvement is not evident.

The low alloy steels undergo complex phenomena attributable to their chemical composition and the effects they undergo prior to the heat treatment after welding. The Cr-Ni-Mo steel being generally supplied in normalized or in normalized-tempered condition, their characteristics after relieving will essentially depend on the initial state (chemical analysis, heat treatment, microstructure), on the temperature and holding time of the tempering treatment.
Already, in the base metal, different effects appear.

- It can be expected that a heat treatment performed at a temperature 30 to 50°C lower than the tempering temperature of the metal will only have little effect on the metallographic structure, the mechanical properties and the impact strength.

- A temperature closer to the tempering temperature, held for a long time, will lead to a structure less acicular by a more pronounced polygonization of the quenching structures; this practice which produces a gradual decrease of the tensile strength with, generally, an improvement of the resistance to brittle fracture, is nevertheless of delicate application because of overtempering risks. Integrating the parameters of the stress relief annealing into the notion of tempering parameter appears, in this respect, to be interesting provided one can be sure of the temperature levels and of the exposure times effectively reached by the steel during its production and also during the successive treatments of manufacturing.

- An irreversible embrittlement of the Cr-Ni-Mo low-alloy steels can occur after a sufficiently long holding time at the treatment temperature. It results in a sharp decrease of the impact strength. This decrease meets the Larson-Miller's law and can appear in the case where several successive stress relief annealings are required during the manufacturing by welding of very thick vessels.

- At a lower heat treatment temperature - generally about 450°C - it can sometimes be observed an other type of embrittlement characterized by an intergranular cracking. This embrittlement however decreases sharply when the temperature increases. It concerns a phenomenon, essentially reversible, which can be damped by a higher holding temperature.

The effects of the heat treatment after welding in the fused metal can result in degradation of the impact strength. In case of Ni additions (2 %) foreseen to increase the impact strength at low temperatures (5), carbide precipitations are produced after stress relief annealing, which are associated to a more or less pronounced presence of a martensitic-austenitic phase. Such effects are damped by the duration of soaking, at the highest treatment temperatures and by a sharper cooling of the piece (6). In other words, attention must be given to low cooling rates but also one can have doubts about the principle of equivalence applied for the process qualification where the effect of several successive heat treatments is simulated by one single treatment of longer duration.

But, it is mainly the heat affected zone which generates the most significant preoccupations.
The reheat cracking is a subject of preoccupation on which IIW has produced an important work of synthesis in the last years (7-8-9).

The phenomenon is defined as an intergranular cracking in the heat affected zone or in the fused steel of a welded joint, being initiated during a heat treatment or during service at a sufficiently high temperature.
The stainless steels, the first ones, have caught the eye on its existence, since the mid-fifthies.

The works, at this time, concerned cases of cracking found in the steam pipes in steel of type AISI 347 niobium-stabilized. Then, it was the turn of the Cr-Mo ferritic steels in the welded assemblies of superheater tubes in the sixties. Afterwards, it has been realized that "so-called" reheat cracks could arise in the zone, affected by the welding, of low-alloy steels used in the manufacturing of pressure vessels as well as in the underlying zone under the stainless steel cladding of such low-alloy steels. Finally, it appeared that the fused metal was not exempt from any danger, particularly in the case of claddings of the 1 1/4 Cr - 1 Mo type.

In general, the risk of reheat cracking of a metal having undergone a welding cycle is attributed to the following mechanism.

During the welding of a seam, the thermal cycle, applied
to the HAZ of the base metal or to the weld metal being
deemed sensitive, produces at the grain boundaries a
segregation of a solution supersatured in carbon and in
impurities. The same zone is subject, during the
depositing of the next seam, to a thermal cycle of lower
amplitude which, in a temperature range from 450°C to
AC_1, induces an intergranular embrittlement by strain
hardening.

During the stress relief annealing, the capability of the
concerned zone to withstand, possibly weakened by a
tempering embrittlement, is by far lower than the one of
the grains, particularly in steels comprising Cr. The
present residual stresses induce, in them, expansions
incompatible with the capability of deformation of the
concerned zones and cause a fracture even before they are
eliminated.

2.4. Effects on the safety from fatigue

The presence of fatigue cracks in a welded structure
consitutes a danger that must be avoided to a maximum.
The residual stresses influence the fatigue behaviour of the
welded assemblies because the local stresses result from the
superposition of the welding residual stresses and of the
variable stresses in service.

---

(5) J.E.M. Braid, J.A. Gianetto - The Effects of PWHT on the
     Toughness of Shielded Metal Arc Weld Metals for Use in Canadian
     Offshore Structure Fabrication - Doc.IX.1585.86/X.1101.86.

(6) D.Cartoud, S.DEBIEZ, Institut de Soudure - Novembre 1984.

(7) A.Dhooge, A.VINCKIER - Stress Relief Cracking - Status Report -
     Doc.IX.1250.82.

(8) I.Hrivnak et Al - Mathemactical Evaluation of Steel
     Susceptibility for Reheat Cracking - Doc.IX.1346.85.

(9) A.Dhooge, A. Vinckier - Reheat Cracking - A Revieuw of Recent
     Studies - Doc.IX.1373.85.

In the endurance range, for the welded and not stress relieved
assemblies, the variable stresses actually present are
characterized by values – maximum Smax and minimum Smin given
by the following expessions :

$$Smax = Re$$

and $\quad$ Smin : Re - dS

where Re is the yield strenght and dS, the stress range
in service. If the ductibility of the material, where a
fatigue crack can possibly appear, is sufficient so as the
plastification, due to the first stress cycle, develops
without cracking, the variable service stresses can be
defined by the single stress intensity value dS.

In the endurance range, it has been experimentally shown
that the relief of the welding residual stresses allows an
increase of the endurance limit which is 20 % at the maximum
and very generally ranging from 10 to 15 %. A significant
diminution, or even a suppresion of the notch effect is, in
this respect, by far more efficient than a stress relief
treatment. In the case of the oligocyclic fatigue, the
unrelieved residual stresses play a small roll because they
are redistributed during the first cycles of plastic
deformation.
One of the dominating factors of the resistance to cracking
becomes the capability of the material to undergo cyclic
plastic deformations each of which deprives the material of
a part of its ductility reserve. This ductility reserve can,
in some cases, be strongly improved by a heat treatment of
metallurgical restoration judiciously applied.

2.5. Effects on the corrosion resistance

The manufacturing of welded structure has shown that some
types of corrosion exist which are specific to the welding.
In this way, when the circumstances are unfavourable, an
important selective attack of the base metal can sometimes
be observed in the vicinity of the weld seams.

Among the main factors governing this phenomenon, one can
mention :

- the nature of the base metal.
- The nature of the weld metal.
- The nature of the corrosive agent, its concentration, its
  temperature, its agitation.
- The macroscopic as well as microscopic heterogeneity of
  the materials in presence (defects, grain
  boundaries,voids, dislocations...).
- The welding conditions, since the preparation of the
  parts, welding sequence, etc... till the preheating and
  the final stress relief annealing.

- The stress state.
- The surface finish.

Besides the main factors governing this phenomenon, the macroscopic and microscopic stresses (the first ones are generated by the welding and the external loadings, the second ones originate in the metal structure itself) can strongly influence the importance of the selective corrosion of the base metal in the vicinity of the welded joints. The stress concentration at the surface defect locations, even of small importance, accelerates the process. Combining their effects with the corrosion itself, the residual stresses can generate the phenomenon of corrosion cracking so redoubtable as far its effects, particularly in the case of high yield strength steels. Here again, the stress relief annealing, when feasible, allows to avoid the growth of a dangerous stress corrosion. A particular case where the stress relief treatment reveals oneself particularly efficient is the case of the sensivity to hydrogen cracking, the so-called delayed cracking. The effect of the postheating immediately after welding is to accelerate the diffusion of hydrogen out of the dangerous structures. The effect of the high temperature is to perform a tempering of such sensitive martensitic structures.

2.6. Effects on the global stability

The boilers, pressure vessels or tubular constructions are rarely subject to instability phenomena, because of the type of applied loads and of the stiffness of the parts. In this way, the existence of residual stresses has, generally, no importance on the calculation of their components.
However, it can exist some circumstances where the global stability of a welded equipment poses a problem. In the constructions having middle or high slenderness, when the stresses in service are added locally to residual stresses of the same sign, the yield strength of the material can be reached prematurely and the corresponding local plastic deformations tend to reduce the stiffness of the element. Here, it turns out a reduction of the instability load of the element which must be taken into account by using the buckling curves for the buckling and warping phenomena, or by a specific calculation method, for the calculation of compressed flanges of the casings. The annealing after welding can suppress the residual stresses which are, locally, of the same sign than the ones in service and so to supply an efficient remedy for the decrease of instability loads associated to the plastic deformations. The efficiency of this remedy is, in some cases, explicitly given in the calculation standards. However, the performed annealing must be such that it does not generate differential expansions due to an excessive thermal gradient during the heat-up, because the resulting distorsions could be more detrimental than the residual stresses we would try to eliminate.

We are faced with this problem particularly in the case of
treating structures comprising pieces of very different
thickness.
However, on a general basis, it can be said that the effect
of the annealing on the global stability is not always the
same and that each particular case asks for a appropriate
study.

***
*

3. ANALYSIS OF THE PRESCRIPTIONS OF THE CODES AND OF THE REGULATIONS

The International Institute of Welding has, several times,
examined the regulations and Codes in force concerning the stress
relief annealing (10-11-13). The roll that the IIW plays in the
international standardization is, this must be mentioned,
determinant. This roll has been recently reaffirmed by the ISO,
in taking a position establishing the IIW as official agency
entitled to preapir the projects of international standardization.
Given the wide variety of the expressed points of view in the
different codes and regulations in force, it is not surprising
that the numerous authors have applied oneself to compare their
prescriptions according to the applications, the steel grades and
the aimed objectives (12).

It is also not surprising that countries member of the IIW tried
to conciliate their points of view, relying upon the
metallurgical considerations which, more and more, have been made
mandatory by the use of new families of steel.
A preliminary remark asserts oneself. A code consitutes an
homogenous entity, to be used globally. It is difficult to
extract one part from it, or to envisage parallelisms, omitting
the general context of the document as well as the provisions of
other articles.

Although the examination of the fields of application foreseen
by the codes reveals a wide variety of points of view, it appears
rather clearly that the stress relief heat treatments are
mandatory in the case of equipments undergoing risk of stress
corrosion, or risk of brittle fracture or fatigue failure in
hydrotest or service conditions.

The analysis leads to keep four essential criteria of comparison :

1) The allowable maximum thicknesses without heat treatment.
2) The heat treatment temperature.
3) The minimum holding times at maximum temperature.
4) The maximum heating and cooling rates, the maximum loading
   and minimum unloading temperatures.

The analysis also leads to distinguish the C. or C-Mn steels, the Cr-Mo low-alloy steels and the Ni alloy steels and foreseen for applications at low temperature.

3.1. Critical thickness

The wall thickness is a derminant factor in the prescriptions of the codes for the application, to the pressure vessels, of heat treatments after welding. Indeed, below a certain tickness, the treatment may be omitted, whereas it is mandatory above a certain limit.

The examination comparing different codes reveals a wide dispersion in the determination of that limit tickness, whatever the steel type may be. One can wonder about the significance of a "go-no go" value. But one may indeed also ask oneself about the scientific or technical bases which have presided over the choice of such arbitrary values. In table 1 are gathered together the requirements of some codes applicable to quality steels for pressure vessels.

The C and C-Mn steels may not be subject to a heat treatment after welding when the wall thickness is lower than a value that, according to the codes, ranges from 20 up to 50 mm. Higher values may however be accepted without treatment if it is demonstrated that there is no risk of brittle fracture, and that the steels are fine grain steels.

A tendency appears to aline the limit values with a thickness of about 35 mm, under which the heat treatment would be (optional, facultative).
Concerning the Nb and V microalloyed steels, a working group within the IIW is getting on the definition of reflection (ways, leads) (13). It is sure that new decision factors, speciallu the heat input of welding, will have to intervene in the definition of the limit of thickness up to which a heat treatment is not required.

The Mo, Cr-Mo, Cr-Mo-V low-alloy steels are also subject to prescriptions, very different according to the codes.
If it is observed that, in general, the maximum limit thickness is reduced gradually as the alloying content of the steel increases, it must be recognized that the widest condusion prevails for the C-Mo steels. Here again, a tendency appears to aline the values with unified tchickness, above which the heat treatment is mandatory and below which it is facultative.

TABLE I

STEELS	ASME VIII/2	BS 5500	AD HP 7 Re < 370	AD HP 7 430 > Re > 370	ISO DIS 2694	NBN F11-001	Rules NL	NF 32-105 Rm <560 Mpa	NF 32-105 Rm <700 Mpa	DIN 17155	IIS Rec
C et CMn	32 38	35 40	30 38 50	30	30 38 50	20 30	32	40	30	30	35
0,35 Mo	16	20	30-38-50	30	30-38-50	30	32				20
0,5 Mo	16	20	30-38-50	30	20	30	0				20
1 Cr - 0,5 Mo	16	0			15	15	0				0
2,25 Cr - 1Mo	0	0			0	0	0				0
Cr Mo V	—				0	0	0				15

3.2. Temperature of the heat treatment

The thermal cycle of the heat treatment after welding is governed in the codes by :

- The maximum loading temperature.
- The maximum heating rate.
- the holding time.
- The maximum cooling rate.
- The maximum unloading temperature.

The holding temperature is considered as the leading factor to obtain a good stress relief.
The upper limit of the temperature is governed by the resistance to deformation of the non supported portions. The holding time at this temperature and the contribution of the heating and cooling phases being taken into account, it must be admitted that it is also the holding temperature which is the driving factor concerning the metallurgical effects of the stress relief treatment.

We always mean by temperature the one that the treated material has to reach, and not the temperature of the furnace atmosphere. In function of the geometry of the pieces to be treated and of the geometry of these furnaces, the know-how of the practician allows to reach the optimum conditions.

The thermal regime of a construction or of an assembly subject to a heat treatment in a furnace is not uniform at each of its points, even if the furnace would have on homogeneous distribution of the temperature in the time and in the space. The thinnest portions of the assembly tend to follow the imposed temperature regime. The thickest portions, on the contrary, are characterized by a important thermal inertia, either during the heating period, or during the cooling period.
Moreover, these same portions may be subject to thermal gradients between their surface and their core.

---

(10) Opportunités du recours à des traitements thermiques de relaxation - Doc.IX.1059.77/X/913.78.
(11) Traitements après soudage des appareils à pression en aciers - Etat de la question - Doc.IIS.798.84, ex Doc.XI.420.84.
(12) Traitement thermique des constructions chaudronnées - colloque Fabrication-Soudage-Contrôle - Paris, Octobre 1983.
(13) WG Influence of Stress Relief Heat Treatment. Statut report Doc IX.1359-85.

Let us consider the simplified case of two separate plates of large dimensions and of thicknesses e_1 and e_2 subject to a cycle of heating, holding at a temperature T and cooling. During the heating, the difference in the instantaneous temperatures between the plate cores increases with the difference in thickness and the rate of heating the furnace. Those differences can be important and lead to compression thermal stresses in the thinnest bracings of an actual assembly (in which however the phenomenon is damped by the thermal conductivity acting via the connection between the thin bracings and the thick bracings).

When the bracing or the thinnest plate has reached the maximum temperature of the treatment, this temperature must be maintained a sufficiently long time to allow the thickest bracings to reach in their turn the wished temperature level.

This indispensable holding time is function of the loading conditions and of the heating rate of the furnace, and of the relative thicknesses of the pieces to be treated. If it is foreseen that the heat treatment will be performed at a high temperature, at which a nearly complete relieving of the residual stresses occurs in a very short time, it is advisable to evaluate the duration of holding which corresponds to a complete homogenizing of the temperatures. This enables to evaluate the dangers of irreversible embrittlement which are likely to intervene, and the risks to see this duration to correspond with the one of a minimum of restoration of the embrittled zones.

If it is foreseen that the heat treatment is performed at lower temperature, but with a longer holding time, the duration of temperature homogenizing will not be highly reduced, but its relative influence on the long range metallurgical degradation will be weaker. During the cooling in the furnace, the thermal gradients between the most massive portion and the thinnest bracings are low. When it is proceeded to withdrawal at, for instance, 300°C, the thinnest bracings are cooling down sharply and therefore are subject to tension. If the thermal gradients are important, these bracings can reach, in some of their weak points, a stress level higher than the yield strength at the temperature at which occur the maxima of temperature differences.
This results in a plastic deformation and, at the end of the cooling the coming out of residual stresses. Moreover, the stress states during cooling are capable of causing cracks starting from defects or microcraks in the vicinity of which an irreversible embrittlement or a recrystallization may have occured. Cases of cracking in thick pieces during a stress relief annealing may certainly be attributed (beside the reheat cracking) to excessive temperature differences either during the heating, or more likely during the cooling.

Hence, it is advisable to proceed to continuous temperature measurements by thermocouples on the product itself and especially in the zones between which the largest temperature differences would occur during the heating and the cooling; the thermocouples must be effectively protected. The measurement in course of heating allows to appreciate the maximum temperature difference reached in transient phase, and the duration that must be met at maximum temperature to reach the required degree of relieving, or, when the case occurs, the most appropriate restoration peak.

The measure in course of cooling gives an indication about the maximum thermal gradients obtained and about the corresponding stresses; and, therefore, about the localization of the zones where crackings can be suspected. However, these measures supply only indications about the temperature at the surface of the products, and, in addition, the thermal gradients between the skin and the core of these products must be taken into account.
Such measurements are usefully completed by a prior estimation of the most unfavourable conditions of the treatment. Such an estimation is possible by mathematical way, knowing the relative thicknesses concerned, and the general programme of the treatment. Especially, it can be very well envisaged to dispose of abaci giving for the most unfavourable case, the regimes of heating, of homogenizing and of cooling in the furnace, and the adequate temperatures of loading which correspond to allowable values of the transient thermal stresses.

Coming back to the comparative examination of the codes, it can be observed that these ones are characterized by the largest scatter in the temperatures to be reached.

The C and C-Mn steels generally present a stress relieving ranging from 80 to 90 % for treatments ranging from 550 to 575°C. Therefore it is not surprising that the recent codes recommend fairly low treatment temperatures consistent with to these results (ISO, CODAP). It is also in this way that IIW took position as early as 1977.

Nevertheless, in general, the holding temperatures are ranging from 530 to 650°C as shown in the table II.
The Yougoslavian code recommends that the temperature be lower than 560°C. A temperature range comprised between 530 and 580°C is recommended :

- by the AD Merkblatt according to SEW089 for the steel with a yield strength lower than 370 Mpa, and for the fine grain steels for thicknesses above 30 mm.

- By the CODAP 80-F2 for the C-Mn Steels and V or Nb microalloyed steels, up to tensile strengths of 530 Mpa, for thicknesses above 40 mm, and up to tensile strengths of 590 Mpa for thicknesses above 30 mm.

- By the DIN 17155 for the STE 460 steels.

In a temperature range slightly higher, one can identify :

- 550°-625°C in the rules of french conception PWR.S1000 applicable to steels of every category with a tensile strength of lower than 500 Mpa and thickness above 35 mm.

- 560°-600°C in the ISO DS 2694.

- 550°-620°C in the DIN 17155 for the ST48-52 steels for thicknesses above 30 mm, the upper limit being reduced down to 600°C for the STE690 steels.

- 580°-620°C in the Belgian standard NBN 629, in the BS5500 for the steels of M0 and M1 categories for thicknesses above 35 mm.

In other respects, the American code ASME VIII recommends, for the P1 steels for thicknesses of 30 mm and above (40 mm if there is a preheat above 95°C), a temperature of 593°C.

The Dutch code imposes a minimum of 600°C, so does the Japanese code for the P1, P2 and P3 steels according to the standard JIS B 8243.

In the range of the highest mini-maxi temperatures, one meets :

- AD Merkblatt according to DIN for the steels ST37 to ST45 : 600°-650°C. The Czechoslovak standard CSN 420284 which recommends the same range, but, interesting fact, which takes into account the types of deposited metals, and lets raise up to 620°C the lower limit when one uses fused alloyed metal (for instance Mo).

- The Soviet codes OP 02CS-66 and RTM-1S-73 which impose a range of 630°-660°C and 650°-680°C respectively.

Owing to the considerations developped at the beginning of the lecture, one understands that the tendency to present, on the market, fine grain steels and steels more sensitive on the metallurgical field, incites to lower the mini-maxi temperatures of the treatment, even if the permanence of a certain level of residual stresses in conceded. However, it is fairly rare that one cares about the heat input of welding, or about the incidence of the fused metal.

TABLE II

CODE	COUNTRY	T° C
CODAP 80F2	F	530° - 580°C
DIS 2694	ISO	560° - 600°C
BS 5500	UK	580° - 620°C
ASME VIII	USA	593°C
AD M DIN	RFA	600° - 650°C
+ SEW 089	RFA	530° - 580°C
JIS B 8243	J	600°C
RULES	NL	600°C
NBN 629	B	580° - 620°C
CODE	Yu	< 560°C
CS 420284	Cs	600° - 650°C
OP-02CS-66	URSS	630° - 660°C
Doc X-861-77	IIS	540° - 580°C

The Cr-Mo and Cr-Mo-V low-alloy steels are in general heat treated after welding, essentially in order to improve the toughness of the fused metal and of the heat affected zones. The examination of some codes shows, here again, fairly large discordances, but which vanish when the Cr content increases (3 % Cr or 5 % Cr).

The tendency is to the increase of the minimum temperature of the treatment with regard to ASME VIII, chich recommends 593°C for the steels 1/2 Cr-1/2 Mo, Cr - 1/2 Mo and 1 1/4 Cr - 1/2 Mo-Si and 675°C for the steels 2 1/4 Cr-1Mo and 3 cr-1Mo. The CODAP recommends respectively 630°-680°C and 670°-710°C, and the ISO 620°-660°C for the steels 1/2 Cr-1/2 Mo and 1 Cr-1/2 Mo and 625°-750°C for the steels 2 1/4 Cr - 1 Mo, what constitutes an enormous range, to be compared with the narrow range of 600°-650°C of the AD Merkblatt, for the same steel.

The C-Mo steels also feature a remarkable scatter in the ranges of maxi-mini temperatures :

- 593°C minimum for the ASME
- 580°-620°C for the ISO/DIS 2694 and the CODAP
- 600°-650°C for the NBN F 11-011 and the AD Merkblatt
- 650°-680°C for the BS 5000.

As for the Ni steels, if everyone recommends the same treatment, for the steels having a Ni content < 3,5 %, as for the unalloyed steels, the recommendations relative to the 3,5 % Ni steels are not homogeneous :

- 550°-610°C for the ANNCC (Sd 5/1) if more than 15 mm
- 580°-620°C for the BS 5500 thickness optional
- 594°-635°C for the ASME VIII-2 (P9b) if more than 16 mm.

The steel with 5 % and 9 % Ni present a risk of reversible embrittlement due to tempering, in such a way that the codes only foresee a heat treatment above a thickness of 50 mm (Italy 40 mm for the 9 Ni and 30 mm for the 5 Ni, but derogations are possible).

The IIW proposes not to proceed to a heat treatment after welding for the steels with 5 % and 9 % Ni, and to assimilate the steels of which Ni content is < 3,5 % to C-Mn steels for low temperatures.

3.3. Minimum holding time at maximum temperature

There is a reciprocal relation between the maximum temperature necessary for relieving the stresses and the duration of exposure at this temperature. In this respect, one uses the Hollomon-Jaffé parameter and the parameter derived from the diffusion laws (14), which allow, in other respects, to assimilate an actual thermal cycle to an equivalent rectangular cycle, and to judiciously use the heating and the cooling periods in order to reduce the exposure duration at maximum temperature, when the code allows it. The prescriptions of the codes and of the standards generally associate the exposure duration with the thickness and the steel grade.

For the C or C-Mn steel, it is observed that several codes (CODAP, AD Merkblatt DIN ans SEW, AFNOR S 1000, NBN F 11-011, ISO DIS 2694, etc) agree to adopt a duration of 2 min/mm in a thickness range comprised between the minimum thickness requiring a treatment and about 60 mm.

---

(14) T. Maynier, M. Toitot, J. Dollet, P. Bastien Mem. Sc. Revue de Metallurgie n° 7/8Q - 1972.

For the lowest thicknesses, the codes often impose a minimum holding time (20 to 60 min.), when the treatment is necessary. Above 60 mm, some codes reduce proportionally the holding time. The code BN 5500 imposes longer holding values (2,5 min/mm), but admits the possibility to integrate the heat-up and cool-down operations.
The lowest data correspond to the highest temperatures ranges required by the codes (ADN - DIN, CS 420284).

The microalloyed steels do not require specific prescriptions. As for the low-alloy steels, they are subject to prescriptions of the same order, but the minimum treatment holding times at the required temperatures are increased for some codes and according to the steel grade.

It will be observed, as an anecdote, the disparity between the codes BS 5500 and CODAP, which for the steel 2 1/4 Cr 1 Mo and, for the same holding time above 72 mm, impose minimum durations of respectively 30 min and 180 min for thicknesses lower than 15 mm.

3.4. Heat-up and cool-down

During the heat-up, the first phase of heating at about 200°-400°C presents a particular importance.
Indeed, it is the temperature range in course of which the internal stresses due to the differential expansions are added to the residual stresses in the transformation microstructures which are not yet in tempered state (martensite, bainite). Therefore, it is
essential to make sure that the heating rate be low, at least up to 300°-400°C, particularly in case of thick walls and when the danger of sensitive microstructures and important differential expansions increases.

The maximum heating rates imposed by the codes are fairly homogeneous between 25 and 100 mm; they are expressed in °C/h in the form $v = 5550/e$ (e in mm) or 200 to $220/t$ (t in ").

The discordance appears below 25 mm :

- max 200°C/h for ISO DIS 2649
- max 220°C/h for CODAP 80-F2, BS 5500 and JIS Z 3700-80
- max not determined for NBN, ASME,...

and to a smaller extent for the allowed minimum rates :

- min 55°C/h for ISO DIS 2649, BS 5500, NBN F 11-001,
- min 50°C/h for JIS Z 3700-80,
- min 38°C/h for ASME VIII.

It must be mentioned that the codes do not contain
prescriptions about the heating rate from the room
temperature up to the maximum furnace temperature during the
furnace loading, which is mostly limited to 300°-400°C.

It must also be mentioned that the maximum temperatures of the
treatment being comprised between 540°C and 650°C, the
necessary durations to reach the precribed temperatures may
be fairly different, with the economic problems which can be
imagined.

Consequently, it is reasonable to take into account as much
as possible, as the BS 5500 does, the heating phase of the
piece in the global balance of the treatment.

***
*

4. SURVEY OF THE INDUSTRIAL PRACTICE OF THE STRESS RELIEVING AFTER WELDING.

The stress relief after welding is performed by thermal processes
and by mechanical processes. The thermal processes generally
include three families of treatments :

- The global treatments, through wich the whole structure is
  uniformly heated, each of its points theoretically reaching the
  same temperature at the same time.

- The uniform local treatment, where only the welds or the
  contiguous zones are heated up to a temperature considered
  uniform at every point.

- The progressive local treatments, where the welds and
  contiguous zones are heated by a moving source, what leads to a
  temperature varying at every point.

As for the mechanical processes of stress relieving, they
essentially include :

- The treatment of overstressing, corresponding to a brutal or
  slow plastic déformation of the metal, obtained by explosion or
  by a progressively increasing mechanical loading.

- The vibration treatment leading to a progressive relieving
  of the stresses trough successive mechanical hysteresis cycles.

These processes may be applied globally or locally.

- The surface local treatments resulting from the shot peening
  or the hammering, the effect of which is to redistribute the
  surface stresses.

4.1. Processes of thermal relieving

The treatments either global or by sections are applied in the annealing furnaces which are designed taking into account the following points :

- Accuracy and sufficient regularity of the temperature at every point.
- Adjustement easiness.
- Easiness for loading and unloading.
- Possibility of automatic recording of the temperature.

In general, the heating of the modern furnaces is performed by means of flame (fuel or gas) or of electricity.

- Flame furnaces : heating by burning gas, in general by means of liquid or gaseous fuel burners.
- Electric furnaces : using electric resistances or induction heating. The low-band induction heating has the advantage to heat the piece within the mass, what eliminates some of the thermal gradients. The use of radiant resistances allows to design modular type furnaces, particularly adapted to energy savings.
- Bath furnaces : for particular treatments.

The heating of the piece may be performed in the direct atmosphere of the furnace or in a chamber possibly filled with a determined atmosphere (treatment in atmosphere).

According to the design, one distinguishes :

- The roll-over type furnace, tunnel-shaped, with a door at each end, which allows a continuous treatment.
- The rotary furnace (with circular hearth) : type generally conceived for the treatment of small pieces.
- The bogie hearth furnace : as the roll-over furnace, it can be provided with two doors which allow to perform treatments of pieces the length of which is larger than the length of the furnace (repeated partial treatment). The bogie hearth only serves to facilitate the loading and the unloading of heavy pieces.
- The furnaces of special design such as removable cover furnace or muffle furnace.

Regarding the temperature control, one disposes of pyrometric devices, of such accuracy and sensitivity that they lie far beyond the requirements and tolerances of temperature control of an annealing furnace.

The question of the temperature adjustment inside of the furnace is much more critical. This adjustment may be performed globally or by distinct zones.
The treatment cycle is feasible by manual control or without attendance of the operator (fully automatic furnace).
A well designed automatic furnace allows, in no-load condition, a temperature steadiness with an accuracy of 5° to 10°C about the temperature set-point in any zone of the furnace. In some cases, the thermal inertia of the furnace plays an important rôle regarding the temperature steadiness and the heating and cooling rates. Nevertheless, let us note that the control and the adequate adjustment of the furnace temperature do not ensure the temperature evenness at any point of the piece, especially when this one consists of elements of very different thicknesses. It has been previously shown that differences can occur between the furnace temperature and the piece temperature. This difference is function of the heat-up and cool-down rates, of the ratio between volume or mass to be annealed and the volume or capacity of the furnace and of its thermal insulation. Measuring the temperature by means of thermocouples duly gauged, fitted to the construction itself and properly protected is useful, indeed indispensable, according to the category of equipment.

The design and the quality of the furnace, as well as the know-how related to this furnace that one can get, are significant factors concerning the control and the temperature steadiness.

The automatic recording of the progress of the treatment is aimed at ensuring that the equipment has effectively undergone the foreseen treatment.

Other causes of important preoccupations are :

- The location of the piece in the furnace in order to obtain a regular circulation of the calorific fluxes around the piece.

- The protection of the construction against the direct contact of the heat source or against an excessive oxidation.

- The furnace tightness or the protection of the portions partially coming out of the furnace (limitation of the temperature gradient and of possible deformations).

Finally, it must be taken into account of the possible deformations of the construction during the heat treatment; therefore, it is important to provide adequate supports.
A variant of the global treatments applicable to pressure vessels consists of an arrangement where the equipment is insulated at the outside and heated inside by every means allowing to meet the conditions foreseen in the used codes and applicables to the case of global treatment by external heating.

When the global treatments are applied by successive sections because of the incompatibility of the furnace dimensions with those of the piece, one recommends an overlapping length of about 1,5 m and an insulation of the portions located outside of the furnace to avoid a too sharp thermal gradient. Moreover, those portions may not include internal partitioning baffle, bottom head or tube sheet.

In some cases, one prefers to proceed to global treatments of separate sections followed by their final assembly and by their final local heat treatment.
A width of treatment is required to be at least three times the wall thickness and the insulation is required to have a width of at least six times the highest wall thickness value.

The local heat treatments are essentially applied to the axisymmetric bodies. They are performed by means of flame or electric heating devices and, in their modern version, work by heating the air, by radiance or by induction. The minimum width over which the heated zone has to extend on both sides of the zone to be treated is governed by certain codes, particularly the French code which fix this width to four times the wall thickness. The insulation extends, as for it, on both sides of the zone to be treated over a width of ten times this thickness. Other codes, as the Dutch code does, recommend the localization of the thermocouples of temperature control in the center plane and at defined distances from that plane, as well as the allowable temperature differences between the different points of measurement.
The industrial applications of local heat treatments are numerous and very diversified : they go well beyond the preoccupation of relieving the welding residual stresses and cover the whole temperature range from the preheat or postheat up to the normalizing annealings.
The performances reached by the stress relieving of welding stresses by local heating must be evaluated not only from the relieving point of view. Significant improvements of the toughness of the welded joint, comparable to those produced by global treatment, may be recorded, even if, after all, the welding residual stresses have not been totally eliminated.

***
*

4.2. Mechanical stress relieving process

In the case where it appears to be unuseful, not very practical or inadequate to proceed to a stress relief by thermal means; it can be made use of the plasticity properties of the material at room temperature to eliminate the peaks of residual stresses.

We note three categories of processes :

- The global or local overtressing, applied slowly or sharply, intended to plastify the stressed metal and to create, if necessary, a state of residual compression.

- The treatment by mechanical vibrations, during which the residual stresses are exceeded by alternate stresses which gradually modify their level.

- The surface local treatments resulting from the hammering shot peening or sand blasting, which create a zone of superficial plastic deformation, and associated compressive residual stresses, thus counterbalancing the preexisting tensile stresses.

The mechanical treatments by overstressing are mainly interesting for the equipments subject to pressure. Their justification lies in the verification that, during the pressure increase, the external loads are added to the internal stresses, producing local plastifications. During the pressure decrease, the elastic retention produces, in the plastified zones, residual stresses playing a rôle of favourable prestressing.
The adequate application of that treatment is easy to control because it is determined by the applied pressure.
In other respects, some regulations, like the French regulation, have foreseen to authorize the application of mechanical treatments with imposed stresses in substitution for a stress relief heat treatment, as far as the reached stress at every point of the welded joints is, at least, equal to 85 % of the yield strength at the test temperature. If it is mostly performed by slow pressurization. The overstressing may also be the local result of a mechanical straining produced by the explosion energy of a charge located for instance in the vicinity of the weld joints. Of a fairly particular type, this treatment presents interest only in a limited number of applications.

The stress relief by mechanical vibrations is applied for several years, mainly in order to stabilize the form of machined/welded assemblies. It is generally performed by fitting a variable frequency vibrator on an element of the structure to be stress relieved and adjusting the frequency to the oscillation frequency of that element till reaching the resonance.

This last condition, however, does not seem to be
compulsory (15). It is the cyclic nature of the applied
loading, and the associated Bauschinger effect, which is at
the origin of the stress relief. This explanation, however,
is simplistic and conceal a set of complex phenomena on the
macroscopic and microscopic planes.
To the scale of dislocations, it is likely that the mechanical
vibrations produce desanchorings and movements of the
interstitial atoms, leading to increase their mobility.
It has been observed (16) an improvement of the reduction of
area during the tensile test on specimens, beforehand
subject to vibration, what leads to think that the stress
relief by vibrations, beyond a simple stress relief, also acts
on the capacity of plastic déformation of a metal.
To the macroscopic scale, the metal undergoes a succession of
loading and unloading cycles, which is distinguishable from
the monotonous loading by reaching the plasticity much
earlier. This results in a stress relieving, and is
accompanied by a softening in the case of the ferritic
steels.

The applications of the stress relief by vibrations are
numerous. It can be mentioned, on a non limitative basis (17)
plate heat exchangers, machine casings, support frames of
precision elements, for which the heat treatments would
entail unallowable distorsions.
Considering the precautions for use, the objections which can
be raised about that process with regard to fatigue cracking
or brittle fracture initiation can be easily rejected.
However it must be recognized that the operational methods
of applications are badly defined and the effectiveness of
the stress relieving, which can be verified by a
modification of the current intensity necessary to impose a
vibration to the piece, has only a qualitative character.

As for the treatments by impact, which induce compressive
residual stresses under the zones where a hammering or a
shot peening is applied, they essentially have a local
character, and are mostly used to improve the resistance
to fatigue and to stress corrosion of the welded joints.
The stresses actually obtained are difficult to determine.

---

(15) P. Barbarin et al CETIM information n° 84 - Février 1984,
pp 43-48.

(16) I. Rak et Al, doc IX-1249-82

(17) G.G. Saunders, R.A. Clayton in "Residual Stresses in Welded
construction and their effects". Publ. the Welding Institute
London 15-17 Nov. 1977.

CONCLUSION

With regard to the preceding considerations, it appears that the metallurgical effects, produced by a so-called stress relief annealing, very often exceed the residual stress relief itself, beneficially or detrimentally.

It also appears that if the intended aim of relieving the stresses, or not creating new ones, imposes to the heat treatment a certain number of rules relating to the loading temperature, the heating rate, the maximum temperature of the treatment, the holding time at this temperature of the treatment, the holding time and the cooling rate, these same rules act on the integrity of the base metal, the fused metal and the HAZ.

Finally, it appears that the safe keeping or the reinforcement of the toughness, of the ductility of the welded joints of a construction takes, in most of the cases, precedence over a complete suppression of the residual stresses which would be accompanied by a degradation of these properties.

The International Institute of Welding has to continue to play its rôle in unifying the views about the merits and the appropriate application of the stress relief treatments.

A standardization of the codes is necessary, but it can not be accomplished in a short time. The positions taken by the IIW, in this respect, are clear : to lower the maximum treatment temperature. To reduce the exposure time at the maximum temperature at the risk of not fully relieving the stresses. To integrate the heat-up and cool-down cycles into the effect of the holding at the maximum temperature. To look into the perverse effects of non controlled temperature increases during the furnace loading.
To extend to the fused metal and to the heat affected
zone the estimation of the effect of the numbers recommended by the codes about the base metal. To adequate these numbers as a function of the steel grades implying microallyed steels and thermomechanically treated steels and as a function of their welding circumstances.
To continue the basic analysis of stress relieving.

The field is wide and the goodwill is general.

Let the Commissions proceed in their work !

Mechanical Aspect of Stress Relieving by Heat and Non-heat Treatment

V. A. Vinokurov

(USSR National Welding Committee)

ABSTRACT

The advisability of different methods stress relieving use depends on the kind of residual stress influence on serviceability of welded structures, especially if this influence is connected with metal properties and stress concentration. The examples have been given. On the basis of stress relaxation studies stress relieving regularities and heat treatment ways of optimization have been determined.

The General Substantiation of the Problem

Residual stresses are virtually elastic deformations which possess some potential energy accumulated in a body. The main feature of residual stress relieving is as follows: they can be avoided only through metal plastic deformation. Where, how and when to develop deformation are those important points of the above-mentioned problem. The choice of residual stress relieving method greatly depends on the kind of removed residual stresses negative affect on the weld structure. There are cases when by decreasing residual stress we give rise to negative after-effects which we could avoid by stress relieving. It is the discussion of stress relieving problem that is closely related with that of residual stress effect and welding in general on the service properties of welding structures.

Among many factors which interact with residual stresses the metal property factor must occupy the first place and also their inhomogenity and reaction of the metal on residual stresses.

As residual stress relieving is due to plastic deformation so one of the main factor is also plastic deformations concentrators availability. Discontinuities often take place as concentrators in the welded joints. The cases dealing with failure make us consider concentrators as one of the most important factor of stress affect and stress relieving. It is important to mention economico-technical factor. Producers and byers of welded structures are interested in stress relieving to be effective both from technical and economic point of view.

Dependence of Stress Relieving Method Role from Influential Factors

Raising the problem as a complex one, first it is necessary clearly to conceive of what negative consequences it is to avoid by stress relieving. The effect of welding stresses or the welded structure serviceability is rather diversive and found its representation in many works /1 - 5/ and articles.

In Fig. 1 different cases of negative influence of residual stresses on the properties of welded structures are schematically shown along the horizontal line. Welding stresses also take part in them. Group one (I) includes the cases which practically depend only on the residual stresses. They are: change of size as a result of mechanical treatment, instability in time caused by creeping and stresses relaxation, rigidity decrease at loading due to plastic deformations or local loss of stability, total loss of stability and others. Here the role of metal properties and stress concentration in minimum. The choice of treatment method for stresses relieving for the first group of cases, generally speaking, is not of principal value. It is the level of stresses relieving that is important. Heat treatment can provide the most significant stresses decrease if complicated structures are taken into consideration. Plastic tension along the joint, rolling by narrow rolls with metal plastic deformation in the direction of thickness, blasting treatment are effective for the simplest sheet parts. Vibration provides less significant stresses relieving which is quite sufficient for a number of cases of mechanical treatment to prevent deformations at structures transporting or their loading by not high working stress. Because of less energy consumption, cost and treatment duration, mechanical stress relieving methods for the first group of cases at insignificant requirements on geometrical stability are often more advantageous than heat treatment. In complicated massive structures, when significant stress relieving is required, heat treatment is beyond competition with other methods.

The second group of negative stresses effect includes the examples in which the role of metal properties acquires the same condition along with the working stresses. Stresses become only a necessary factor of negative effect and they are not a sufficient factor. There can be given the following examples: failure due to volume stresses in the metal depth; corrosion damage and failure, depending on the level of residual stresses and metal proper ties; cold cracks (hardening) formation; endurance relieving which at the absence of stresses concentrators depends on the metal yield strength, i.e. on its properties; failure in the creeping mechanism, in particular, in the crack reheated; stress relaxation and change of dimensions due to structural instability of metal properties.

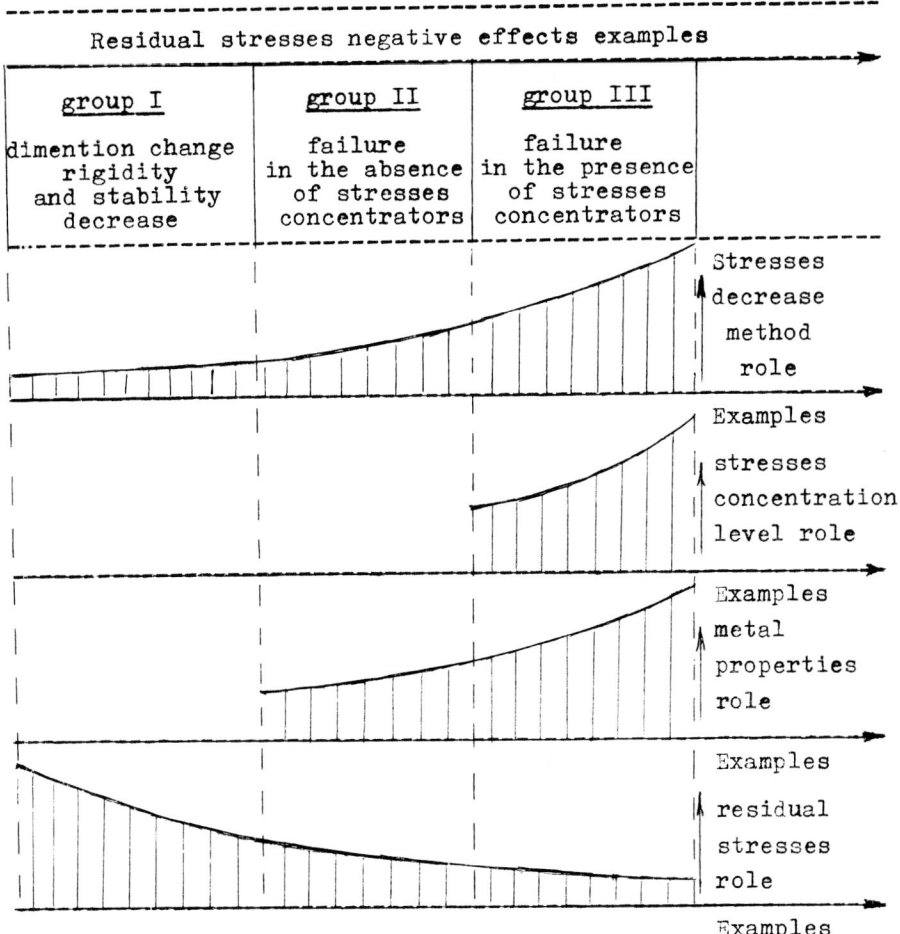

Fig. 1.

The role of treatment method for stress relieving for optimal results sufficiently is increased in the second group. For example, to prevent volume failure, mechanical methods are likely to be optimal and, to prevent failure in the mechanism of creeping and to increase endurance suitably if the structures are not limited by complexity there can be mechanical methods rather than thermal ones. Methods of regulated metal properties at welding, f.e. in case of cold cracks formation, cracks reheated, corrosion failures play a special role. As stresses are only a necessary factor, so their role can be minimized by obtaining the appropriate method properties after welding and in a number of cases not to play special attention to residual stresses. Moreover, in the second group the stress relieving method itself can affect the metal properties. For example, the cold cracks formation (hardening) can be prevented by the fact that after heat treatment the metal properties become such kind that no high level of stresses can any longer cause cold crack rather than at the heat treatment stresses are relieved.

The situation changes more radically if such a strong factor as stress concentration and plastic deformations is added to the residual stresses and metal properties. This is the third group of cases. It would appear that stresses concentration ratio should increase the role of residual stresses as such. In fact very often stresses concentrator changes metal properties in the tip of the notch and strengthens the role of the service stresses so much that the relieving of the residual stresses makes no sense at all. Welded joints and structures strengthening fatigue is a vivid example of the abovementioned. The role of concentrators at high stresses concentration is so great that residual stresses relieving is of not so much importance. Moreover, if stress relieving is carried out by thermal treatment which affects the metal mechanical properties (in this case on metal yield strength in the concentrator tip) then the fatigue strength will decrease.

If we take as an example the case of cold strength in the welded joint with the concentrator, then the choice of stress relieving method will significantly change compared with the fatigue strength. Heat treatment will significantly increase cold strength and mechanical methods of stress relieving will decrease it in a number of cases.

Methods of Producing the Plastic Deformations for Stress Relieving

Let us take the simpliest example of welded plate with residual uniaxial stresses σ_x (Fig. 2). To avoid residual stresses σ_x, it is necessary to form plastic deformations of elongation ε_{xpl}, which are approximately equal to the deformations ε_{xel}. For this purpose there are two ways:
1. Total deformations ε are constant while treatment

$$\varepsilon_{el} + \varepsilon_{pl} = \varepsilon = const. \tag{1}$$

and elastic deformations ε_{el} turn into plastic ε_{pl}. Relaxation of stresses at heat treatment follows due to the above mechanism. Stresses during heat treatment are decreased approximately at the same per cent level both in the region of high and low stresses. The proportional stress decreasing in all the points of the workpiece does not cause any changes of equilibrium forces, deformations have merely no changes. This explains the fact why the distortions of forms and sizes appearing at welding are not removed at heat treatment for stress relieving.

2. Total deformations in the axial direction X are increased, f.e. plates by means of tension. The process of plastic deformation of tension in the zone of $2\varepsilon_{pl}$ width. If plastic deformations of tension in this zone are $\frac{\sigma_{xmax} + \sigma_{pz}}{E}$, then after tension loading removal the residual stresses will decrease up to zero level. During the plate loading along the axis X there

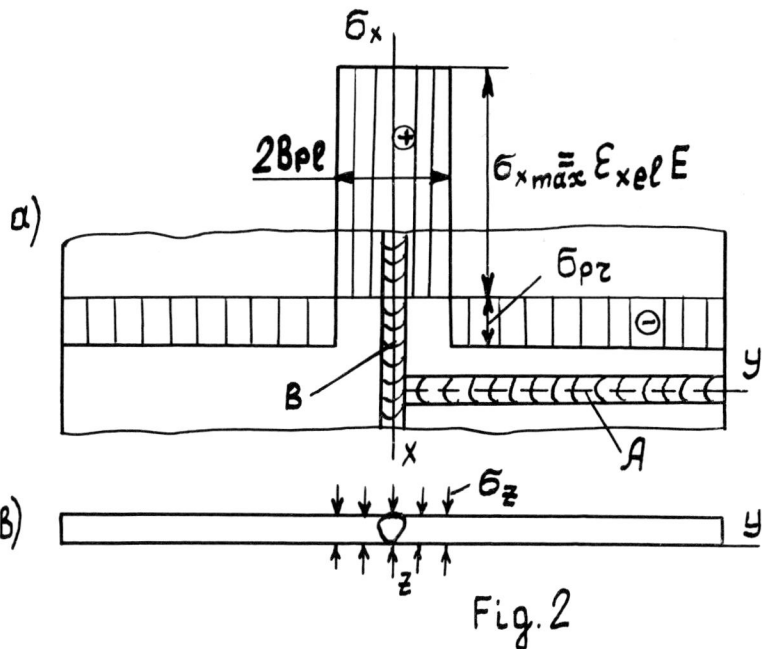

Fig. 2

occurs active plastic deformation in the zone $2\beta_{pl}$. The latter also appears at the use of bending and at the vibration methods of stress relieving. During active plastic deformation the metal properties considerably change, especially at the plastic deformation concentrators availability when their plane is perpendicular to the axis X, f.e. poor penetration in the joint A (Fig. 2,b).

3. Active plastic deformation is produced in the direction of Z axis by the compression forces σ_z (Fig. 2,b). In the direction of X axis there appear passive plastic deformations of elongation necessary for stress relieving σ_x. If the passive plastic deformation of elongation ε_{xpl} in the zone $2\beta_{pl}$ is $\frac{\sigma_{xmax}+\sigma_{pz}}{E}$, then the residual stresses σ_x will decrease up to zero level. The change of metal properties in the stresses concentrators zone whose plane is perpendicular to X axis in this case can be minimal. Due to the abovementioned mechanism there occur stress relieving processes at forging, rolling by narrow rolls, at blasting and partly at vibration and at other methods.

Where does the change of metal properties during the abovementioned three ways of plastic deformations for stress relieving lie? In Fig. 3 of the curve I we can see an ordinary diagram of metal deformation at the coordinates with stress intensities σ_i and deformation intensities ε_i. Curves II and III show changes of maximum stress σ_{max} at the tip of the notch also in the function ε_i for two types of concentrators: II -

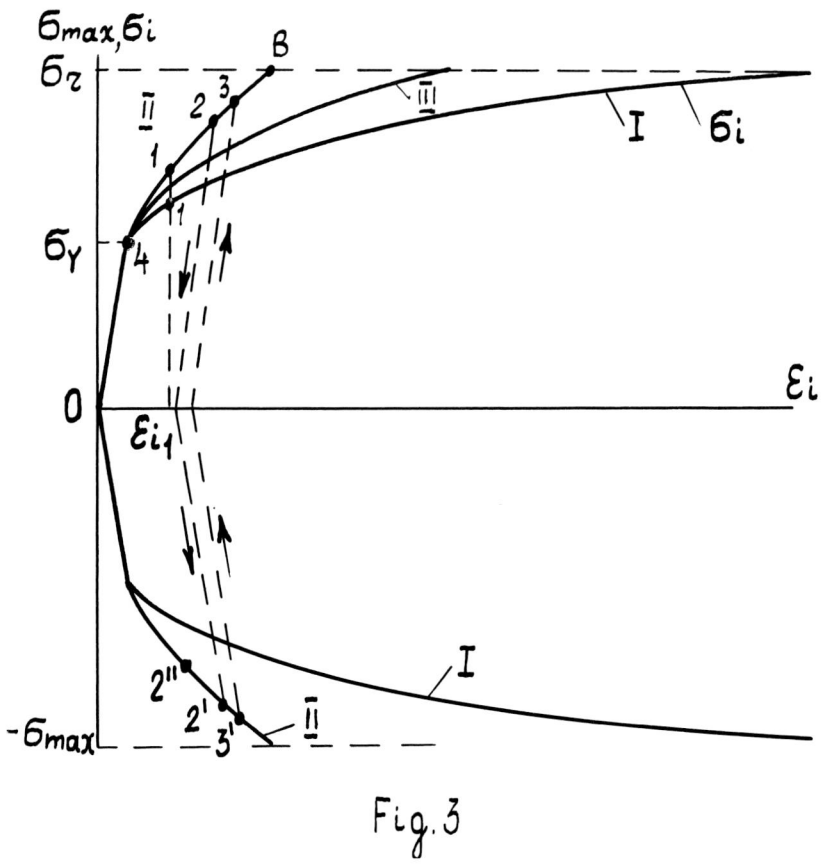

Fig. 3

for the concentrator with the extreme small radius; III-for concentrator with non-zero radius. Stresses concentrator ratio is of essential value. The greater α is, the greater the values of ε_i appear at the same nominal (average) deformation.

Let's imagine that while welding joint δ in Fig. 2 in the process of cooling there appears a certain plastic deformation. With a concentrator perpendicular to x axis, f.e., of poor penetration in the joint A in Fig. 2,a there appears deformation intensity ε_{i1} in it. Correspondingly, we obtain point 1 on the curve II in the diagram. Let's imagine that for residual stress relieving the method of tension is applied, i.e. active plastic deformation was produced and passed from point 1 into point 2 in Fig. 3. At relieving the tension loading applied for stress relieving maximum stress will cover the path 2'' 3' and metal properties due to additional plastic metal deformation at the tip of the notch will be characterized by point 3' on the curve 1'. At service loading maximum stress σ_{max} will pass 3'3 and will approach the point of failure B. In case of stress relieving by compression passive plastic deformation will appear and instead of path 122'3' the stress in the metal will pass from point 1 to point 2'' which is to the left of point 2'. However, in this case we have the decrease of metal plasticity at the tip of the notch (point 2'' is to the right of point 1).

The situation changes basically in case of heat treatment (high temperature tempering). Metal properties ($\sigma_{0.2}$) after tempering will agree with point 4 on the curve I and the residual maximum stresses σ_{max} will be in the vicinity of point 0 . Metal properties change from point 1 on the cirve I to point 4 on the cirve 1 will result in some endurance decrease in case of variable loadings. However, if the joint is loaded at low temperature when the metal is brittle, then, in case of tempering the welded joint plasticity will be significantly great (OB section) than after stress relieving by tension (the section from point 3 up to point B).

Some Stress Relieving Features by Heat Treatment

Heat treatment as a method of stress relieving differs from other methods by many factors although we can note some common features.

Change of metals properties is common and characteristic for all stress relieving methods. These changings can be useful and harmful. And this gives rise to incorrect interpretation of residual stresses role. Good or bad in the properties changing which has been achieved at stress relieving is often associated with the residual stresses role. The example of the above idea are the experiments carried out in tens an of countries by hundreds of scientists, when heat treatment markedly increased the static strength and plasticity of joints at low temperatures. Plastic deformations concentration or one which is added by ageing approached the metal to the failure point B (Fig. 3) so much, that it was insignificant service stress doze given at low temperatures that resulted in brittle failure. High tempering in the majority of steels practically totally restored metals properties after strain hardening or strain ageing, static strength at low temperatures greatly increased /2/ and it seemed that this effect depended on residual stresses relieving. Other methods of stress relieving do not give such picture of metals properties changing which is important for low temperatures.

If mechanical properties of metals are the same in different parts of the welded structure then stresses relieving σ_i (as the temperature increases during heating) in the majority of steels occurs approximately in proportion to the initial level of stresses. As was mentioned above this explains unavoidable welded distortions despite of residual stress relieving at tempering. At other methods of relieving stresses are changed in cases where plastic deformation is produced, and if the produced strain is uniform, then stresses are relieved where they are maximum.

Stresses relieving at tempering during heating practically

does not depend on heating rate, and basically depends on the temperature attained /2/. The role of temperature factor at stresses relaxation is much greater than that of time.

Stresses relaxation in time at constant temperature obeys the law
$$\sigma = \sigma_c \left(1 + \frac{t}{t_o}\right)^{\beta} \qquad (2)$$

i.e. we have straight lines in double logarithm coordinates $\lg \sigma$ - $\lg t$.

In (2) σ_c - stress is at the beginning of holding at constant temperature,
σ - stress changing in time, t - time.

Triaxis stress relaxation in massive structures has some peculiarities. As it is known metal creeping occurs without volume changing. That is why at residual thriaxis uniform tension or compression where volume changing is necessary for stresses changing, stresses relieving can occur only due to stresses redistribution because of plastic deformation at the surface layers, where stresses are doubleaxial. In Fig. 4 we can see the regularity of stresses relieving with different degree of thriaxis stresses volume, 1 - shear, 2 - plainscheme of stresses, 3 - 5 - continuous cylinders and continuous spheres. Residual stresses can exceed 2-2.5 times the uniaxial ones; in this case there is no possibility to get significant stress relieving on the account of holding time. Noneconomic stress relieving by holding time increase was discussed in the work /6/. It was shown that by the temperature increase of 20°C the same effect is attained as we have at holding time increase for 4-10 hours. The Standard /7/ establishes the time of holding for 2-3 hours after temperature flattering in the end of heating.

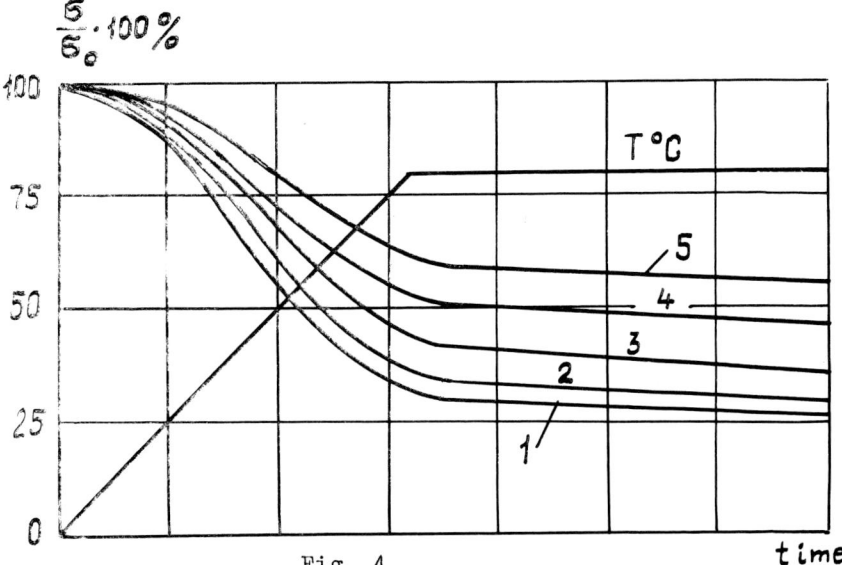

Fig. 4.

The essential feature of stresses changing at tempering is different degree of stresses relaxation in the same steel when it has different structure. In Fig. 5 some data are given. The curves 1 and 2 correspond to the steel of chemical content 0.16 C; 1.06 Si; 1.35 Mn; 0.16 Mo; 0.13 V, which has in the annealing condition $\sigma_{0.2}$ = 480 MPa, and after heat cycle by electroslag welding $\sigma_{0.2}$ = 560 MPa.

Curves 3 and 4 correspond to steel: 0.14 C; 0.21 Si; 0.85 Mn; 4.5 Ni; 0.18 Mo which has in the annealing conditions $\sigma_{0.2}$ = 360 MPa and after electroslag welding $\sigma_{0.2}$ = 830 MPa.

Stress relieving in the heat-affected zone (curves 2 and 4) occurs quicker. It means that if there are residual tension stresses equal in different structure zones in welded joint of transverse direction to the joint axis, then during the relaxation ptocess there will take place the loading of heat-affected zone from the base annealed metal.

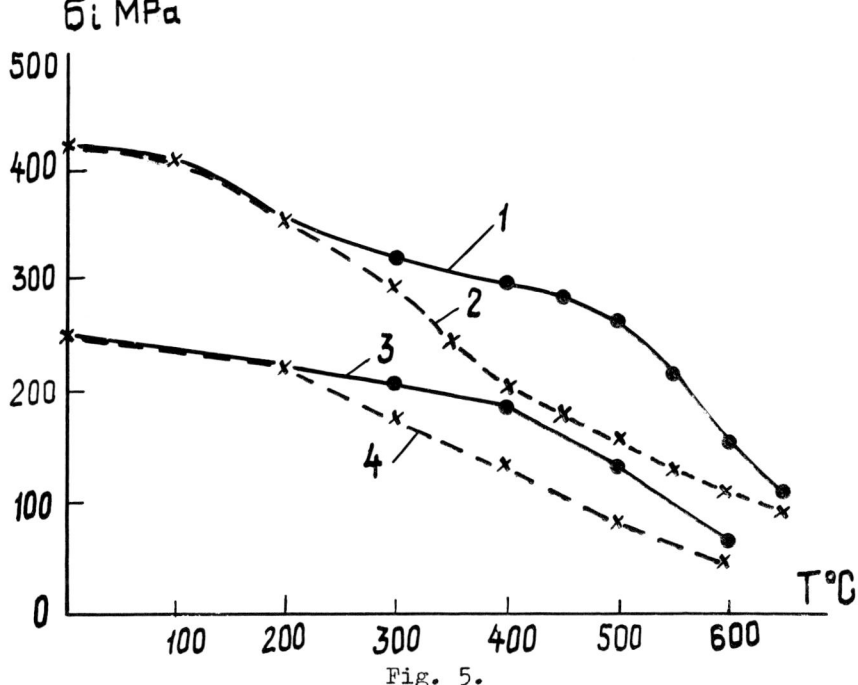

Fig. 5.

The essential feature of stresses changing at tempering is different degree of stresses relaxation in the same steel when it has different structure. In Fig. 5 some data are given. The curves 1 and 2 correspond to the steel of chemical content 0.16 C; 1.06 Si; 1.35 Mn; 0.16 Mo; 0.13 V, which has in the annealing condition $\sigma_{0.2}$ = 480 MPa, and after heat cycle by electroslag welding $\sigma_{0.2}$ = 560 MPa.

Curves 3 and 4 correspond to steel: 0.14 C; 0.21 Si; 0.85 Mn;

4.5 Ni; 0.18 Mo which has in the annealing conditions $\sigma_{0.2}$ = 360 MPa and after electroslag welding $\sigma_{0.2}$ = 830 MPa.

Stresses relieving in the heating affect zone (curves 2 and 4) occurs quicker. It means that if there are residual tension stresses equal in different structure zones in welded joint of transverse direction to the joint axis, then during the relaxation process there will take place the loading of heat affect zone from the base annealed metal. The plastic strains of creeping will be concentrated in less stable heat affect zone which can result in its failure at tempering.

To investigate the possibility of welded joint failure at variable temperature and the inherent stresses availability the method of plastic approach /3/ which means that at first the problem is solved by the method of finite elements as elastoplastic was developed. The solution obtained is then used to carry out thermomechanical tests of the metal at the variable temperature and variable strains obtained from the first solution. The second solution with the use of the mechanical properties obtained at thermomechanical tests which reflect metal creeping and stresses relaxation is carried out. Second solution results allow to carry out proper tests of welded joint to define its strength at heat treatment with regard of inhomogeneity of mechanical properties.

CONCLUSIONS

The application of different residual stresses relieving methods essentially depends on the type of negative stresses affect and on the properties of the metal, stresses concentrators and some economic considerations.

Stresses relieving by heat treatment can be followed by sufficient changes of the metal properties which are sometimes more important for strength than for relieving of stresses.

Stresses intensity relieving σ_i at heat treatment practically does not depend on the heating rate and is basically determined by temperature attained and by an insignificant time of holding.

Stresses relaxation in the investigated steels in the annealed state is less than in that which took place as a result of heat welding cycle; that is why tha heat-affected zone at tempering experiences the increased plastic strains due to the elastic strain transfer out of the base metal.

REFERENCES

Masubuchi K. (1980). Analysis of Welded Structures. Pergamon Press, 642.
Vinokurov V.A. (1973). Tempering of Welded Constructions for Stress Relieving. Mashinostrojenije, 215.
Vinokurov V.A., and A.G.Grigorjants (1984). Welding Strain and Stress Theory. Mashinostrojenije, 280.

Sagalevich V.M. (1974). Methods of welding strain and stress relieving. Mashinostrojenije,248.
Sagalevich V.M., and V.F.Savelyev (1986). Welding joints and constructions stability. Mashinostrojenije, 264.
Vinokurov V.A., Chernykh V.V., and L.A.Skvortsova (1985). Some comments to the discussion of the report "Post weld heat treatment of steel structures", IIW, X-1097-85.
Vinokurov V.A. (1983). Post weld heat treatment of steel structures. IIW, X-1056-83.

Principles of Mechanical Stress-Relief Treatment

I. Hrivňák* and K. A. Yushchenko**

*Welding Research Institute, Bratislava, Czechoslovakia
**Paton's Institute of Electric Welding, Kiev, USSR

ABSTRACT

During welding transient thermal stresses are formed. These stresses interact with the reaction stresses that are caused by external restraint of weldment. The interaction results in residual stresses which are present in a weldment after welding. The most useful techniques of mechanical stress relief treatment are described. It is shown that overloading of welded pressure vessel or welded structure causes considerable relief of internal stresses but does not improve the metallurgical microstructure of weldment. For the dimensional stability as well as for partial stress relief the vibrational stress relief treatment is now widely applied. Peening techniques, both hammer and shot peening are evaluated. Peening reduces subsurface residual stresses and improves the fatigue life of weldment. Similar effect has grinding. Explosive relaxation treatment is also described in the paper.

KEYWORDS

Transient thermal stress; residual stress; stress relaxation treatment of weldments; overloading; vibrational stress relief; hammer peening; shot peening; grinding; explosive relaxation treatment.

INTRODUCTION

During welding the weldment is subjected to thermal changes given by the complex shape of weld thermal cycle. This causes transient thermal stresses which

produce strains in the weld vicinity. Residual stresses are the internal stresses which remain after welding is completed. Residual stresses are of two types (Masubuchi, 1980a, Welding Handbook, 1984): stresses that are produced in welding of non-restrained members and reaction stresses that are caused by external restraint. The level of residual stresses and their distribution depends on the contraction of solid metal during cooling and the yield strength of the weld metal at given temperature. In addition to residual stresses formed by thermal and strain changes, the stresses may arise also from transformation in steel (Linert, 1983, Hrivňák, 1985a). Transformations that occur above $\sim 650°C$ will cause no shrinkage stresses because the yield strength of the steel is still low and so the expansion during $\gamma \rightarrow \alpha$ transformation can be accomodated (e.g. by permanent elongation). However, if austenite transforms to low temperature decomposition products, the structural stress may arise. The martensite is under high compressive stresses, whereas the non-transformed austenite is under high tensile stresses. Residual stresses do not decrease their intensity with time. If a weldment is subjected to loading conditions which introduce certain amount of plastic strain in highly stressed zones, then the maximum stresses in these locations will be reduced. Therefore, it is supposed that application of a small degree of overloading at temperature over the transition temperature will lead to internal stress reduction and redistribution. In general, the residual stresses are decreased by stable or dynamic loading sufficient to cause permanent strain. Residual stresses are usually relieved by post weld heat treatment (PWHT). The PWHT is usually applied with the aim to achieve the following improvements (Status report, 1985f):
- stress relief
- to minimize the susceptibility to crack formation particularly under the conditions which call for high notch toughness
- to improve the dimensional stability
- to decrease the heat affected zone (HAZ) hardness by decomposition of martensite and other supersaturated structures
- to increase the resistance to corrosion, especially to stress corrosion cracking
- to remove cold working.

From this it can be concluded that the stress relief post weld heat treatment is not only expected to relieve the stresses but also to decompose the supersaturated structural decomposition products (martensite, bainite), what can lead to precipitation of special carbides, to decompose the residual austenite and strongly modify the distribution of dislocations. Annihilation processes, polygonization of dislocations

and partially recrystallization are the processes
which are observed in the stress relief heat treated
steel.

The metallurgical and mechanical aspects of stress
relief PWHT were the topic of interest in the previous
two papers presented by R.V.Salkin and Prof. Vinokurov. The
relief of stresses and, in some cases, also improvement of metallurgical structure of a weldment can be achieved also by
mechanical stress relief treatment. However, mechanical treatment cannot replace the PWHT in all consequences. In the next
part of this paper we shall discuss the overloading technique,
vibrational treatment, peening, grinding and explosive treatment.

Fig. 1 shows an example of residual stress distribution in vicinity of the circumferential weld on ∅ 760 mm pipe.

OVERLOADING

As it has been already mentioned, overloading of welded structure results in reduction of residual stresses. Fig. 2 shows
the distribution of stresses in the plane perpendicular to the
weld axis in as-welded condition. If the weld is subjected to
loading (2b) then in a part of the weld internal stresses
caused by external loading will exceed the yield stress of the
material and stresses in this part will relax by permanent
yielding. The result of overloading is shown in Fig. 2c. In
that part of the weld where yielding has proceeded, the level
of residual stresses is reduced. Weld can contain cracks and
other discontinuities and stress raisers. In this way, the
peak residual stresses of stress-raisers are reduced in their
value while the steel is in tough condition and the relief of
stresses is caused by plastic flow. The first application of
load will cause localized yielding at the tips of these defects
and on unloading these regions will be compressed by the
surrounding metal. But a potential cracking hazard may exist
when the steel used is notch-sensitive or has low notch
toughness at the overload temperature. Therefore the temperature at which overloading is applied must lie over the
ductile-brittle transition temperature of any part (including
the HAZ and weld metal) of the loaded structure. It must be
noted that in order to achieve effective relaxation of stresses the loading must produce uniform stress throughout the
weldment, because a non-uniform stress may result in higher
inherent stresses than those caused by welding.

The overloading technique is applied usually to pressure
vessels simple in geometry. In cylindrical vessels the hydrostatic loading produces circumferential stress nearly two
times higher than the longitudinal stress. Therefore relief
of residual stress in circumferential welds is supposed to be
only half of that relieved in longitudinal welds.

The application of overloading is usually connected with the pressure test of a vessel. The temperature of test is increased to 20-40°C. The exact value of this temperature is calculated from the brittle-fracture tests of the welds. Further on it must be pointed out that the success of this treatment depends upon the homogeneity of the welds. In the case of marked heterogeneity caused e.g. by very hard HAZ, the overloading treatment will result in overall decrease of residual stresses but the local stress peaks in the weld vicinity will remain unaltered. Therefore steps must be taken to improve the welding technology and proper filler materials for welding must be selected in order to achieve metallurgically and mechanically as homogeneous welds as possible. The overloading or pressure test usually consists of application of three subsequent pressure cycles, between which the pressure drops nearly to zero. The maximum pressure for treated vessels should be calculated from the membrane stress in the wall. This membrane stress should be close to the guaranteed yield strength of the material used.

Overloading can provide appreciable safety against brittle fracture even at lower temperatures, provided the mode of service loading is the same and if the overload stresses are not reached in service. But if large initial defects occur during the loading a severe strain ageing takes place and the material at the tip of defect can be embrittled for next loadings at lower temperatures (The Welding Institute, 1981; Lezzi and Scanavino, 1984). Overloading has also benefits in improvement of the fatigue strength of welded joints (Alpsten and Tall, 1970). The reason for this consists in reduction of gross residual stresses and introduction of local compressive residual stresses around the regions of stress concentration.

VIBRATIONAL TECHNIQUES

The process of stress relieving by vibration involves inducing a welded structure into one or more resonant or subresonant states, using high force exciters (Claxton and Saunders, 1977, Chmelan et al., 1985b, Welding Research Institute, 1986c). The result of this is the overall elastic straining of the treated structure. Altough the nominal applied strains are elastic, regions or locations of stress concentration give rise to local straining while the resultant stress exceeds the yield stress of the steel concerned. The degree of overall stress relaxation is not too high and it is supposed to be < 40% (Zveginceva, 1968,Zubchenko, 1974b). Polycrystalline materials consist of grains with different crystallographic orientation and alloys may contain solute atoms, precipitates and inclusions. During the vibrational treatment surface macrostresses relax due to redistribution of internal stresses (Procházka et al., 1974a). We assume, the relaxation proceeds on a grain by grain basis and at a rate which is a function of the orientation of the primary slip planes in each grain with respect to the external stress axis. During the vibration the sections that contain tensile residual stresses elongate slightly, while the

sections that contain compressive stresses compress slightly.
As a result of these small strains the residual stresses are
dissipated. Since the level of strains is very small the part
undergoes little macroscopic dimensional change.

The effectiveness of vibrational stress relief (VRS) treatment
depends on the type of structure treated (size, rigidity and
weight). The vibration is attached to the structure at an
appropriate point (Fig. 3) and then it is actuated and scanned
very slowly from zero to its maximum frequency. During this
operation the response of the treated structure is monitored
until the resonance is achieved. Then the vibration is continued for a certain number of cycles (e.g. 5000 - see Fig. 4),
after which the frequency is slowly decreased until some other
resonant frequency is found. After the full frequency range has
been scanned with some dwell at other resonant frequencies, it
is scanned once more at the same rate without dwell. For most
of the welded structures it is usual to pass through two or
three resonant states. There is laso other possibility: to dwell
the vibration at frequencies between the resonant peaks where
no resonance occurs for a long time.

VSR treatment is effective for shape stabilization where a balanced stress state is more important than the reduction of
residual stresses. It works effectively for a wide range of
welded and cast structures, is cheaper than the heat treatment
and in some cases can be even more effective. The use of VSR
is to be avoided in situations where brittle fracture is a serious risk.

The basic equipment for application of the VSR treatment consists of a vibrator which is driven by a motor, controlled
from a console. The vibration frequency ranges between 80 to
200 Hz. The vibratory unit is attached on the upper surface of
the weldment. It is important to find a proper location for
attachment of this unit. Diagram 4 shows the relationships
between the vibration parameters for light structures (1),
structures from aluminium or stainless austenitic steels (2),
high-strength steels (3) and heavy section structures (4).

PEENING

Peening consists in light hammering the weld and/or the
surrounding parent metal in order to relieve the stresses present and to consolidate the structure of metal (Madox, 1985c).
It may be applied while the weld is still hot or after the
weld has cooled. Peening has been used for over 40 years, but
the code requirements governing this procedure were based on
opinion and experiences rather than on scientific data. There
has been no practical method for measuring the effect of peening. Peening can be realized by hammering the surface (hammer
peening) or by directing a high velocity stream of particles
on its surface (shot peening). Peening has a number of effects,
some of which are beneficial and other detrimental. Peening
causes yielding and to some extent plastic flow of the surface
and subsurface layers of the material to which it is applied.

By this it reduces the residual stresses and may introduce compressive stresses at the surface. In some cases peening may give rise to formation of other stresses, or small cracks (in the case of low ductility of treated metal) or to the occurance of strain ageing. Care must be taken in peening hot metal whether slag particles are not entrapped under the surface. Tests on deposited weld metals showed that peening of the last layer may result in a pronounced loss of notch toughness. This effect is most pronounced when weld metal is peened in the blue, brittle temperature range and when the peening is too severe. This adverse effect of peening weld layer can be eliminated by a subsequent covering deposit. Peening may also result in a pronounced hardness increase. Peening can lead also to recrystallization of the part of weld metal when subsequent weld bead is deposited. Peening can also enhance the corrosion attack of the treated surface. Therefore peening technique should only be applied to joints which can be adequately protected from corrosion.

The ASME Code Section VIII - Unfired Pressure Vessels prescribes a method for checking the effectiveness of peening procedure in an actual welding operation. According to this code punch marks should be made on opposite sides of the weld. The distance between these marks should be kept within 0.8 mm accuracy. Peening is performed during the welding of the bead with the first measurement being made executed so as to hold the deviation in distance between the punch marks to a minimum until the weld has been built up to a depth of 31.75 mm to 38.1 mm, the same degree of peening may be applied safely to the remainder of the weld.

Peening is also known as a method to avoid the fatigue failure of a particular weld section. Peening can introduce compressive residual stresses into the regions where fatigue cracking might initiate. Although this method has been employed in a number of studies it is not included in any fatigue design codes. Peening operation can actually treat the sharp discontinuity at the toe and thus reduces the notch acuity. Peening can be more effective for improvement of the fatigue life than grinding or remelting the toe. Maddox (1985c) reviewed the results of investigations to examine the effect of peening on the fatigue strength of joints. He concluded, this method can be practically implemented.

For hammer peening pneumatic or electric hammers are used. These hammers are usually heavy and therefore a remarkable physical effort is needed on the part of the operator to control their positions. This can lead to an unadequate peening, which is reflected by the depth of straining achieved.

Fig. 5 and 6 show the effect of proper and improper peening of transverse non-load carrying fillet welds. Therefore the optimum depth of penetration of the peened surfaces should be established. Maddox reports 0.6 mm as the optimum depth for mild steel fillet welds and 0.3 mm depth of straining for Al-Zn-Mg alloy fillet welds. Therefore the pneumatic peening hammers should be calibrated. The energy for blow generally tends to increase with higher air pressure and with heavier riding loads.

Peening is not suitable for short or intermittent fillet welds because the local weld shape tends to be less favourable.

Controlled shot peening has been found to be effective for increasing the fatigue strength of butt and fillet welds in structural steels. For controlling the intensity of peening the Almen Strip technique has been developed at The Welding Institute Abington. This is a strip of steel which is shot peened on one surface under the selected conditions. Then the curvature induced on the surface is measured (by the height of the arc) and is proportioned to the intensity of peening.

One of the uses of peening is to control the distortion. It is necessary to study carefully the dimensional changes as some sections of the weldment can posses the compressive and the other tensile stresses.

Peening may redistribute the residual stresses, but cannot be considered as a stress-relieving operation unless the last pass is peened, which may be undesirable because of the loss in notch toughness. Peening should be used when the bead or layer is not too thick. For deposits 6 mm or more thick the peening is of doubtful value.

The effects of shot peening on the residual stress distribution can be exerted as follows:
- the compressive residual stresses do not achieve their maximum level at the surface,
- the maximum level of compressive residual stresses appears at a distinct distance below the surface,
- the depth of the maximum level of residual stresses with increasing shot intensity produced either with a larger shot diameter or an enhanced shot velocity, or a higher coverage,
- an enhanced shot velocity and a higher coverage increase also the maximum level of residual stresses.

GRINDING

Grinding is believed as a way of removing the intrusions at the toes of fillet welds. These intrusions can produce reduction in fatigue performance, particularly in high strength steels. It is to be expected that if such intrusions are removed, an appreciable increase in fatigue strength would result. To remove material at the weld toe to a depth of about 0.25 mm the grinding must be carried out very carefully. The machining marks of the grinding should be parallel to the direction of applied stress in order not to act as other potential initiation sites for fatigue cracks.

EXPLOSIVE TREATMENT

Explosive treatment (Kudinov, 1976, 1986a, Petushkov, 1985d, 1980b, 1986b, 1985, 1987) is a very simple and effective method of mechanical relaxation of residual stresses. Explosive treatment is based upon the action of very short but very intensive stress-strain cycle which is introduced into the treated surface by means of explosive charge. Explosive charge can have a shape of solid line (wire) or solid strip. The quantity of explosive charge and its precise shape and lay-out are precisely calculated. The shock wave reaching the treated surface penetrates into the treated material and results in stress-strain wave propagating over the whole structure. It is strongly suspected that this stress-strain cycle applied to a welded structure containing residual stresses produce highly localized yielding, so that a small amount of localized plastic flow very quickly relieves the residual stresses in the treated area. Nowadays, a set of programs is available which permits a very thorough knowledge about physical processes taking place during explosive treatment. The maximum depth of the strain wave penetration into the material is 10-12 mm.

The explosive treatment results in many improvements. Firstly it improves the dimensional stability of weldments by removing the stresses remaining in the structure.

The explosive treatment was shown to eliminate or remarkably decrease the sensitivity of weldment to stress corrosion cracking. Apart from the corrosive environment, the high level residual stresses in the vicinity of welded joints are responsible for exceeding the threshold of cracking conditions and reduction of these stresses is sufficient to avoid the difficulty. Fig. 7 describes the results of stress corrosion cracking tests in butt welds made in St3 mild steel of various plate thicknesses: 1 - $6 \div 8$ mm; 2 - $10 \div 14$ mm; 3 - $16 \div 20$ mm; 4 - $24 \div 30$ mm; tested in boiling aqueous solution of nitrates. The welds were subjected to explosive treatments which resulted in various decrease of residual stresses. The dependence between time to fracture (hrs) and ratio of residual stress R_r to strength R_m, R_r/R_m is shown in the figure.

Explosive treatment improves also the fatigue properties of weldments. The improvements are caused not only by relaxation of internal stresses but also by introduction of small portion of compressive stresses into the treated steel surface. Fig. 8 shows an example of fatigue life curves for butt weld in alloyed steel tested at $R_\sigma = -1$ [8] in as welded condition -1 and after explosive treatment -2. The explosive treated weld shows remarkable increase in fatigue strength (+ 300MPa).

By explosive treatment it is also possible to control the propagation rate of a fatigue crack, as it is shown in Fig. 9. This figure represents the relationship between the total length of fatigue crack and number of applied cycles ($R_\sigma = 0$) in butt weld of 15ChSND steel in as-welded condition (1) and after explosive treatment (2).

Explosive treatment results also in improvements to brittle fracture resistance. Fig. 10 shows the results of brittle fracture tests of welds in structural steel. Fig. 10 shows the dependence between the fracture stress and test temperature. Curve 1 is the result of base material tests. Curve 4 represents the temperature dependence of yield stress for the material tested. Curve 2 gives the result obtained on non-treated welded joint. Comparing to base material there is a remarkable decrease in fracture stress below $\sim -40^{\circ}C$, which indicates high sensitivity to brittle fracture. After applying an explosive relaxation treatment the temperature dependence of fracture stress is more even. As it can be concluded from Fig. 10 the explosive treatment can result in remarkable improvement of brittle fracture behaviour of the weldment.

CONCLUSIONS

The paper deals with some, mostly used mechanical relaxation treatments. It can be concluded that in engineering practice various mechanical treatments are employed. For pressure vessels, simple in shape, especially for spherical pressure vessels (e.g. storage tanks for liquefied gases) the application of overloading cycles during pressure testing, can fully replace the stress relief heat treatment if the weld quality fulfills the requirements. Overloading is also used in welded structures (e.g. bridges) where heat treatment would be impractical.

Vibrational relaxation treatment is used for simple or complex shapes where the main requirement is to ensure the dimensional stability of the structure. This treatment is very simple but noisy, saving time and energy and its proper application can lead also to remarkable decrease of residual stresses.

Explosive treatment shows another example with ever extending applicability. Explosive treatment can be applied namely to circumferential welds, pipelines and other structures. This treatment results in improving the corrosion behaviour, fatigue life and brittle fracture resistance of a weldment.

Peening and grinding are applied mainly for improving the fatigue life of the structure due to the fact, that they introduce compressive stresses into the surface and subsurface area.

The discussed modes of mechanical relaxation treatment cannot replace fully the stress-relief heat treatment. Where the needs for relaxation treatment do not include decrease of internal (residual) stresses only, but call for improving the metallurgical structure of weld, mechanical relaxation treatments cannot replace the heat treatment.

REFERENCES

Alpsten, G.A., and Tall, L. (1970). Welding J., 93s-105s.
Claxton, R.A., and Saunders, G.G. (1977). Vibratory stress relief. IIW Doc. X-846-77.
Hrivňák, I. (1985a). Int. J. Pres.Ves. and Piping, 20, 223-237.
Chmelar, I. et al. (1985b). Zváranie, 34, No.9, 281-285.
Kudinov, V.M. et al. (1976). Avt. Svarka, No.1, 46-49.
Kudinov, V.M., and Petushkov, V.G. (1986a). Svar.proizvodstvo No.7, 1-4.
Linnert, G.E. (1983). Welding Metallurgy, AWS, New York.
Lezzi, F., and Scanavino, S. (1984). Rivista Ital.della Saldatura, 36, No.3, 139-153
Madox, S.J., (1985c). Metal Construction, 17, No.4, 220-226.
Masubuchi, K., (1980a). Analysis of Welded Structures, Oxford, Pergamon Press, XI.
Petushkov, V.G., and Kudinov, V.M. (1985d). Avt.svarka, No. 7, 1-7.
Petushkov, V.G. et al. (1980b). Avt.svarka, No. 8, 11-13.
Petushkov, V.G. et al. (1986b). Avt.svarka, No. 4, 5.
Petushkov, V.G., and Kasatkin, S.B. (1980c). Avt.svarka, No. 6, 11-12.
Petushkov, V.G. (1985e). Avt.svarka, No. 4, 1-4.
Petushkov, V.G., Fadenko, Y.I., and Pervoy, V.M. (1987) The explosion treatment of welded structures - Report for this lecture, Kiev.
Prohászka, J. et al. (1974a). Vibration induced internal stress relief. Techn. univ. Budapest, May.
Status report (1985f). Influence of Stress Relief Heat Treatment, IIW Doc. IX-1359-85 (X-1088-85).
The Welding Institute (1981) Residual Stresses and Their Effect. Abington.
Welding Research Institute (1986c). Vibration treatment of weldments. Bratislava.
Welding Handbook (1984) 4 - Metals and Their Weldability, AWS
Zubtchenko, O.J., (1974b), Avt.svarka, No. 9.
Zveginceva, K.V., (1968), Svar.proizvodstvo, No. 11.

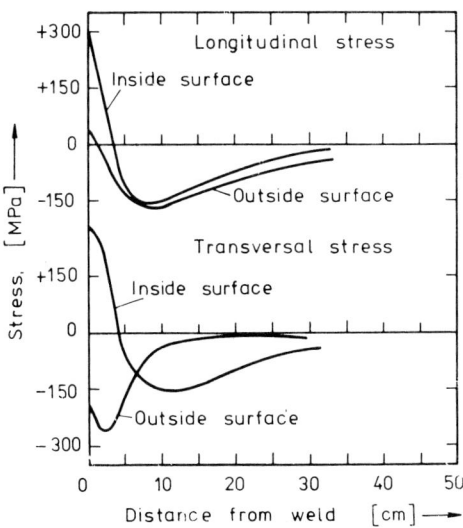

1. **Distribution of surface residual stresses in vicinity of the circumferential weld on ∅ 760 mm pipes with 11 mm wall thickness**

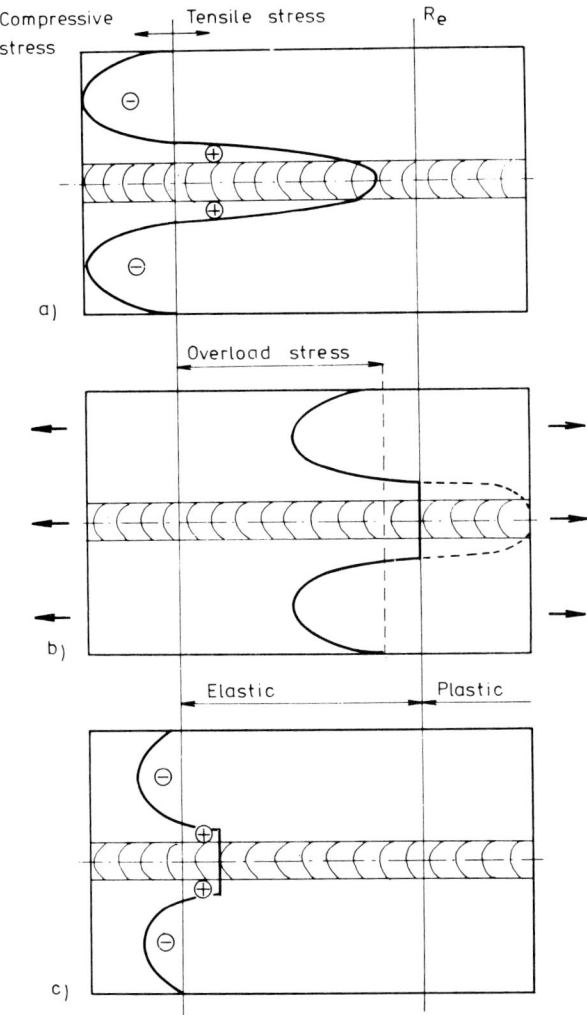

Fig. 2. The effect of overloading on mechanisms of mechanical stress relief in welded joints.

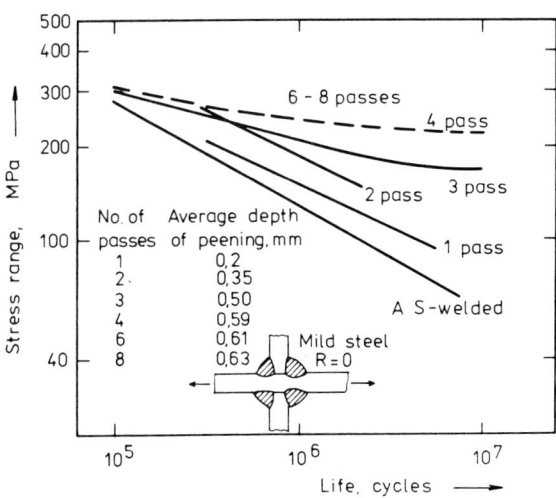

Fig. 3. Effect of number of passes on peening depth and fatigue strength of hammer-peened transverse non-load carrying fillet welds

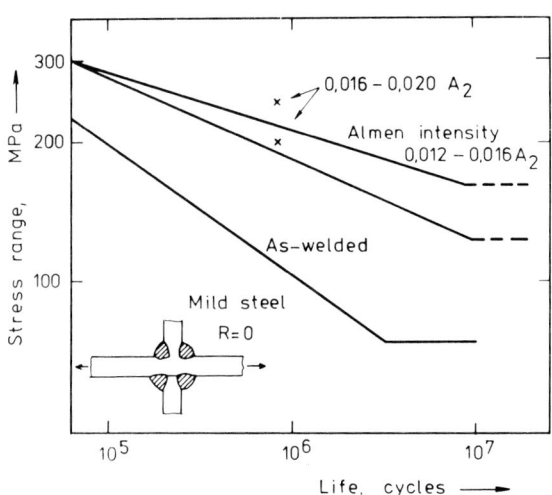

Fig. 4. The results of fatigue test obtained from shot peened transverse non-load carrying fillet welds.

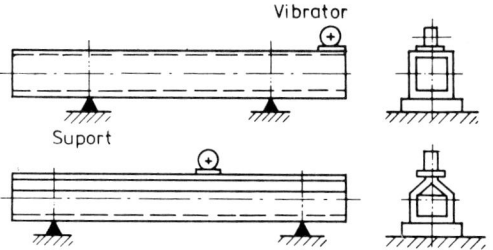

Fig. 5. Vibrational treatment of welded structure.

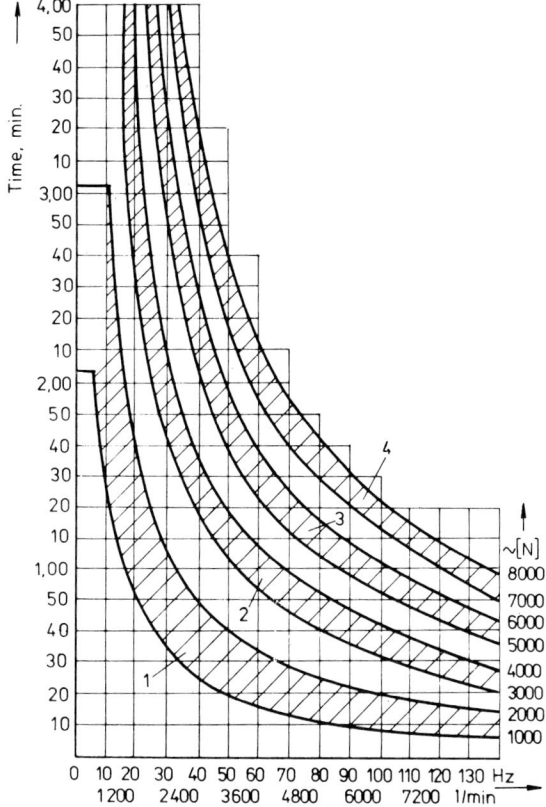

Fig. 6. Nomogram between numbers of cycles by vibration, N for one resonance state, vibration frequency and time of treatment

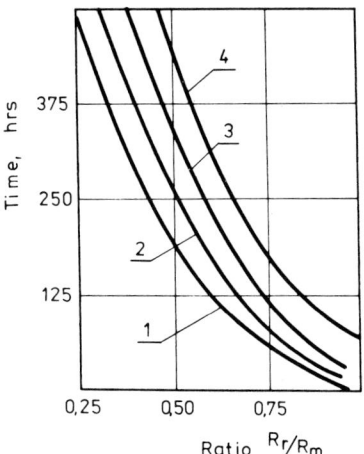

Fig. 7. Results of stress corrosion cracking tests in nitrate environment of explosive treated welds of various thicknesses.

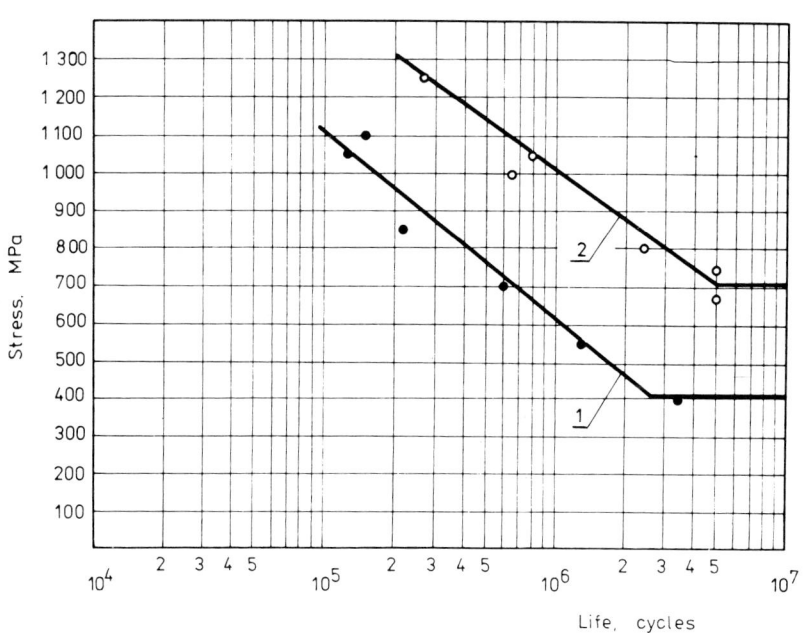

Fig. 8. The results of fatigue test obtained on as-welded (1) and explosive treated (2) weldments.

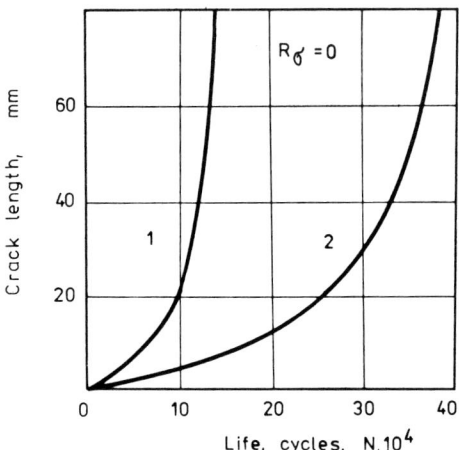

Fig. 9. Growth of fatigue crack on non-treated (1) and explosively treated (2) welds in 15ChSND steel.

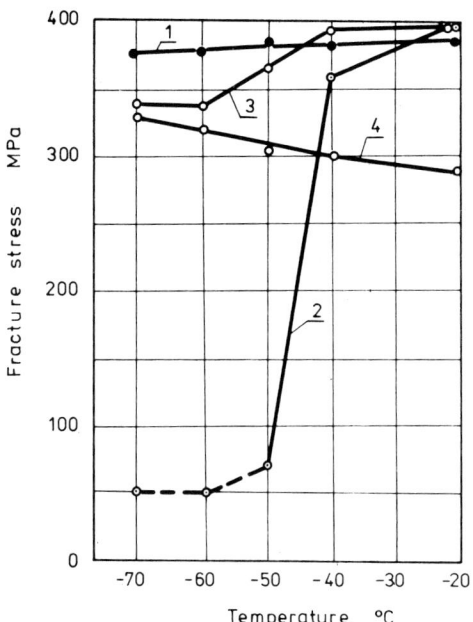

Fig. 10. The brittle fracture tests of untreated (2) and explosiv treated (3) welds. Notice the temperature dependence of base metal (1) and its yield stress (4).

SESSION I

METALLURGY OF STRESS-RELIEVING HEAT TREATMENTS AND MECHANICAL CONSEQUENCES/METALLURGIE DES TRAITEMENTS THERMIQUES DE RELAXATION ET CONSEQUENCES MECANIQUES

I.1	The effect of stress relief heat treatment of welded joints on changes in structural and mechanical characteristics (J. Bosansky, L. Mraz — Czechoslovakia/Tchécoslovaquie)	29
I.2	Properties of dissimilar Cr-Mo steel joints as a basis for the selection of the PWHT-temperature (J.-J. Chene, M. E. J. Shakeshaft — Switzerland/Suisse)	37
I.3	Prevision de l'evolution des caractéristiques mécaniques de la zone affectée et du métal fondu lors du détensionnement (Ph. Bourges, Ph. Maynier, R. Blondeau — France)	53
I.4	Prediction of HAZ Hardness after PWHT (M. Okumura, N. Yurioka, T. Kasuya and J.-J. Cotton — Japan/Japon)	61
I.5	Effets secondaires des traitements de relaxation sur la ténacité de la zone fondue de soudures multipasses d'aciers non alliés ou faiblement alliés (S. Debiez — France)	69
I.6	Influence of post weld treatment on the fracture toughness in welding procedure qualification and component welds (J. G. Blauel, W. Burget — FRG/RFA)	79
I.7	Effects of heat input and stress-relief annealing on the toughness of the HAZ in the welding of the boiler steels type HI, HII, 17 Mn4 and 19 Mn5 (S. Anik, K. Tulbentçi, A. Dikicioglu — Turkey/Turquie)	91
I.8	Heat treatment and welded structures dimensions stability (V. M. Sagalevich, V. F. Savelyev — USSR/URSS)	99
I.9	The improvement of fatigue resistance of structure welded joints by heat treatment (V. I. Trufiakov, P. P. Mikheev, Yu. F. Kudryavtsev — USSR/URSS	101

I.1

The Effect of Stress Relief Heat Treatment of Welded Joints on Changes in Structural and Mechanical Characteristics

J. Bošansky, Ľ, Mráz

Welding Research Institute, Bratislava, Czechoslovakia

ABSTRACT

The contribution deals with a direct proof of precipitation of the coherent niobium carbonitride (NbX) in the weld metal which causes plasticity drop after the stress relief heat treatment. The precipitation was observed in the dark field of electron microscope. The coherence of precipitate and base matrix was proved by streaks on the diffraction pattern. The results of study were applied for optimalization of stress relief heat treatment temperature. It was found out that the annealing temperatures within 550 - 580 °C temperature range have more favourable effect on notch toughness of weld metal than the temperatures within 630 - 650 °C range. Besides the precipitation also the effect of different configuration of dislocations on the notch toughness was evaluated.

KEYWORDS

Stress relief heat treatment; Nb-microalloyed steel; welded joint; electron microscopy; coherent precipitates; stress relief heat treatment temperature; configuration of dislocations; plasticity; weld metal.

INTRODUCTION

It is generally known that the stress relief heat treatment of welded joints of Nb-microaaloyed steels deteriorates the notch toughness. Though it was supposed that this notch toughness drop was caused by the precipitation effects of niobium carbonitride (NbX) no direct proofs were observed. A review of works dealing with the effects of niobium on the structure and properties of weld metal (WM) compiled Dolby in 1980. He has modified a model (Garland and Kirkwood, 1975) considering not only the effect of Nb on the yield point of WM through lowering the transformation temperature and he has also supplemented it with the effect of wire-flux combinations.

Some authors in their work (Watson et al 1981 a; Watson et al 1981 b) on the basis of extensive study of different filler metals and fluxes have come to conclusion that the notch toughness of WM depended on the amount of acicular ferrite and the yield point. They discovered unfavourable effect of side plate ferrite. The structure is affected also by the oxygen content. The effect of Nb and V on WM properties was studied also by Japanese researchers (Stiga et al, 1977). The effect of Nb is more pronounced than that of V. They have found out that due to stress relief heat treatment the transition temperature increased by 25 - 50 ^{0}C depending on the filler metal type. Lowering of transition temperature in spite of stress relief heat treatment was obtained when Ti-B alloyed wire was applied and it was explained by increased proportion of acicular ferite and lowered hardness. According to those authors B enhances NbX precipitation which increases after the stress relief heat treatment. However, they did not provide any direct evidence on NbX precipitation.

The favourable effect of Ti - B was reported also by (Mori and Howa, 1980, 1982) which was caused by preference of acicular ferrite on detriment of proeutectoid ferrite. Unfavourable effect of Nb on the WM properties after the stress relief heat treatment reported also (Hannerz, 1975). NbX precipitation in steel was studied also by (Ohmori, 1975) who has found out that after quenching and tempering NbX preferencially precipitates at the boundaries of laths and dislocations. After normalizing the interphase precipitation was observed.

METHODS AND RESULTS

The effect of different temperature and annealing time on NbX precipitation we studied at University Lulea in Sweden. The steels welded by submerged arc process contained Nb within 0.006 - 0.29 % what means that after dilution the Nb content in WM ranged within 0.002 - 0,117 %. The WM without Mo and those containing 0.2 % Mo were also studied. Due to methodic reasons the NbX precipitation was studied in the WM with higher Nb content. For observing of precipitates the dark field of electron microscope and prepared thin foils were applied. The method and test results are more in detail presented in (Bošanský, J. at al 1977).

The hardness course in Fig. 1 shows that precipitation hardening occurs. Degree of hardening depends on Nb content in the WM. Molybdenum shifts the maximum of hardness towards longer times what means that it hinders Nb diffusion after precipitation. Higher temperature shifts the maximum hardness of WM towards shorter times. Maximum hardness was attained during 2 - 10 hrs annealing.

By electron microscopy analysis on thin foils from the zones of maximum hardness of weld metal the dispersion NbX precipitates as shown in Fig. 2 were observed. The precipitates were

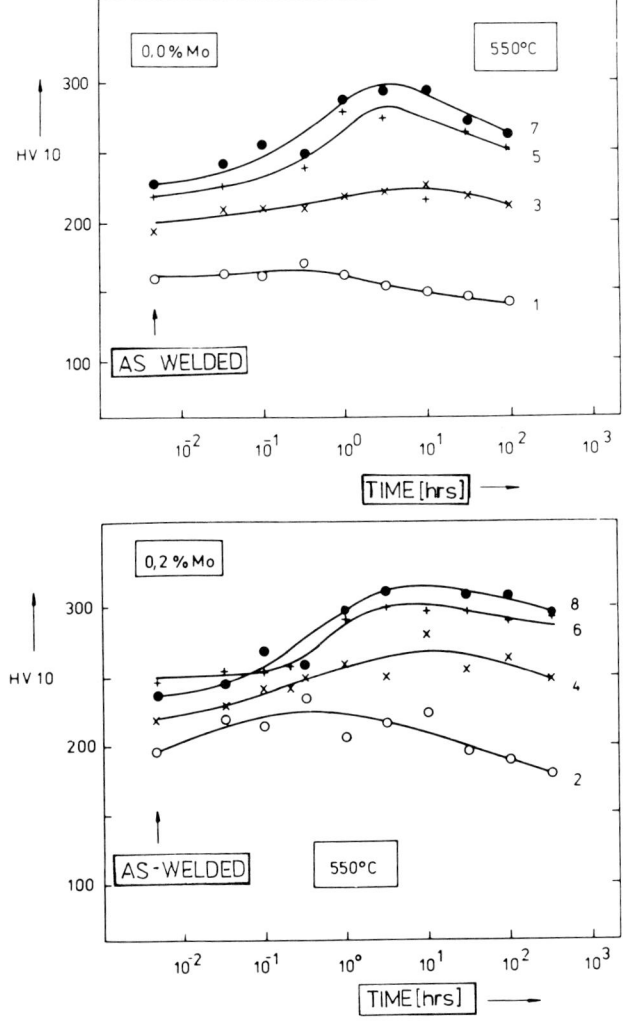

Fig. 1 Hardness courses in WM in dependence on annealing time at 550 °C temperature

located either on dislocations and partially as the interphase precipitation in acicular ferrite. From the diffraction spectrum shown in Fig. 2 it can be concluded that the diffraction traces from the base matrix contain the streaks what means that the NbX precipitates are coherent. The streaks were observed in both $\langle 100 \rangle$ directions in the acicular ferrite. In the polygonal ferrite the streaks were observed only in one $\langle 100 \rangle$ direction as a result of interphase precipitation. At greater magnification the decomposition of particles into three parts

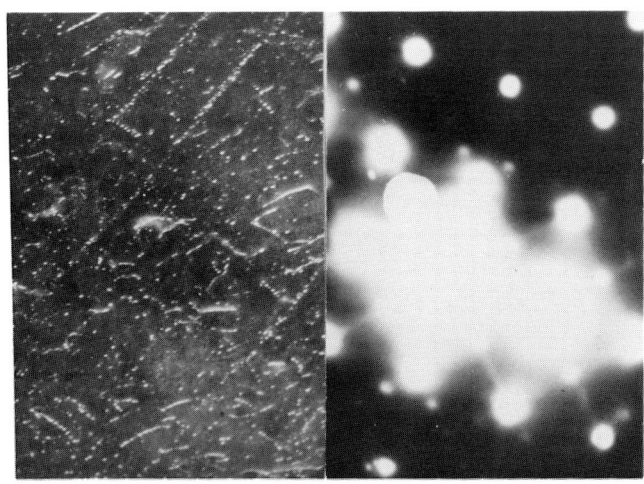

Fig. 2 NbX precipitates in WM after annealing
in the range of maximum hardness with
the streaks in the diffraction pattern

was observed. This phenomenon can be explained from the thermodynamics by the additional elastic energy caused by cooling from the precipitation temperature. The stresses at 1000 MPa level were calculated. The orientational relationship between the NbX and α-Fe obeys the relationship suggested by Baker and Nutting. The size of the smallest precipitates was approximately 1 nm.

As concluded in the IIW document (Hrivňák, 1975) the standards in some countries have already specified relatively low temperatures of the stress relief heat treatment. These are, for example IIW-X-867-77 - 540-580 °C; CODAP - 550-600 °C; ISO/DIS 2694 - 560-600 °C and BS 5500 - 580-620 °C. The Czechoslovak Standard ČSN 42 0284 specifies for C-Mn steels and their welded joints 600-650 °C temperatures and for the alloyed WM it specifies 620-650 °C temperatures of stress relief heat treatment.

The results from the study of precipitation as well as from our works dealing with the effect of configuration of dislocations on the change in plasticity of welded joints (Bošanský and Mráz, 1985, Bošanský,1986) were utilized for optimalization of stress relief heat treatment temperatures. If the NbX precipitation should not cause the notch toughness drop the temperature level and time of the stress relief heat treatment must be selected in order to prevent the formation of coherent or dispersion precipitation. It means that the stress relief heat treatment must be performed before attaining the maximum hardness of the WM. For 550 °C temperature it is approximately up to 10 hrs. This was proved experimentally on SA welded joints of Nb-microalloyed steel ČSN 11503. The results are processed in Fig. 3. Based on this figure it can be concluded that the welded joints annealed below 600 °C temperature exert lower transition temperature than the joints annealed above 600 °C. The weld metal centre exerted in general higher transition temperature than in surface vicinity.

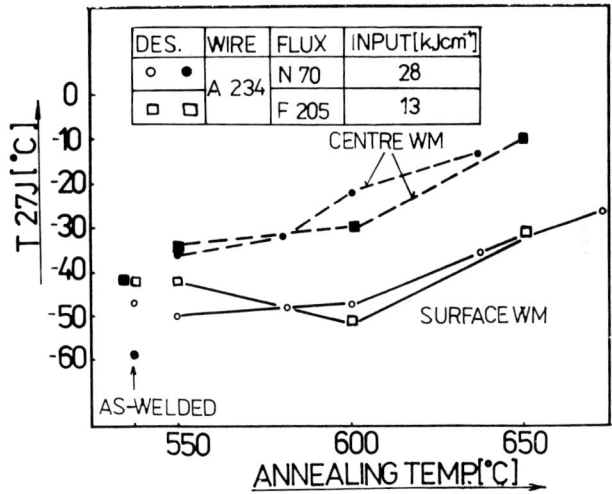

Fig. 3 The effect of annealing temperature on the transition temperature

As it follows from Table 1 the annealing temperatures 550 °C and 580 °C respectively were sufficient for the lowering of residual stresses.

TABLE 1 The Effect of Annealing Temperature on the Drop of Residual Stresses

Condition	Stress Rx MPa	Stress Ry MPa
as-welded	+ 143.3	+ 129.7
PWHT at 550 °C	- 18.0	- 40.8
PWHT at 580 °C	- 3.35	- 20.2

The structural changes after stress relief heat treatment were evaluated from the aspects of configuration of dislocations and hardness measurements in the polygonal as well as in the acicular ferrite. Microhardness decreased with increasing annealing temperature and was always lower than that measured in as-welded condition. From Fig. 4 it can be concluded that at

+ 20 °C testing temperature with the decreasing hardness the notch toughness increases. However, at - 20 °C testing temperature with decreasing hardness also the notch toughness decreases. The study of dislocations has shown that with increasing annealing temperature the average dislocation density decreased and their polygonization became more marked. Based on this finding as well as on our previous works it can be stated that besides the precipitation the notch toughness after stress relief heat treatment is affected also by the change in the dislocation density.

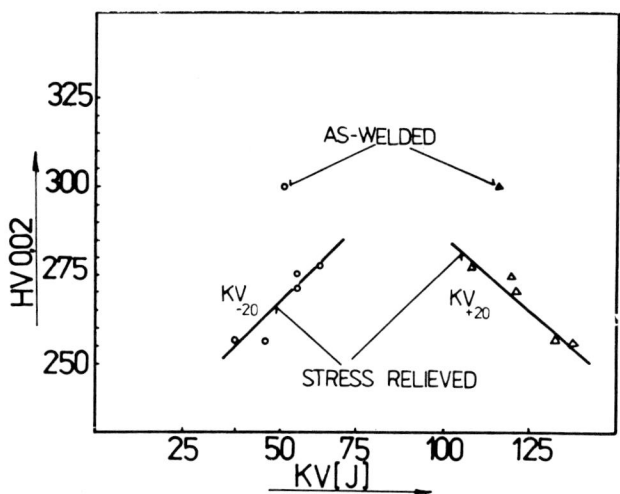

Fig. 4 The effect of microhardness on notch toughness of WM at + 20 and - 20 °C.

As we have shown in the previous works due to 500 - 650 °C annealing temperatures the dislocation network and subgrain boundaries are formed as the result of dynamic recovery. This type of configuration of dislocations contains the dislocation barrier type a ⟨100⟩. This configuration of dislocations has roughly doubled effect on the changes in strength and plastic properties as the same dislocation density formed after straining at ambient temperature without the dislocation barriers type a ⟨100⟩.

Finally, it can be concluded that the notch toughness and also other mechanical characteristics after the stress relief heat treatment are affected not only by the precipitation of carbides and carbonitrides but also by the change in density and configuration of dislocations. The unfavourable configuration of dislocations containing the dislocation barrier type a⟨100⟩ can decrease the notch toughness of welded joint in spite of hardness drop following from the overall decrease in average dislocation density. In optimalization of the stress relief heat treatment temperature not only the precipitation but also the change in configuration of dislocations must be considered.

REFERENCES

Bošanský, J., Porter, D. A., Astrom, H., Easterling, K. E. (1977). Scand. J. Metallurgy No. 6, p. 125.
Bošanský, J., Mráz, L. (1985). Kovové materiály, 23, No. 3, p. 341.
Bošanský, J. (1986). IIW Document-IXB-133-86.
Dolby, R. E. (1980). IIW Document-IX-1175-80.
Garland, J. G. and Kirkwood, P. R. (1975). Metal Construction 7, No. 6, p. 320.
Hannerz, N. E. (1975). Welding Journal, 54, p. 162.
Hrivňák, I. (1985). IIW Document-IX-1359-85.
Mori, N. and Howa et al (1980). IIW Document-IX-1158-80.
Mori, N. and Howa et al (1982). IIW Document-IX-1229-82.
Ohmori, Y. (1975). Trans. Iron Steel Inst. Japan, 15, p. 194.
Stiga, A. et al (1977). IIW Document-IX-1049-77.
Watson, M. N. et al (1981a). Weld. Metal Fabrication, March, p. 101.
Watson, M. N. et al (1981b). Weld. Metal Fabrication, April, p. 161.

Properties of Dissimilar Cr-Mo-Steel Joints as a Basis for the Selection of the PWHT-temperature

J.-J. Chene* and M. E. J. Shakeshaft**

*Technical University of Lausanne and Sulzer Bros. Ltd.
**Sulzer Bros. Ltd., Winterthur, Switzerland

ABSTRACT

The prime object of any necessary PWHT should be to ensure weld integrity and low levels of internal stress. It is essential to achieve a satisfactory combination of microstructures which will fulfill the service requirements. Low alloy Cr-Mo-steels are well known in the design of boiler plant and it is our practice to strive for a controlled chromium content across the joint to hinder carbon migration. Thus the properties of the filler metal - in this case E CrMo1 B20 - become of major importance to the selection of the PWHT conditions.
This investigation concerns joints made between the DIN steels 10 CrMo 9 10 and 15 Mo 3 resp. 15 NiCuMoNb 5. It is shown that the most suitable temperature range for the PWHT of such joints is 660...690°C

KEYWORDS

Cr-Mo-Steels; Carbon Migration; PWHT; Fracture Toughness; Tensile and Hardness Properties.

INTRODUCTION

If a PWHT is considered necessary or desirable it is important to review all the factors influencing a satisfactory life of the entire construction under the service conditions. Indeed, it may then be shown that a PWHT should either not be performed at all or only between strict temperature limits. Of paramount importance is the behaviour of the joint as a whole and not simply the characteristics of the base materials to be joined.
Factors such as workpiece geometry, welding process and conditions, design details of the total construction and selection of the base materials and filler all play their part and should be appraised as a whole, together with any stipulated or recommended PWHT-conditions published by the pertinent Authorities.
The degree by which internal stress - both beneficial and detrimental depending upon the service loading - can be relieved is a function

of the yield behaviour of the total welded contruction and each single weldment at the selected PWHT temperature. In order to ensure a satisfactory microstructure across the joint it is often necessary to select a PWHT-temperature outside the normal range of simple stress-relief. The complexities of the microstructure across the weld influence the macro-behaviour of the joint and subsequently that of the complete construction under review. A weldment of higher strength than that of the base materials will tend to support its neighbouring regions under elastic loading. A base material of lower strength will in turn support the welded zone under plastic conditions. In the same way, micromechanisms of mutual support are possible within the varied structures across the weld. Once this is achieved it becomes possible to produce high strength joints of high toughness.
The special problem of high service temperatures and perhaps high temperatures during the PWHT influences the choice of filler metal which in turn once again influences the choice of PWHT-temperature. Our practice is to select the filler metal to obtain a suitable chromium gradient across the weld and thus hinder carbon migration during the PWHT which is generally specified because of the wall-thickness employed in our valves and vessels. Two systems were selected on this basis and investigated as a function of PWHT-temperature in order to ascertain an optimum range. Common to each system was the steel 10 CrMo 9 10 of about 2.25 % Cr and the filler metal E CrMo1 B20 of about 1.10 % Cr. The other base material was either 15 Mo 3 of usually less than 0.25 % Cr or the steel 15 NiCuMoNb 5 which usually exhibits chromium levels below 0.30 %. Thus the selected filler metal had an intermediate chromium-content in terms of the base materials.

BASE MATERIALS

(A) 4 Tubes, 10 CrMo 9 10: ⌀ 267 x ⌀ 220 x 250 mm*
    4 Tubes, 15 Mo 3:   ⌀ 267 x ⌀ 220 x 250 mm*
(B) 4 Tubes, 10 CrMo 9 10: ⌀ 309 x ⌀ 245 x 250 mm**
    4 Tubes, 15 NiCuMoNb 5: ⌀ 309 x ⌀ 245 x 250 mm**
*wall thickness: 24 mm
**wall thickness: 32 mm

Identification System

10 CrMo 9 10 - 15 Mo 3:   A1, A2, A3, A4
10 CrMo 9 10 - 15 NiCuMoNb 5: B1, B2, B3, B4
Four PWHT-temperatures were selected (1...4) and thus each specimen identified:
        e.g. A1/1...A1/4
             B4/1...B4/4
Smaller test specimens were further identified in numerical fashion (see Fig. 1).

Chemical Composition

Random samples were spectrographically examined and the results are presented in Table 1. (Base and Filler materials).

FILLER METAL

Carbon Migration in dissimilar Joints with different Carbon content of the Base Metals

TABLE 1 Chemical Compositions (%)

	15 Mo 3		10 CrMo 9 10			15 NiCuMoNb 5		E CrMo1 B20	
	Sample	DIN*	Sample	Sample	DIN*	Sample	DIN*	Sample	DIN*
C	0.175	0.16	0.089	0.123	0.11	0.134	≦0.17	0.053	<0.07
P	0.0113	≦0.035	0.0195	-	≦0.040	0.0064	≦0.035	0.015	<0.02
S	0.0248	≦0.030	0.0129	-	≦0.040	0.0086	≦0.035	0.011	<0.02
Si	0.249	0.20	0.189	-	≦0.50	0.350	0.35	0.541	0.50
Mn	0.627	0.60	0.450	-	0.55	0.950	1.00	0.908	1.00
Cr	0.042	≦0.25	2.320	2.230	2.25	0.281	≦0.30	1.16	1.10
Ni	0.059	-	0.034	-	-	1.170	1.15	0.045	-
Mo	0.310	0.30	0.964	-	1.05	0.346	0.30	0.452	0.45
V	-	-	0.003	-	-	-	-	0.017	-
Cu	0.107	-	0.024	-	-	~0.62	0.70	0.031	-
Nb	-	-	-	-	-	-	~0.02	-	-
Al	0.014	-	0.019	-	-	0.011	-	0.012	-
Ti	-	-	0.0019	-	-	0.0015	-	0.0029	-

*Mean Value resp. Typical Value.
The particularly low S and P contents of the 15 NiCuMoNb 5 will be noted; any possible low toughness value can therefore not be attributed to these elements.

It is comon practice in creep resistant steels to balance the chromium and carbon contents such that a steel with higher chromium has generally a lower carbon content; a steel with relatively high carbon level exhibits therefore a generally lower chromium content. When such steels are joined it is possible to observe carbon migration from the low-chromium to the high-chromium steel. Such migration has been noted within twenty minutes' holding time at temperatures above 700°C and it seems that a threshold temperature of around 600°C exists for many low-alloyed ferritic steels. This must therefore be considered during the selection of the filler metal and PWHT conditions. Carbon migration weakens the joint due to the formation of a coarse ferrite zone at the transition line and it seems advisable to hinder this by reducing the chromium-gradient across the joint. This is most simply achieved by selecting a filler metal of intermediate chromium content.
The major consequence of this selection of a low-carbon Cr-Mo filler metal is a weld metal zone of relatively high strength and low ductility

in the as-welded condition. Thus the choice of PWHT-temperature will be governed to a large extent by the characteristics of the filler metal and not directly by those of the base materials to be joined.

TEST-COUPON PREPARATION

Single-U-Groove

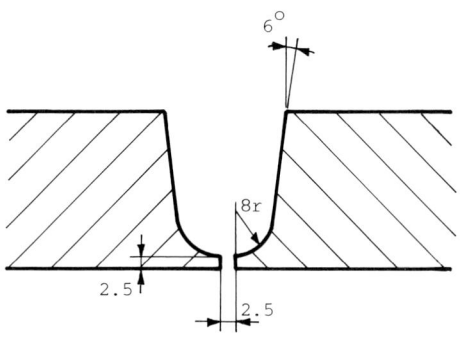

Welding Parameters

Procedure: TIG+E
Filler Metal: TIG; DCMS-IG Ø 2.5 mm
               E;   E CrMol B20 Ø 2.5-5.0 mm
Preheat: 300°C; max. Pass-Temperature 450°C
Backing gas: up to 4th. Pass; Argon
Cooling: in still air

Heat Treatment

Joints A1, B1: 560±2°C / $2^{+1/4}_{0}$ h / Furnace to 300°C
       A2, B2: 620±2°C   "   "
       A3, B3: 690±3°C   "   "
       A4, B4: 730±3°C   "   "

SPECIMEN PREPARATION

Tensile and Impact Specimens according to DIN 50120 resp. 50115 were prepared (see Fig. 1). The impact specimens were etched in order to ensure exact notching in relation to the weld metal and HAZ regions (see Fig. 2). Pieces for hardness-traverse measurements were also prepared (see Fig. 3).

TESTS

The tensile specimens were tested at RT while the impact tests were conducted at temperatures between -20°C and RT. Each specimen was suitably identified. Hardness traverses were also made (HV 10) on further samples of each joint.

RESULTS

Tensile Tests

As expected, each specimen fractured in the base material of lower tensile strength; A-specimens in 15 Mo 3 and B-specimens in 10 CrMo 9 10. The tensile properties are exhibited in Fig. 4 and a typical photo of the specimens is given in Fig. 5.
It may thus be assumed that the UTS and yield strength of the joint 10 CrMo 9 10 - 15 Mo 3 become markedly reduced at PWHT-temperatures between 560 and 620 °C while no further important loss occurs from 620 to 730 °C. The joint 10 CrMo 9 10 - 15 NiCuMoNb 5 showed a relative intensitivity to PWHT-temperature over the tested range.

Impact Tests

The results are displayed in Figs. 6, 7. Due to the large scatter definite conclusions are difficult to make, but the following tendencies seem valid:

- impact strength increases with increasing PWHT-temperature up to a maximum and then decreases
- the influence of test temperature between $-20^{\circ}C$ and RT is negligible
- impact values of 100 J or more may only be expected in connection with a PWHT-temperature of $690^{\circ}C$
- steel 15 NiCuMoNb 5 exhibits lower toughness values than 10 CrMo 9 10 and 15 Mo 3
- it seems that low toughness may be expected at the fusion line; this is specially valid for the steel 15 NiCuMoNb 5
- the lowest toughness values were registered in the weld metal zone.

Hardness Traverses

The hardness traverses (HV 10) may be seen in Fig. 8. It will be noted that the steel 15 NiCuMoNb 5 is associated generally with high hardness values although the lowest toughness values are associated with the weld metal zone of generally intermediate hardness. It may therefore not be automatically assumed that high hardness is directly associated with low toughness.

TABLE 2   Hardness Peaks and Toughness

PWHT Temp. ($^{\circ}C$)	Peak Hardness (HV)		Toughness Value Impact Energy (J)	
	Weld Metal	15 NiCuMoNb 5	Weld Metal	15 NiCuMoNb 5 Fusion Line
560	250...300	ca. 330	-	ca. 60
620	ca. 260	ca. 330	ca. 25	ca. 65
690	ca. 230	ca. 280	ca. 125	ca. 140
730	ca. 220	ca. 290	ca. 120	ca. 50

DISCUSSION

The simple tensile test can be used to demonstrate the principle of mutual support at the macro-level. Despite discrete zones of high local hardness and others of low toughness, each tensile specimen fractured

within a zone where higher deformation was possible. The presence of high-strength material, hardness peaks and brittle zones had no influence upon the behaviour of the tensile specimen as a whole. Each specimen was characterised by the behaviour under load of that part of the joint which was able to exhibit a necessary degree of deformation.
The most dangerous combination of properties is when a material of low yield strength also exhibits low toughness i.e. when the material is not allowed to deform. The partner of higher strength carries the applied load in an elastic manner while the material of lower yield strength endeavours to deform plastically. If this deformation is hindered, fracture will occur in a brittle manner at relatively low applied loads.
Weld defects reduce the mechanical properties of a joint dependent upon defect size, shape and position. Crack-like defects are the most dangerous since they cause very high stress concentration. Such defects are most likely in the deposited weld metal and in the HAZ. If deformation is hindered around such a defect a brittle fracture is possible. A sensible precaution is to ensure that sufficient deformation can occur in these sensitive areas.
Microstructural strengthening can also reduce the deformation capacity of a material. Microstructures having a high concentration of precipitates but a low dislocation density tend to show a definite yield point and such materials deform readily. Structures of high precipitation and dislocation densities are of particularly low fracture toughness since deformation is extremely limited and an unstable crack propagation is fostered. In the case where high hardness is attributable solely to high dislocation density, a low toughness may not be automatically assumed. If, however, such a microstructure were heat treated so that precipitation effects could compound the prevailing high level of dislocation density, the danger of brittle fracture would be increased. Certain materials cannot exhibit such dangerous combinations of dislocation and precipitation density in the as-welded condition and therefore can exhibit sufficient ductility either without PWHT or at relatively low PWHT-temperatures where only the low level of internal stress has to be further reduced because of certain service conditions. In terms of fracture toughness, however, it has been shown, for example, that for the steel 17 MnCrMo 33 the beneficial reduction of internal stress at 560°C was counteracted by such severe embrittlement that the PWHT proved highly undesirable [1].
A welded joint exhibits a variety of micro-structures and it is their degree of interplay which dictates the observed behaviour under load. A stable crack will seek out zones of lower strength i.e. of higher deformation capability in order to propagate. Only in this way can the available energy be adequately dissipated. An unstable crack is associated with minimum deformation and will therefore select zones where the least deformation energy is required at the given deformation speed. Such cracks can only be blunted or deviated by neighbouring zones of higher toughness. The internationally recognised ISO-V-specimen is incapable of describing such interplay with sufficient accuracy; the presence of this aggregate behaviour is reflected, however, in the large scatter in measured impact resistance (see Figs. 6, 7). It would seem prudent to use the lowest measured values to determine the most suitable range of PWHT-temperature.
A consideration of the diagrams in Figs. 6, 7 yields the following:
- the fusion line (10 CrMo 9 10) exhibits adequate toughness over the whole PWHT-temperature range, while 690°C seems advisable.
- the fusion line (15 Mo 3) is associated with high but progressively reduced toughness with increasing PWHT-temperature

- the fusion line (15 NiCuMoNb 5) is of relatively low toughness and reasonable values can only be expected after a heat treatment around 690°C
- the HAZ (10 CrMo 9 10) is of high toughness throughout the range but temperatures of 690...730°C seem indicated.
- the HAZ (15 Mo 3) exhibits high values throughout with perhaps an advisable limit of 690°C
- the HAZ (15 NiCuMoNb 5) is also of high toughness with an indication that a PWHT-temperature of around 620°C is associated with the highest general values.
- the weld metal zone is clearly of low toughness afer treatments in the range 560...620°C. Values in excess of 50 J can only be expected at PWHT-temperatures in excess of approx. 630°C while the 100 J limit requires a treatment at 660°C. A suitable range seems to lie between 660 and 730°C.

An appraisal of all the measured values and their influence upon the selection of a suitable range of PWHT-temperature leads to the conclusion that the properties of the weld metal zone are paramount. The final tensile properties of the base materials thus become of secondary importance to the main object of ensuring joint integrity as a whole, bearing in mind that the creep properties at the service temperature are of prime importance. In this present work these properties happen to be in excess of the minima required by DIN, but even if the finally selected PWHT had resulted in tensile and yield strengths below the minimum requirements, it would still have been maintained that joint integrity is of more importance. In such a case the lowest PWHT-temperature in the recommended range would be deemed adequate.

The presence of hardness peaks must be considered in relation to a possible embrittlement of martensitic and bainitic regions by hydrogen. It is normal practice to assume no embrittlement at hardness levels below $\sim$HV 350. It remains, of course, possible that the hardness traverse failed to show a small area of above-average hardness and it would seem prudent to set a limiting peak hardness of HV 300...330 as a selection criterion for the PWHT-temperature. On this basis it can be assumed that a PWHT at temperatures below approximately 650°C should not be employed.

CONCLUSIONS

It is recommended that the PWHT for these joints be performed within the range 660 to 690°C.

Provided that the minimum tensile properties can be guaranteed for the particular Heat, it seems possible to tolerate temperatures up to 720...730°C.

PWHT-temperatures above 690°C are associated with a reduction in toughness although the minimum requirements according to DIN seem capable of being fulfilled.

The presence of hardness peaks of around HV 280...300 in the HAZ of these joints has no direct bearing on the selection of the PWHT-temperature.

It has been shown that mutual support in zones of varying metallurgical and mechanical properties is possible. The macro-behaviour of a joint seems related to the active volumes of "brittle" and "tough" regions within the various zones of the joint. The PWHT should guarantee that sufficient deformation at both the micro and macro level can take place.

REFERENCE

[1] Braun, O. Thesis Nr. 578, 1985, University of Lausanne, Switzerland.

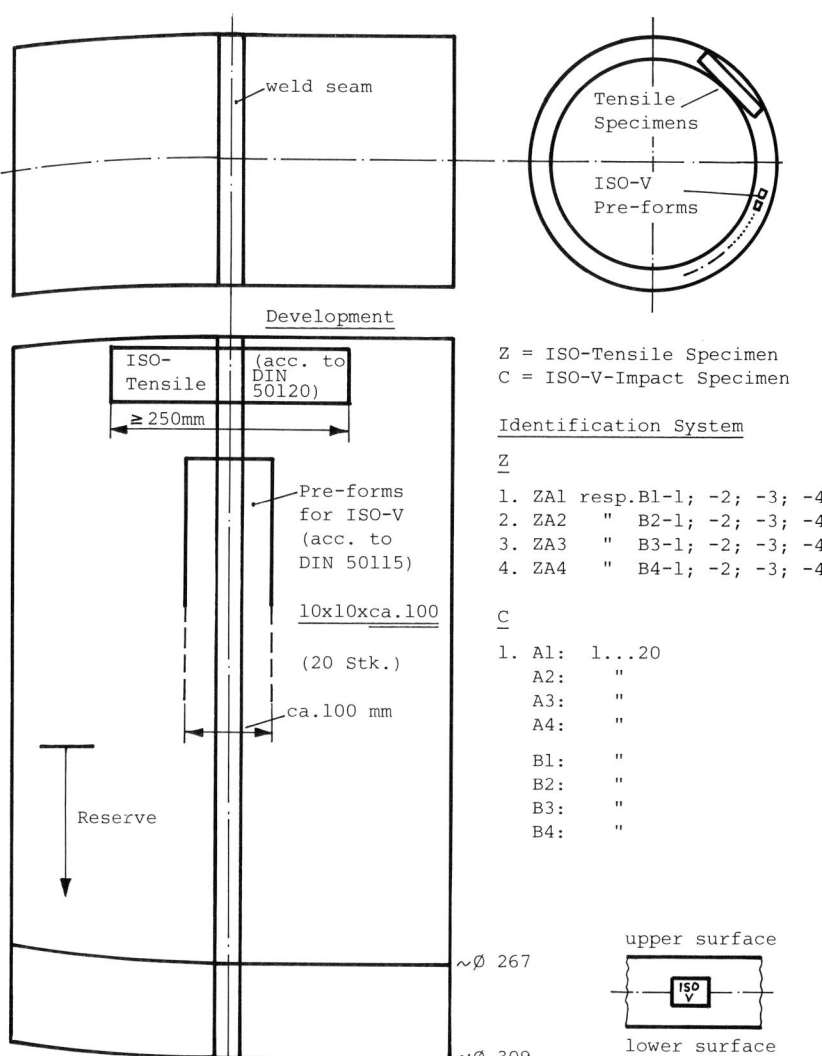

Fig. 1. Specimen Plan and Identification.

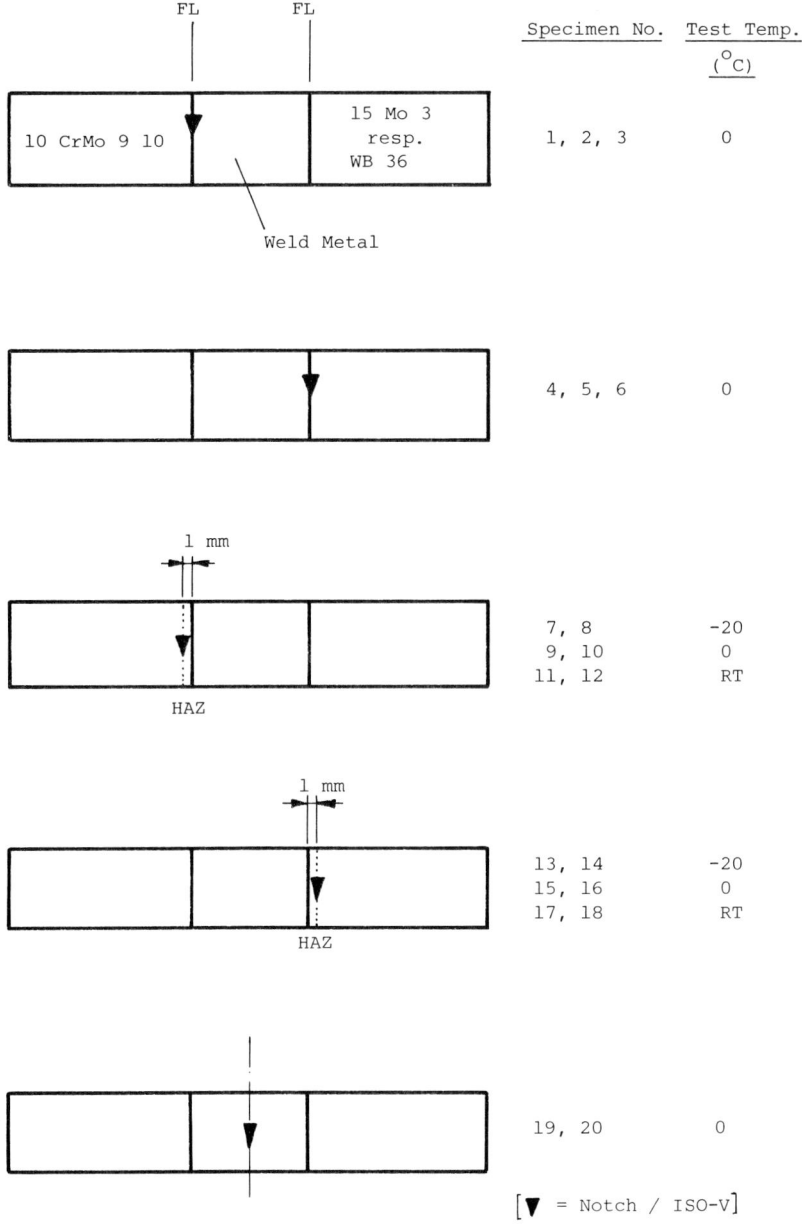

Fig. 2. Notch positions for the Impact Specimens

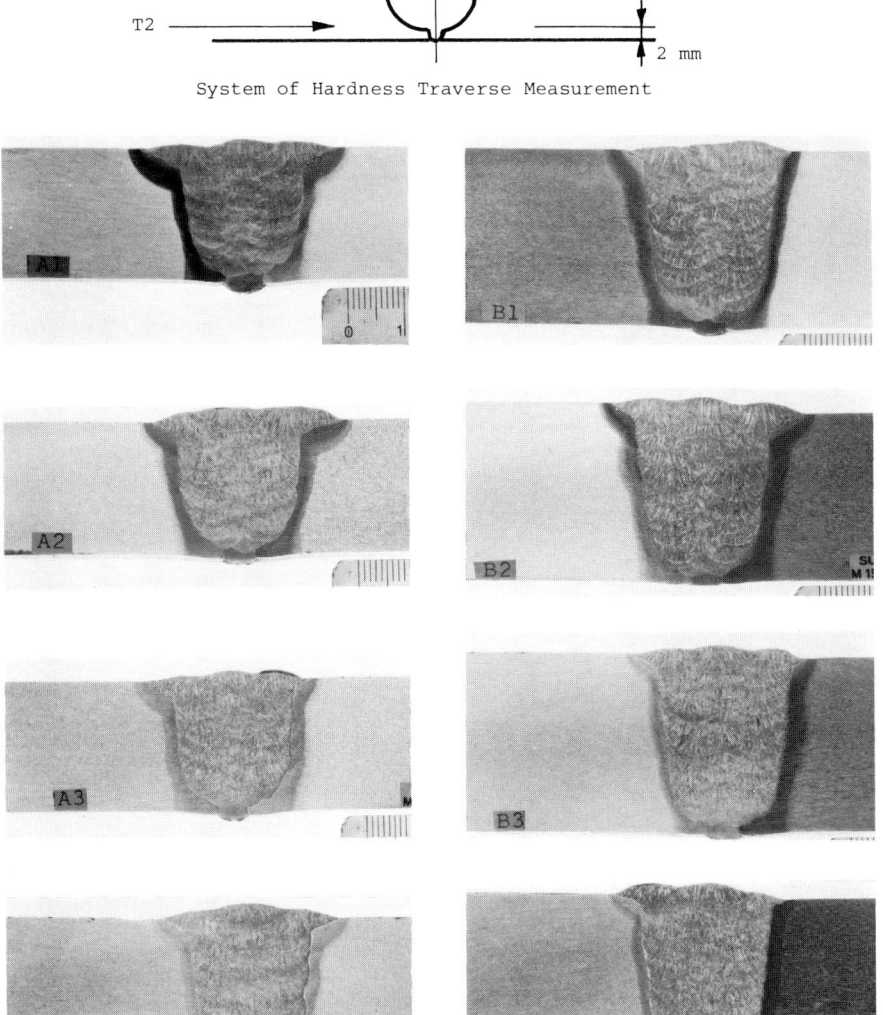

Fig. 3. Macrophotographs of the Joints

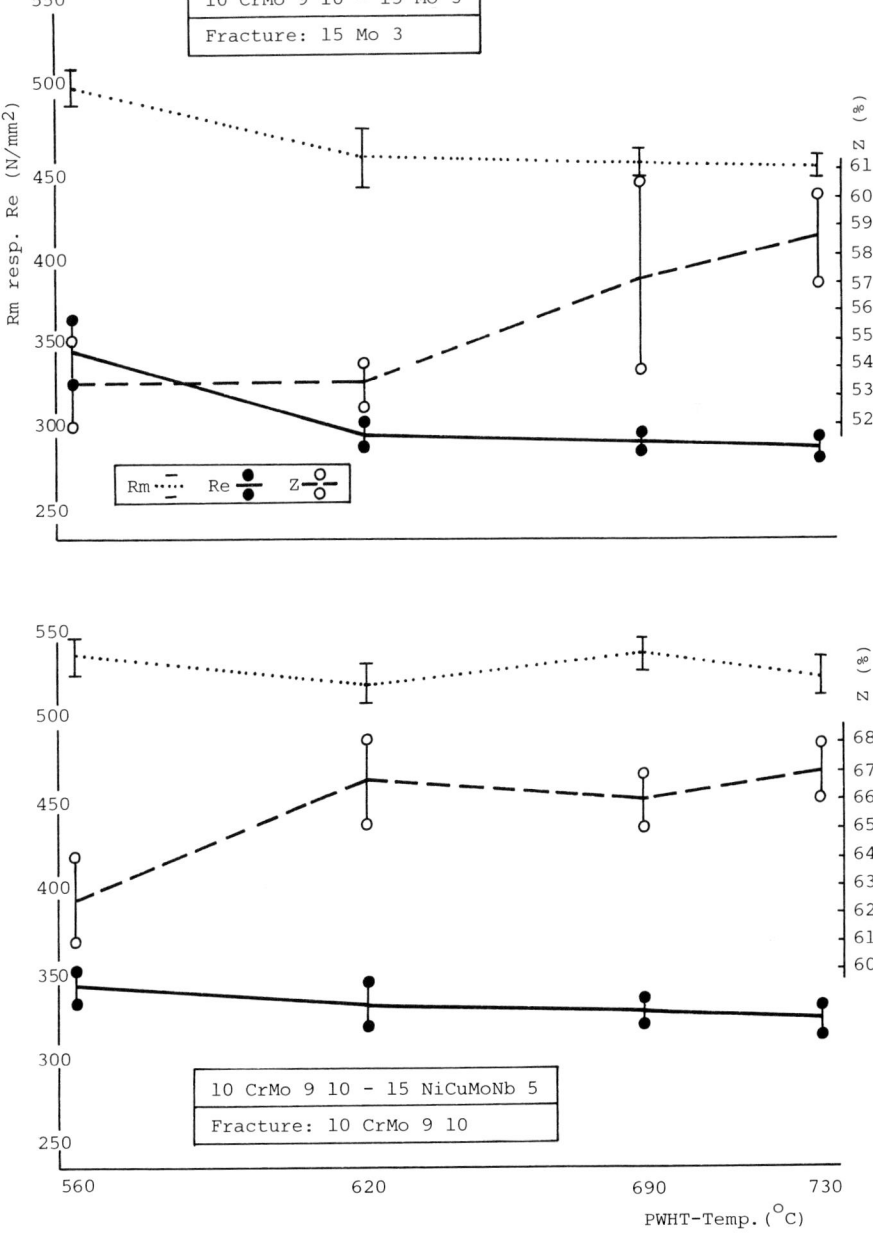

Fig. 4. Tensile Test Results (RT)

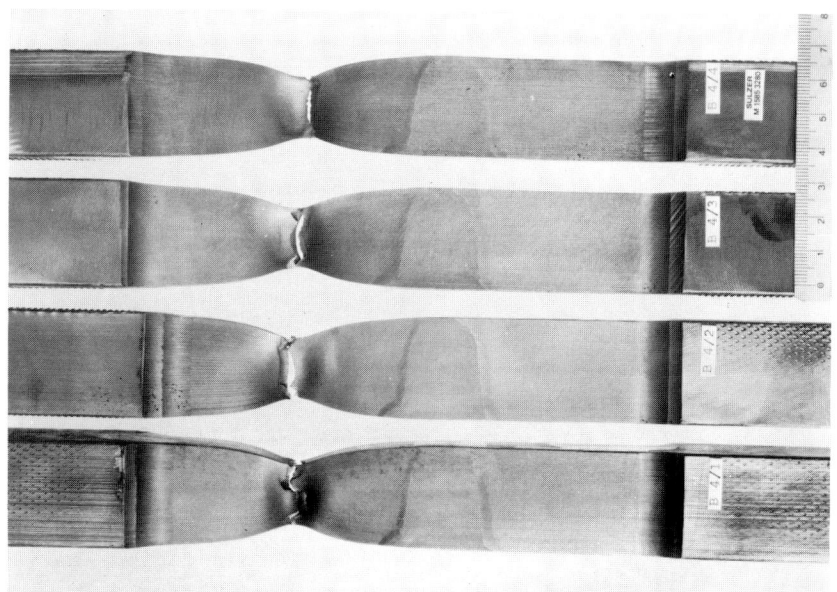

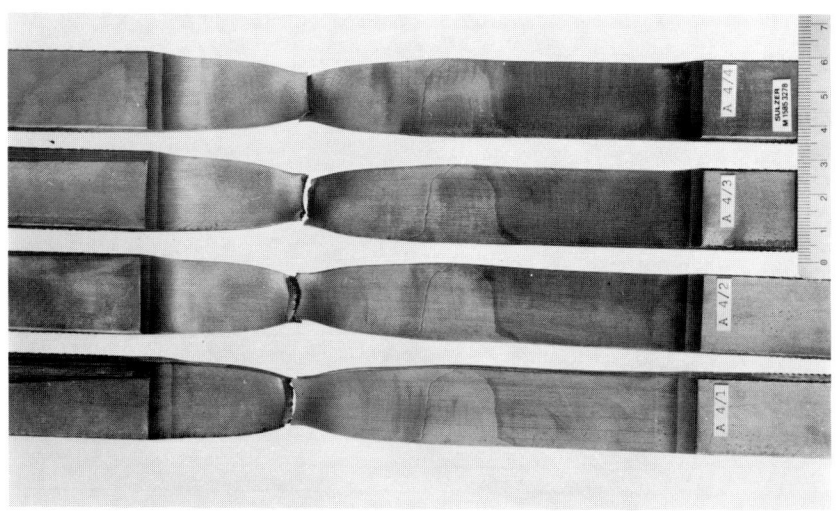

Fig. 5. Fracture zones; Tensile Specimens.

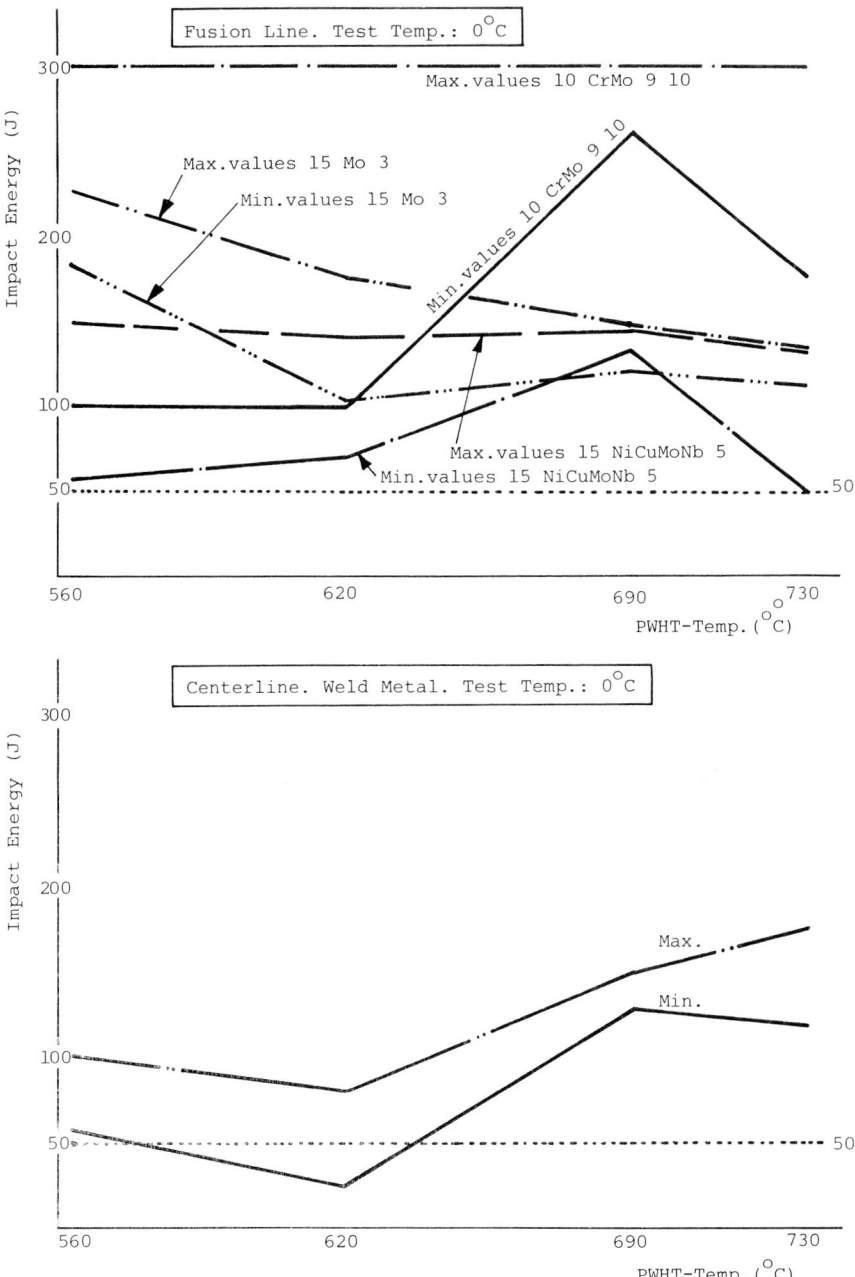

Fig. 6. Impact Energy Scatterbands as a Function of PWHT-Temp.

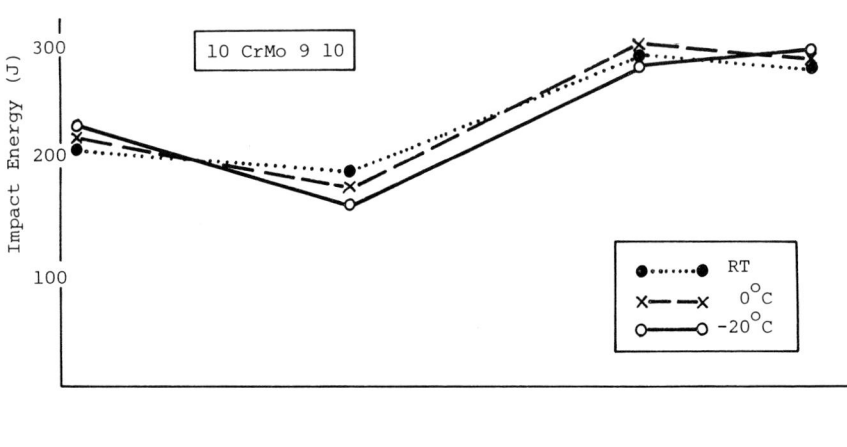
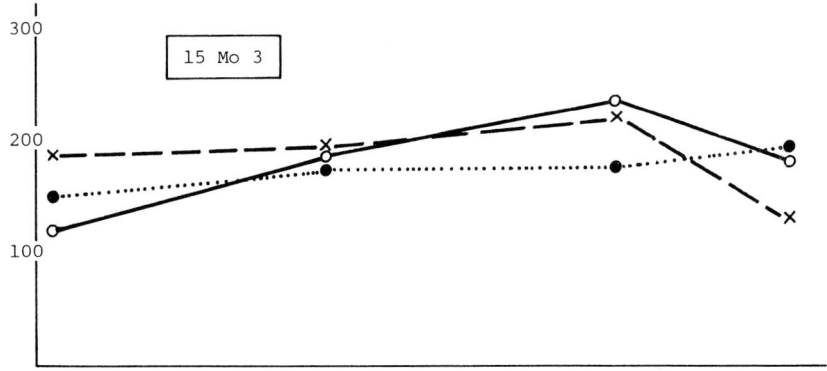
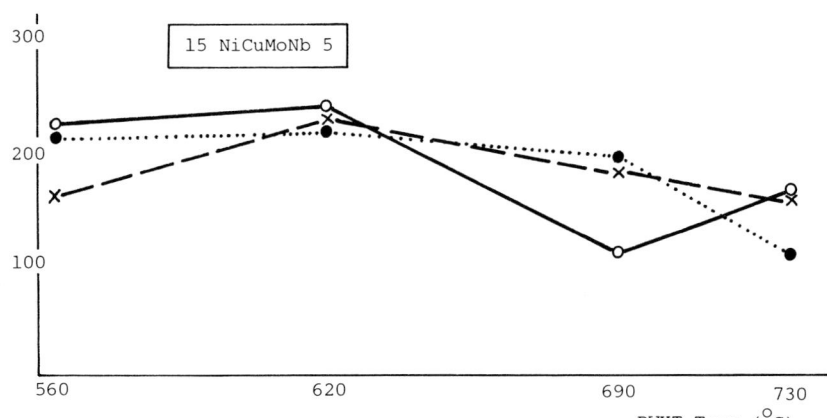

Fig. 7. Impact Energy Minima. HAZ; FL + 1 mm

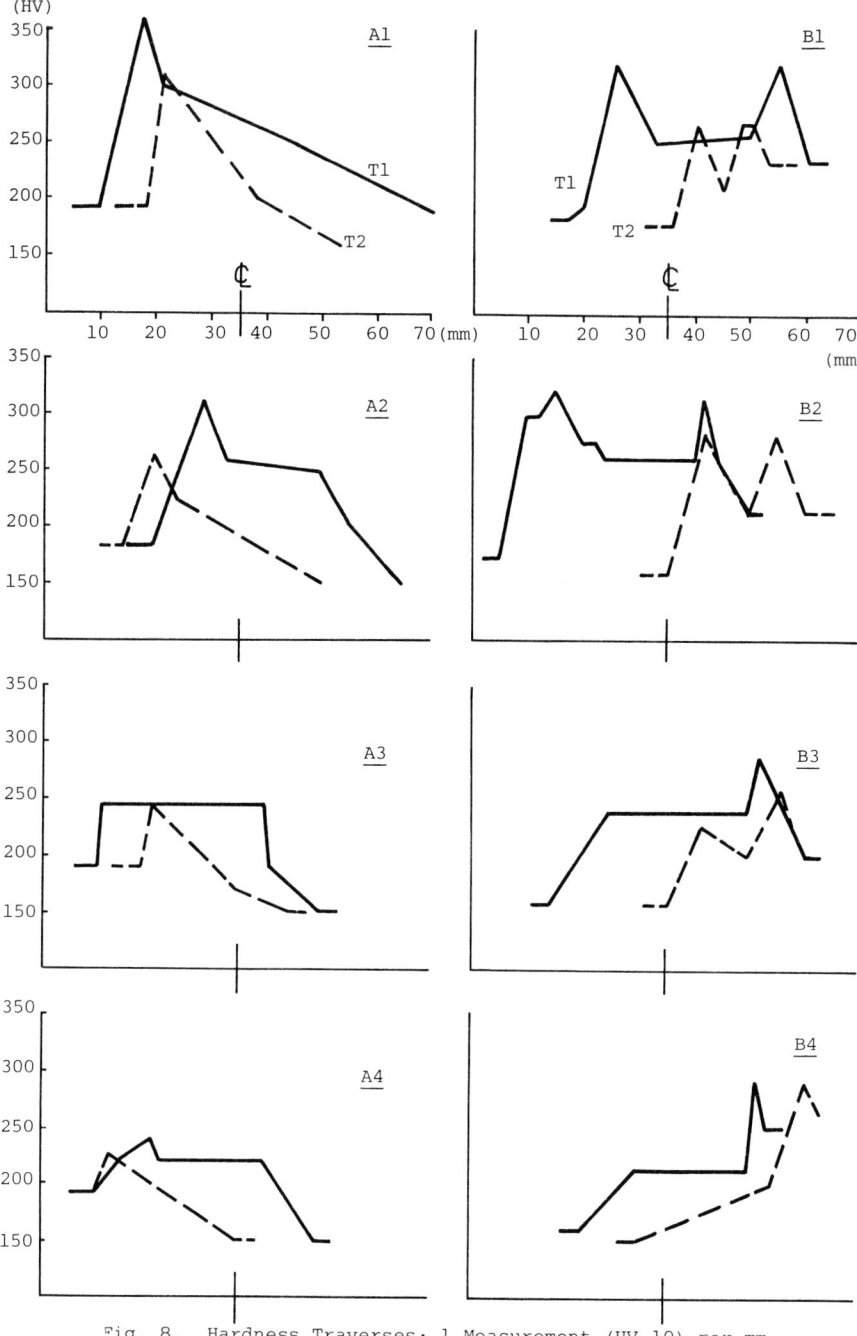

Fig. 8. Hardness Traverses; 1 Measurement (HV 10) per mm.

I.3

Prevision de L'evolution des Caracteristiques Mecaniques de la Zone Affectee et du Metal Fondu Lors du Detensionnement

Ph. Bourges, Ph. Maynier and R. Blondeau

Creusot-Loire Industrie, Centre de Recherches des Matériaux du Creusot, Le Creusot, France

RESUME

La détermination du comportement lors du détensionnement des zones affectées et du métal fondu est souvent difficile à entreprendre de façon systématique. Dans ce but une méthode prévisionnelle a été développée permettant de connaître l'évolution des caractéristiques au cours des traitements thermiques après soudage.

Conformément au système PREVERT déjà développé pour les matériaux de base, la méthode est axée sur la connaissance des constituants de la microstructure obtenue à partir des diagrammes en refroidissement continu et des conditions de soudage. A partir du comportement des différents constituants lors du revenu, il est possible de prédire le comportement global.

Les effets de recristallisation et de réaffectation observés en soudage multipasses sont pris en compte.

Les caractéristiques de dureté, de traction, ainsi que la résilience maximale sont prédites par le modèle. Une comparaison est faite avec les mesures expérimentales.

MOTS CLES :

Soudage - Détensionnement - Caractéristiques mécaniques - Prévision

INTRODUCTION

L'un des problèmes essentiels de l'ingénieur soudeur est de choisir les matériaux et les conditions opératoires qui lui permettront de satisfaire les spécifications de son client.

Si le choix du matériau de base est souvent imposé par des considérations autres que le soudage, le matériau d'apport peut souvent être choisi assez librement. Par ailleurs, les conditions de soudage, généralement très liées au choix du procédé, sont elles aussi de la responsabilité de l'ingénieur soudeur.

D'autre part, le traitement thermique après soudage influe significativement sur les caractéristiques des matériaux.

L'objet du présent exposé est de montrer qu'il est possible par le calcul de prévoir les caractéristiques des différentes zones soudées et donc de réaliser un choix préalable à tout essai ou bien de déterminer l'évolution du comportement d'un matériau donné, en particulier lors du détensionnement.

Cette méthode de calcul concernant les zones soudées s'intègre dans le système général de prédiction PREVERT qui s'intéresse plus globalement aux caractéristiques des aciers au carbone et faiblement alliés, soudés ou non.

APPROCHE METALLURGIQUE

Cas Général

Le système PREVERT est basé sur la connaissance des microstructures obtenues et de leur comportement lors des traitements thermiques.

La prédiction des microstructures est liée à la détermination du diagramme de transformations en refroidissement continu en fonction de la composition chimique et des conditions d'austénitisation (Maynier, 1972 ; Blondeau, 1975). Plus précisément, dix vitesses critiques de refroidissement sont ainsi calculées délimitant les différentes structures et/ou les différentes proportions de structures.

Ces vitesses critiques dépendent de la composition chimique et de l'austénitisation subie par le matériau. Cette austénitisation peut être caractérisée par un paramètre Pa d'équivalence entre le temps et la température, en faisant comme Maynier (1972) l'hypothèse que la diffusion est le mécanisme contrôlant en particulier le grossissement du grain austénitique.

$$Pa = (1/T - R/\Delta H \ \text{Ln} \ t/t_o)^{-1} \qquad [1]$$

où T température en K, t temps, to unité de temps, R constante des gaz parfaits, ΔH énergie d'activation du phénomène.

Il est ainsi possible de déterminer l'effet des austénitisations provoquées par des cycles thermiques extrêmement variés

Des modèles permettent de déterminer la vitesse de refroidissement en fonction des conditions étudiées (cycle thermique, géométrie de la pièce). Il est alors possible de comparer la vitesse obtenue aux vitesses critiques et de déterminer la microstructure par interpolation linéaire (Maynier, 1972).

Des formules permettent de calculer les caractéristiques (Hv, Re, Rm, A %, KCV) des différents constituants de la microstructure. Par combinaison proportionnelle aux pourcentages respectifs de structures, on obtient les caractéristiques de la structure désirée.

Revenu

Le problème du détensionnement peut se traduire en fait par l'introduction de l'adoucissement dans les formules de prédiction de la dureté des structures.

Comme précédemment concernant l'austénitisation, on peut définir un paramètre d'équivalence P entre le temps et la température en faisant comme seule hypothèse que la diffusion contrôle les mécanismes de revenu. Ce paramètre a exactement la même expression théorique que le paramètre Pa.

Comme le montre Pont (1970) et comme reporté sur la figure 1, le Molybdène provoque une variation importante de l'énergie d'activation qui passe de 57,5 kcal/mole pour les aciers sans Molybdène à 100 kcal/mole pour les aciers ayant une teneur supérieure à 0,03 % de Molybdène. Cette dernière valeur n'évolue plus quelle que soit la teneur en cet élément ou en autres éléments carburigènes. Remarquons que la valeur de 57,5 kcal/mole correspond à l'autodiffusion du fer α.

L'invariance de la chaleur d'activation permet d'affirmer la validité du modèle pour tous les cas de traitements de revenus

Une des conséquences de ce résultat est que l'adoucissement au revenu, quelle que soit la structure est une fonction linéaire de l'inverse du paramètre de revenu. On remarque la convergence des courbes pour les revenus importants et l'existence d'un point de cassure (Pc) qui montre pour les revenus à haute température la prédominance de la coalescence sur la précipitation des carbures. La valeur critique ainsi définie peut s'exprimer en fonction de la composition chimique (teneurs en %) :

$1/Pc = 10^{-3}(1,365 - 0,205\ C + 0,233\ Mo + 0,135\ V)$ [2]

A partir de ces considérations, les formules donnant la dureté de la martensite, de la bainite et de la ferrite perlite en fonction de la composition chimique et du paramètre P ont été établies, de même pour la résistance et la limite d'élasticité.

Comme précédemment les caractéristiques réelles sont obtenues par interpolation linéaire.

PREDICTON DES CARACTERISTIQUES DES SOUDURES

Dans le cas des soudures, il existe un certain nombre de spécificités qui doivent être précisées.

Austénitisation

Il est particulièrement intéressant de relier l'austénitisation subie aux conditions de soudage.

Suivant en cela Christensen (1965) et Rykaline (1961) nous pouvons exprimer le temps passé au-dessus d'une température T. Il est alors possible par intégration de la formule [1] de calculer l'austénitisation équivalente à un cycle de soudage.

$$Pa = 1/T_M - R/\Delta H \ln(\alpha E/\lambda (T_M - To))^{-1} \quad [3]$$

où T_M température maximale atteinte (en K), $\alpha = 0,03$ dans le cas des aciers au carbone et faiblement alliés, λ conductibilité thermique, To température de préchauffage (en K), E énergie de soudage (kJ/cm).

Par ce moyen, il est alors possible de déterminer les vitesses critiques tel que précisé précédemment. Cette formulation est d'application générale sauf dans le cas du métal fondu brut de solidification comme précisé par ailleurs.

Vitesse de Refroidissement

Adams (1958) a intégré les équations de diffusion de la chaleur dans le cas du soudage à partir des hypothèses de Rosenthal (1941). Il est ainsi possible de calculer les conditions de refroidissement à partir de l'énergie de soudage et de la géométrie des pièces à souder en tenant compte du mode de refroidissement comme l'a montré Brisson (1972).

Cas du Métal Fondu

Le cas particulier du métal fondu nécessite quelques considérations particulières spécialement en ce qui concerne le soudage multipasses.

Le métal fondu multipasses apparaît formé de différentes zones :

a) Zones brutes de solidification n'ayant pas subi ultérieurement de cycles thermiques au-dessus du point de cassure difini par Pc.

b) Zones réausténitisées par un pic de température au dessus de AC_3.

c) Zones a) ou b) ayant subi ultérieurement un traitement intercritique

d) Zones a ou b ayant subi ultérieurement un traitement de revenu au-dessus du point de cassure.

Les caractéristiques du métal fondu sont une combinaison des caractéristiques des différentes zones. Les proportions de ces dernières peuvent être définies à partir des conditions de soudage en utilisant les formules d'Adams (1958) relatives à la géométrie des isothermes en soudage.

Il est important de noter comme le fait Chaillet (1976) que la trempabilité des zones a) est nettement plus faible que celle des zones b). L'explication réside dans la solidification du métal fondu en phase delta. Il existe en effet un retard à la transformation delta-gamma qui diminue l'austénitisation réellement subie. Le réchauffage ultérieur ne conduisant pas (ou très rarement) à l'apparition de phase delta, l'austénitisation subie alors est comparable à celle d'une zone affectée, significativement plus importante.

Suivant les différentes zones constitutives on applique les formules de caractéristiques valables à l'état brut ou à l'état revenu.

POSSIBILITES DE PREVISION

Actuellement le système PREVERT permet de déterminer la dureté, la résistance à rupture, la limite d'élasticité, l'allongement à rupture dans le métal fondu. Pour des raisons évidentes seule la dureté est prévue dans la zone affectée (Vallon, 1987). Toutes ces caractéristiques peuvent être obtenues avant et après traitement thermique. Ce dernier peut être un simple revenu ou bien comporter un traitement d'auténitisation par exemple une normalisation ou bien une opération de trempe et revenu. Les figures 2 et 3 mettent en évidence les performances de ce système de prévision. Il convient de remarquer que la comparaison relative permet d'accroître considérablement la précision de la prévision.

La résilience ne fait pas encore l'objet de prédictions. Toutefois à partir des formulations existant en métal de base (Maynier, 1976, 1977) l'application à la zone affectée apparaît possible en déterminant un cycle thermique satisfaisant, en particulier pour le soudage multipasses (le problème est analogue à celui de la simulation thermique). En métal fondu, une formulation de la résilience au palier ductile est aujourd'hui disponible.

$$Em = 16 \times 10^4/Rm - 12 \times 10^{-3} [O] - 5,5 \times 10^{-3} [S] - 2 \qquad [4]$$

avec Rm en MPa

[O] [S] en ppm

Cette expression qui met clairement en évidence l'effet des teneurs en oxygène (et accessoirement soufre) présente une excellente corrélation calcul mesure (figure 4) sur environ 60 points de mesure.

Il est important de préciser le domaine d'application en ce qui concerne la composition chimique. Le tableau ci-dessous le précise.

	Zone affecté	Métal fondu
C %	0,05-0,5	0,02-0,15
Si %	< 1	< 0,9
Mn %	< 2	< 2
Ni %	< 4	< 2,5
Cr %	< 3	< 2,5
Mo %	< 1	< 1
V %	< 0,2	< 0,1
Cu %	< 0,5	< 0,5
Al %	0,01-0,05	< 0,05

Dans tous les cas il est nécessaire que la somme Mn + Ni + Cr + Mo soit inférieure à 5 %.

APPLICATIONS

Les applications du système PREVERT sont extrêmement variées. Nous n'en citerons ici que quelques exemples représentatifs.

Le problème de la dureté sous cordon se pose fréquemment. Vallon (1987) présente les résultats obtenus à l'état brut de soudage sur une plage de compositions chimiques beaucoup plus étendue que celle évoquée précédemment. Toutefois il convient de ne pas oublier que l'influence du traitement thermique peut être recherchée et par exemple on peut essayer ainsi de déterminer le traitement thermique adapté à l'obtention d'une dureté souhaitée (inférieure à une certaine valeur ou l'inverse).

Dans le cas du métal fondu, il peut être intéressant de sélectionner préalablement différents produits d'apport en fonction des impositions et des conditions de réalisation (y compris le traitement thermique). Ceci peut par exemple se concrétiser dans la recherche d'un produit d'apport satisfaisant aux impositions après traitement de qualité type normalisation ou trempe-revenu (problème de fonds soudés emboutis à chaud par exemple).

Le système de calcul PREVERT est disponible sur serveur télématique. Ce processus permet de garantir à l'utilisateur une remise à jour des formulations au fur et à mesure des améliorations apportées.

Conclusions

Le système PREVERT permet de déterminer les caractéristiques mécaniques des zones soudées avec une précision satisfaisante. En particulier il permet de déterminer l'influence des traitements thermiques de postsoudage sur ces caractéristiques.

REFERENCES

Adams, C.M., (1958). Welding journal, 20 (5), p 210 s.
Blondeau, R., Ph. Maynier et J. Dollet (1975). Prévision de la dureté et de la résistance des aciers au carbone et faiblement alliés d'après leur composition et leur traitement thermique. Revue de Métallurgie, (11), pp 759-769.
Brisson, J., Ph. Maynier, J. Dollet et P. Bastien, (1972). Investigations of underbead hardness in carbon and low alloy steels. Welding Journal, pp 208 s - 220 s.
Chaillet, J.M., F. Chevet, P. Bocquet et J. Dollet, (1976). Prediction of the microstructure and tensile properties of weld metal deposits. In A.B. Rothwell et J. Malcolm Gray (Ed.), Welding of HSLA (microalloyed) structural steels. ASM pp 298 - 321.
Christensen, N., V. de L. Davies et K. Gjer-Mundsen, (1965). British Welding Journal, pp 54 - 75.
Maynier, Ph., M. TOITOT et J. DOLLET, (1972). Application des lois de la diffusion aux traitements thermiques des aciers. Revue de Métallurgie, Août 1972, pp 502 - 511.
Maynier, Ph., F. Olive et J. Dollet, "Prévision de la forme de la courbe de transition Charpy V des aciers au carbone et faiblement alliés", Revue de Métallurgie, 73 (12), pp 765-772.
Maynier, Ph., J. Lagrange, M. Palmier, A. Ponsot et J. Dollet (1977). "Contribution à l'étude de la cinétique de la fragilisation de revenu réversible des aciers au carbone et faiblement alliés", Mémoires Scientifiques de la Revue de Métallurgie, 74 (12), pp 757-763.

Pont, G., Ph. Maynier, J. Dollet et P. Bastien, (1970). Contribution à l'étude de l'influence du Molybdène sur l'énergie d'activation de l'adoucissement au revenu. <u>Mémoires Scientifiques de la Revue de Métallurgie, 67 (10)</u>, pp 629 - 636.

Rosenthal, D., (1941). Mathematical Theory of heat distribution during welding and cutting. <u>Welding Journal, 43 (5)</u>, pp 220 s - 225 s.

Rikaline, N.N., (1961). Soudage et Techniques Connexes, <u>15 (1/2)</u>, pp5-38.

Vallon, B., Ph. Bourges, Ph. Maynier et R. Blondeau, (1987). Prévision de la dureté sous cordon des aciers au carbone et faiblement alliés pour les soudures monopasse. Document IIS-IIW IX B, 137-87.

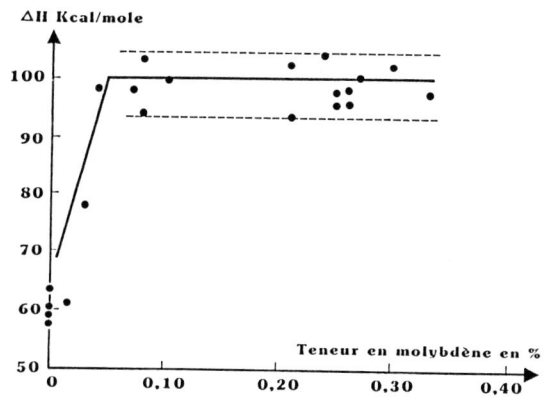

INFLUENCE DU MOLYBDENE
SUR L'ENERGIE D'ACTIVATION AU REVENU

Figure 1

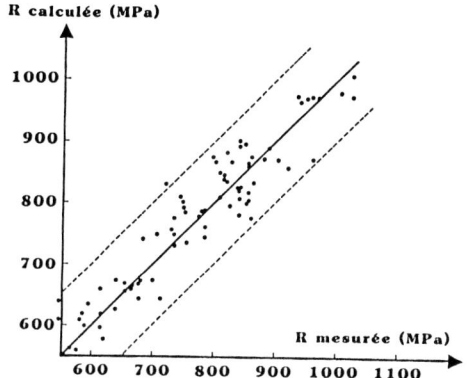

RESISTANCE A LA RUPTURE - ETAT BRUT DE SOUDAGE
CORRELATION CALCUL-MESURE

Figure 2

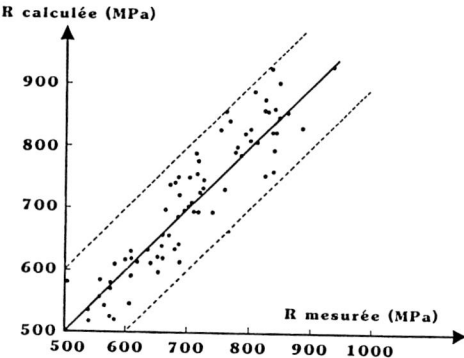

RESISTANCE A LA RUPTURE - ETAT DETENSIONNE
CORRELATION CALCUL-MESURE

Figure 3

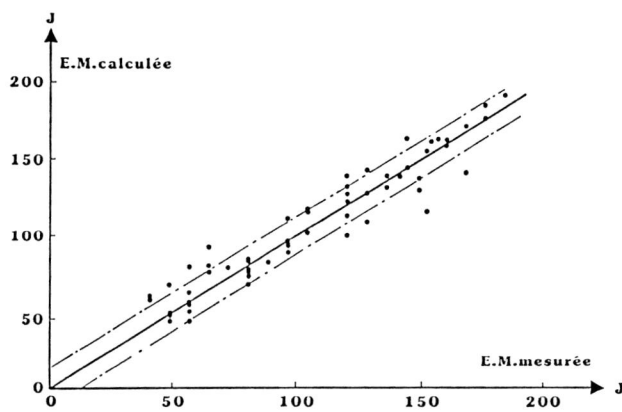

ENERGIE MAXIMALE DE RESILIENCE
CORRELATION CALCUL-MESURE

Figure 4

I.4

Prediction of HAZ Hardness after PWHT

M. Okumura*, N. Yurioka*, T. Kasuya* and H. J. Cotton**

*R & D Labos. II, Nippon Steel Corporation, Sagamihara, Kanagawa-Pref. Japan
**Fracture Department, The Welding Institute, Abington Hall, Cambridge, U.K.

ABSTRACT

In this experiment, the change in hardness at heat affected zones (HAZ) of low alloy high strength steels after post weld heat treatment (PWHT) was investigated. The hardness change was found to be determined by the decomposition of martensite at HAZ and the precipitates of Nb, V and Mo. Based on the experiments, a formula to predict the HAZ hardness after PWHT was proposed.

KEYWORDS

Hardness; heat affected zone; low alloy steel; post weld heat treatment; precipitation hardening; martensite.

INTRODUCTION

Linepipes and pressure vessels for sour gas service are required to have a certain level of reduced hardness at their HAZ (Heat-affected zone) primarily for avoiding stress corrosion cracking during service. The required level of hardness tends to be lower. Some steels cannot meet this requirement in an as-welded conition, therefore post weld heat treatment (PWHT) has to be employed to reduce HAZ hardness. However, the hardness unexpectedly increases after PWHT in some steels, under some conditions.

The prediction of HAZ hardness after PWHT is desired not only for selecting the adequate PWHT condition but also for designing the suitable chemical compositions of steels for sour gas service.

EXPERIMENTS

Laboratory-melt steels were used in order to investigate the effect of V and Nb, which are known as precipitation-hardening elements. Table 1 shows their chemical compositions. The maximum level of V and Nb was 0.10%. The test steels were melted in a 50kg capacity vacuum furnace. They were reheated at 1250°C for one hour and hot-rolled into 20mm thick plates. Then the plates were normalized at 900°C for 20 mins. Commercial basic oxygen converter (BOC) pressure vessel steels were used to examine the validity of a formula proposed to predict the HAZ hardness after PWHT. Their chemical compositions are shown in Table 2.

TABLE 1   Chemical Composition of Laboratory-melt Steel

Symbol	C	Si	Mn	P	S	V	Nb
NV-1	0.090	0.40	1.42	0.007	0.005	0.003	0.003
NV-2	0.088	0.40	1.41	0.006	0.005	0.030	0.003
NV-3	0.088	0.40	1.41	0.006	0.005	0.096	0.003
NV-4	0.093	0.41	1.45	0.006	0.007	0.003	0.012
NV-5	0.093	0.41	1.43	0.006	0.006	0.003	0.066
NV-6	0.088	0.41	1.43	0.006	0.006	0.003	0.110

TABLE 2   Chemical Composition of BOC steel

Symbol	C	Si	Mn	P	S	Cu	Ni	Cr	Mo	V	Nb
A299	0.245	0.26	1.41	0.017	0.007	0.105	0.116	0.150	0.142	<.010	<.010
SB49	0.263	0.25	0.77	0.012	0.006	<.010	0.107	0.124	<.010	<.010	<.005
SB49M	0.144	0.17	0.71	0.007	0.005	0.119	0.056	0.062	0.562	<.010	<.005
A302B	0.176	0.27	1.40	0.006	0.003	<.010	0.189	0.116	0.462	<.010	<.005
A203D	0.105	0.24	0.56	0.007	0.002	<.010	3.328	0.035	<.010	<.010	<.005
Ducol	0.155	0.24	1.32	0.003	0.002	0.066	0.529	0.504	0.224	0.051	<.005

The HAZ hardness is examined generally by the bead-on-plate testing. However, the laboratory-melt steels were not sufficient in quantity to conduct this testing. Instead, the present test employed a method similar to the Implant testing which used a small round bar machined from the laboratory-melt steel. A supporting plate made of mild steel was 20mm thick, 250mm wide and 300 mm long. The Implant-type testing was conducted by a shielded metal arc welding (SMAW) process. The welding heat input was at four different levels, i.e., 0.8, 1.7, 3.0 and 4.5kJ/mm. The welding cooling time between 800 and 500°C, T8/5, were 3.5, 7.3, 15.8 and 29.2s corresponding to each heat input.

The HAZ hardness for commercial BOC steels was examined by a bead-on-plate test with the tapered thickness from 10 to 30mm. Four different heat inputs were employed. The resultant T8/5 ranged between 2.4 and 63.5s.

The extent of this heat treatment is expressed generally as TP (temper parameter), which is as follows:

$$TP = T(20 + \log t)/ 10^3 \qquad (1)$$

where, T (K) is temperature and t (hr) is duration of PWHT. In the present experiment, PWHT was conducted at temperatures from 550 to 650°C and for duration from 1 to 10hrs. As a result, TP ranged between 16.5 and 19.4.

The Vicker's hardnesses were measured along the line parallel to the plate surface and tangent to the weld metal boundary at the point of maximum penetration. This tangential method is according to JIS Z3101. The load level in the Vickers test was 5kg.

RESULTS

Fig. 1 shows a change in the HAZ hardness against T8/5 for the laboratory-melt steels in as-welded condition and post-PWHT conditions. HAZ hardness was seen to decrease after PWHT, and the reduction was more extensive in the shorter time side in T8/5, in which HAZ microstructure is more martensitic. The reduction in post-PWHT HAZ hardness was suppressed to a varying degree, depending on the content of Nb and V ,because of precipitation hardening caused by Nb and V during PWHT.

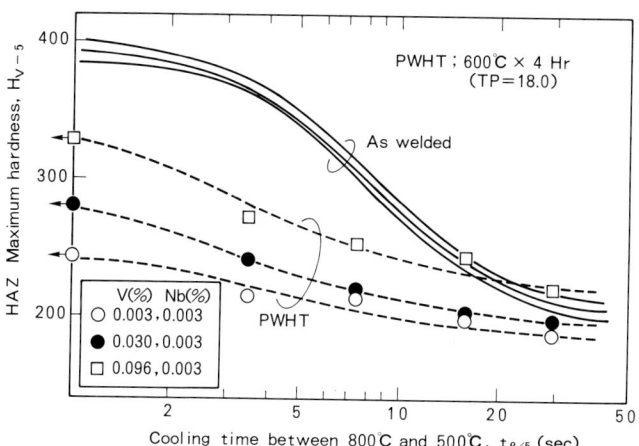

Fig. 1. HAZ hardness change of laboratory-melt steel

DISCUSSION

Basic Equation to Predict HAZ Hardness

The following equations (Yurioka, 1987) predicts the martensite volume and maximum hardness at HAZ of carbon steels and low-alloy steels (C<0.3%, Si<1.2%, Mn<2%, Cu<0.9%, Ni<5%, Cr<1%, Mo<2%, B<2ppm):

$$Hv = (H_M + H_B)/2 - (H_M - H_B) \arctan(x)/2.20 \quad (2)$$

$$x(\text{radian}) = 4 \log\{(T8/5)/T_M\}/\log\{T_B/T_M\} - 2 \quad (3)$$

where, H_M (hardness of 100% martensite HAZ) = 884 C + 294 (4)
H_B (hardness of 0% martensite HAZ) = 197 CEII + 117 (5)
T_M (critical T8/5 for 100% martensite)= exp(10.6 CEI -4.8) (6)
T_B (critical T8/5 for 0% martensite)= exp(6.2 CEIII +0.74) (7)

CEI=C +Si/24 +Mn/6 +Cu/15 +Ni/12 +Cr/8 +Mo/4 (8)
CEII=C +Si/24 +Mn/5 +Cu/10 +Ni/18 +Cr/5 +Mo/2.5 +V/5 +Nb/3 (9)
CEIII=C +Mn/3.6 +Cu/20 +Ni/9 +Cr/5 +Mo/4 (10)

Since the hardness in an as-welded condition, $(Hv)_{AW}$,can be predicted by Eq. 2, the post-PWHT hardness, $(Hv)_{PWHT}$, can be predicted once the reduction in HAZ hardness after PWHT, ΔHv is known:

$$(Hv)_{PWHT} = (Hv)_{AW} - \Delta Hv \quad (11)$$

As indicated in Fig. 1, ΔHv was assumed to depend on the volumetric ratio of martensitic structure in HAZ, M, which is expressed as (Yurioka,1987):
$$M = 0.5 - 0.455 \arctan(x) \quad (12)$$

In Fig. 2, ΔHv for steels NV1 and A299 were plotted against M, which was calculated by substituting the chemical compositions into Eq. 12. The preferable linear relation like these was recognized in all the steels. ΔHv can thus be given as a function M as:
$$\Delta Hv = A M + B \quad (13)$$

The coefficients A and B were considered to be a function of the chemical composition and the temper parameter, TP. If no interactive effect of each element on the coefficients is assumed, A and B may be described as follows:

$$A = a + \sum_{i=1}^{n} f_i(TP) [C]_i^{mi} \quad ; \quad B = b + \sum_{i=1}^{n} g_i(TP) [C]_i^{ni} \quad (14)$$

where $[C]_i$ is the weight percent of ith element.

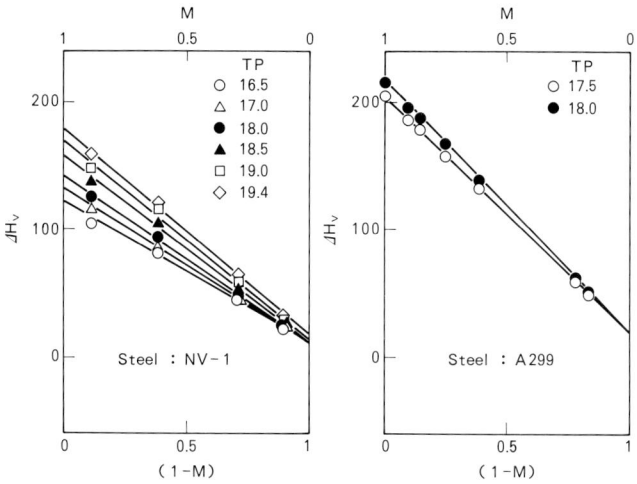

Fig. 2. Change in post-PWHT hardness and martensite volume

Precipitation Hardening of V and Nb

The effects of V and Nb on the post-PWHT HAZ hardness were examined. Fig. 3 shows the relation of ΔHv to square roots of the contents of V and Nb, under various levels of M, when TP was 18.0. Preferable linear relations were observed as shown in Fig. 3, and this linear relation was also observed under the other TP conditions. The negative value ΔHv in Fig. 3 means that HAZ hardness increased after PWHT. Nb was found to be a stronger precipitation-hardening element than V.

The following relation holds against an arbitrarily chosen element, X, under constant TP and constant concentrations of all elements other than X:

$$\Delta Hv = f_x(TP)C_x^{mx} M + g_x(TP)C_x^{nx} + \{a + \sum_{i \neq x} f_i(TP)[C]_i^{mi}\}M + b + \sum_{i \neq x} g_i(TP)[C]_i^{ni}$$

$$= f_X(TP) \, C_X^{m_X} M + g_X(TP) \, C_X^{n_X} + A_X M + B_X \tag{15}$$

where A_X and B_X are constants.

The linear relation of ΔHv to $V^{1/2}$ and $Nb^{1/2}$, found in Fig. 3, leads to derive the following relation:
$$m_X = n_X = 1/2 \quad \text{for } X=V \text{ or } X=Nb \tag{16}$$

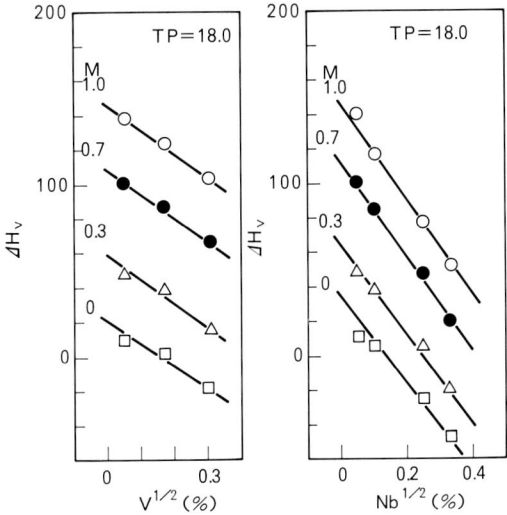

Fig. 3. Post-PWHT hardness change and contents of V and Nb

TABLE 3 Coefficients for Post-PWHT Hardness Change

Element(X)	T.P.	$f_X(TP)$	$g_X(TP)$	A_X	B_X
	16.5	0	-100	105	13
	17.0	-10	-110	117	12
X = V	18.0	-5	-130	125	15
	18.5	0	-140	136	17
	19.0	0	-125	148	18
	19.4	5	-105	160	17
	16.5	0	-225	98	16
	17.0	-10	-240	110	18
X = Nb	18.0	-15	-260	111	22
	18.5	-15	-270	122	27
	19.0	-10	-250	138	27
	19.4	-10	-225	146	28

The coefficient $g_X(TP)$ in Eq. 15 can be given from the gradient of the line in Fig. 3 at M=0. Also, $f_X(TP) + g_X(TP)$ can be given from that at M=1, and then $f_X(TP)$ is determined. The coefficients A_X and B_X were determined in the same manner from the value of ΔH_v at M=1 and M=0, when $C_X = 0$. Table 3 shows the coefficients thus determined, when X=V and X=Nb. As shown in Table 3, $f_X(TP)$ is negligibly small in both case X=V and X=Nb. $f_X(TP)$ was considered to be practically zero. This means that the precipitation hardening of Nb and V had no relation to the martensite volume in HAZ.

$g_X(TP)$ is considered to describe the resistance to hardness reduction per unit concentration of the element X. In other words, its negative value, $-g_X(TP)$, is a factor representing the extent of precipitation hardening. $g_X(TP)$s were plotted against TP, as shown in Fig. 4 and the following relations were found:

$$g_V(TP) = 18\ (TP-18.0)^2 - 138\ ;\ g_{Nb}(TP) = 20\ (TP-18.0)^2 - 268 \quad (17)$$

It was found that the precipitation hardening ($-g_X(TP)$) caused by V as well as Nb appeared at the highest when TP was 18.

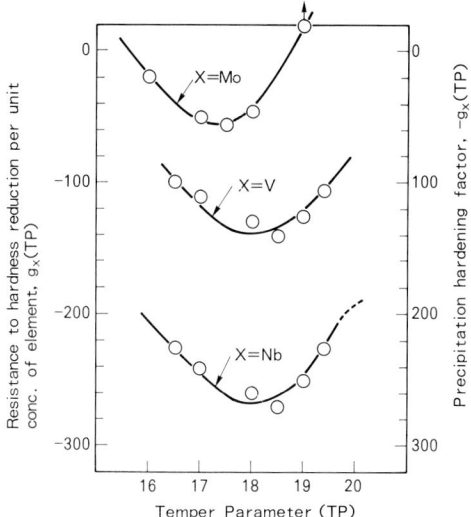

Fig. 4. PWHT hardening due to precipitates of V, Nb and Mo

Hardness of Matrix after PWHT

$(A_X M + B_X)$ in Eq. 15 is assumed to represent the reduction of matrix hardness exclusive of the precipitation hardening.

$$(\Delta Hv)_{matrix} = A_X M + B_X \quad (18)$$

When PWHT is excessively performed, martensitic structures are completely tempered and they are decomposed to those consisting of ferrite and pearlite. If the hardness of this sturcture is expressed by H_F. The maximum value of the coefficients A_X and B_X were estimated as follows:

M=1: $\lim_{TP\to\infty} (\Delta Hv)_{matrix} = H_M - H_F = (A_x + B_x)_{max}$

M=0: $\lim_{TP\to\infty} (\Delta Hv)_{matrix} = H_B - H_F = (B_x)_{max}$

Therefore, $(A_x)_{max} = H_M - H_B \quad (19)$

Based on the results by Kihara (1968), H_F was assumed as follows:

$$H_F = 204 \text{ CEII} + 83 \tag{20}$$

The extreme value of the coefficients, (A_x)max and (B_x)max, is a function of chemical compositions, as seen in Eq. 19. In order to find the relation of A_x and B_x, in Table 3, to the temper parameter, $A_x - (A_x)$max and $B_x - (B_x)$max were plotted against TP, when X = V and X = Nb, as shown in Fig. 5. The relations observed in Fig. 5 give the following equations:

$$(A_x)_{average} = (884 \text{ C} -197 \text{ CEII} +177) + 16.5(\text{TP}-21.5)$$
$$(B_x)_{average} = 26 -7 \text{ CEII} \qquad (16 < \text{TP} < 20) \tag{21}$$

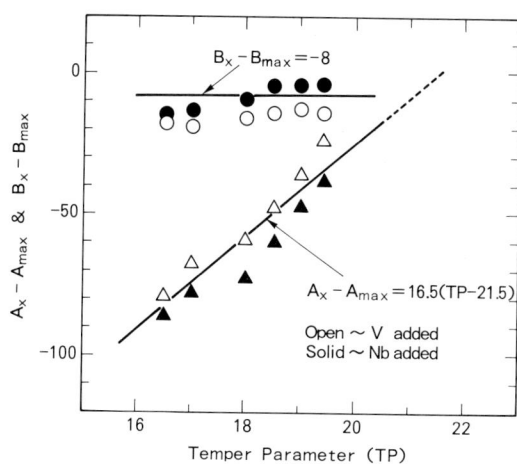

Fig. 5. Post-PWHT hardness change in matrix and TP

Consideration of Effect of Cr and Mo

Cr and Mo are known as precipitation-hardening elements as important as Nb and V. From the experiments by Irvine(1960), precipitation hardening caused by Mo was determined as:

$$g_{Mo}(\text{TP}) = 25(\text{TP}-17.3) -55 \qquad (\text{Mo} < 1\%) \tag{22}$$

Irvine (1960) stated that appreciable precipitation hardening was not observed when steel contains less than 0.5% Cr. This statement was verified in the present experiment, as described later.

Validity of Prediction of PWHT Hardness

By substituting Eqs. 17, 22 and 21 into Eq. 15, the reduction of HAZ hardness after PWHT from that as-welded, ΔHv, is given as:

$$\Delta\text{Hv} = \{884\text{C} +177 -197\text{CEII} +16.5(\text{TP} -21.5)\} \text{ M} + \{18(\text{TP}-18.0)^2 -138\} \text{ V}^{1/2}$$
$$+\{20(\text{TP}-18.0)^2 -268\} \text{ Nb}^{1/2} +\{25(\text{TP}-17.3)^2 -55\} \text{ Mo}^{1/2} -7.0\text{CEII} +26 \tag{23}$$

where, M is given in Eq. 12.

The hardness predicted by Eq. 23 was compared with those measured on the BOC steels whose compositions are shown in Table 2. The results of A299 steel are shown in Fig. 6. Preferable coincidence between the prediction and measurement was recognized for all the BOC steels.

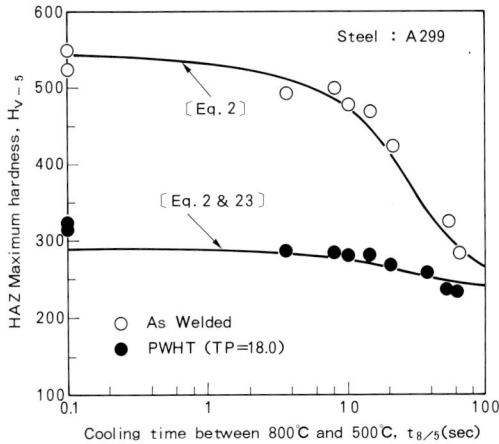

Fig. 6. Comparison of measured HAZ hardness with predicted one

Eq. 23, which does not include the precipitation-hardening effect of Cr, was valid for the BOC steels. This fact means that no precipitation hardening was found in steels with a Cr content of less than 0.5%. However, further investigation on secondary hardening will be necessary for heat resistant steels which normally contain higher Cr.

CONCLUSION

1. A change in the post-PWHT HAZ hardness is determined by precipitation hardnening and decomposition of martensitic structrues.

2. Nb, V and Mo are major elements causing precipitation hardening during PWHT, while no appreciable hardening is found in steels with less than 0.5% Cr.

3. Precipitation hardening by V and Nb appears at the maximum when the temper parameter was 18.

4. Eqs. 2 and 23 preferably predict the HAZ hardness after PWHT as a function of the steel compositions and temper parameter.

REFERENCES

Irvine, K, J. (1960). J. I. S. I., 194, 137-153
Kihara, H. (1968). Welding of high-strength steel, 34, Sanpo Inc. Tokyo
Yurioka, N and M. Okumura (1987). to be published in J. Japan Welding Eng. Soc.

Effets Secondaires des Traitements de Relaxation sur la Tenacite de la Zone Fondue de Soudures Multipasses D'aciers non Allies ou Faiblement Allies

S. Debiez

Institute de Soudure, Paris

RESUME

Une analyse des effets du traitement de relaxation sur la ténacité de la zone fondue est menée sur quatre dépôts de nuances C-Mn, C-Mn-Mo, C-Mn-Ni, en tenant compte des facteurs initiaux : macrostructure, microstructure et composition chimique. On observe l'évolution de la courbe de résilience en fonction des données du traitement, température, durée du palier, vitesse de refroidissement. L'analyse distingue les effets à l'échauffement et durant le palier des effets au refroidissement, elle révèle des influences opposées dont la résultante provoque, soit une évolution aller et retour de la ténacité au cours du traitement avec un bilan positif négatif ou nul, soit un glissement continu défavorable.

MOTS CLES

Traitement thermique de relaxation - Ténacité des zones fondues des soudures.

INTRODUCTION

Le présent exposé concerne le traitement thermique de relaxation des constructions en acier à moyenne résistance travaillant à des températures moyennement basses ou faiblement élevées, essentiellement appliqué pour réduire les contraintes propres produites par le soudage, et la ténacité des zones fondues des soudures des nuances appropriées à ces constructions à savoir :
- C-Mn,
- C-Mn 0,5 % Mo,
- C-Mn 2 % Ni.

L'intention est généreuse puisque le traitement est prescrit pour augmenter la sécurité des constructions. Toutefois, il produit également des effets secondaires qui modifient les propriétés mécaniques de traction et de ténacité notamment, qu'il importe de connaître pour effectuer le bilan final de l'efficacité du traitement à réduire le risque de rupture fragile. Rappelons que le comportement à rupture d'une structure est déterminé par les lois de la mécanique de la rupture qui s'appuient, d'une part, sur le facteur d'intensité de contrainte en rapport avec le champ local des contraintes, et d'autre part, sur la ténacité du matériau. Si le traitement venait à réduire nettement la ténacité des assemblages, son opportunité déjà contestée par certains mécaniciens qui minimisent l'influence réelle des contraintes résiduelles de soudage, pourrait être remise en question. Par contre, si le traitement est un moyen pour améliorer la ténacité défaillante de l'assemblage ou par trop inférieure au métal de base, son utilité se trouverait renforcée. L'effet d'adoucissement généralement produit par le traitement, qui se traduit par une faible diminution des caractéristiques de traction étant assez bien connu et prévisible, nos observations portent essentiellement sur l'évolution de la ténacité évaluée par l'essai le plus simple et le plus courant, à savoir l'essai de résilience, ou plutôt la courbe KCV = f(T) permettant de déterminer une température de transition conventionnelle, par exemple la température TK5 correspondant au niveau 5 daJ/cm^2.

On remarquera qu'en prélevant l'éprouvette de résilience dans la zone fondue d'une soudure multipasse par nature très hétérogène, on fait là une utilisation très particulière de l'essai pour laquelle il n'est pas spécialement adapté. De plus, la valeur de la résilience peut varier selon la position du prélèvement.

La ténacité est déterminée à l'état brut de soudage par trois facteurs principaux qui sont :
- la composition chimique y compris les micro-éléments, les impuretés et l'oxygène ;
- la structure macro et micrographique dont la définition est très complexe pour une soudure multipasse, elle-même déterminée par la composition chimique et le mode opératoire de soudage ;
- le degré de vieillissement tenso-thermique pour les soudures épaisses.

Sans modifier apparemment ces données, le traitement de relaxation effectué à une température au plus égale à 650° peut mettre en jeu des phénomènes fragilisants ou adoucissants tels que ceux qui sont indiqués à la figure 1, capables de provoquer un déplacement latéral de la courbe de résilience qui peut être favorable ou défavorable à la ténacité. Ces phénomènes peuvent produire des effets opposés au même endroit ou à des endroits différents de la zone fondue en raison de l'hétérogénéité de la structure. Certains phénomènes se manifestent durant la montée en température et le palier alors que d'autres n'interviennent que durant le refroidissement. Les influences sont donc multiples et complexes, de ce fait, selon les cas, le bilan final du traitement peut être largement positif ou négatif voire nul.

Notre propos est de décrire les effets du traitement de relaxation en observant les déplacements de la courbe de résilience, en fonction de la structure métallurgique et de la nuance de la zone fondue, de la teneur en Phosphore et des données du traitement. Ces observations ont été tirées d'expé-

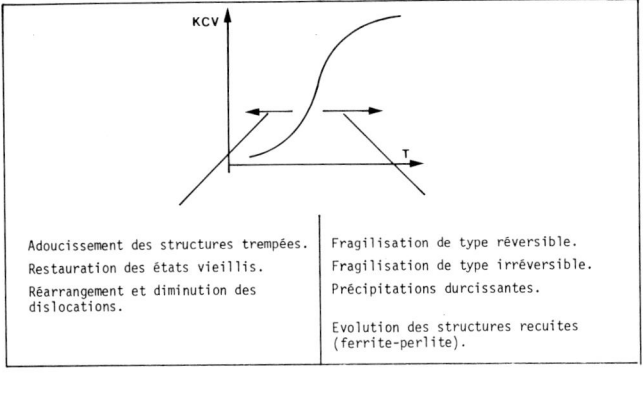

Figure 1 :
Effets secondaires du traitement de relaxation sur la ténacité.

rimentations pour lesquelles on a exigé le respect rigoureux des modes opératoires de soudage afin de figer les structures et l'architecture macrographique de la zone fondue, puis, le tracé du prélèvement de chaque éprouvette sur une tranche de soudure après préparation macrographique, pour réduire le plus possible les causes de dispersion.
Pour mieux comprendre les déplacements de la courbe, on s'est efforcé de distinguer les effets du palier de ceux qui se sont produits durant le refroidissement en procédant à une double détermination d'abord à l'issue du palier après arrêt par trempe à l'eau, puis après le refroidissement normal à la vitesse donnée.

COMPORTEMENT EN FONCTION DE LA STRUCTURE ET DE LA NUANCE

Les observations doivent être faites en considérant la nuance caractérisée par la composition chimique principale, la structure macrographique prise en compte par la section de l'éprouvette de résilience en mesurant les pourcentages de la section occupés par les structures basaltique, transformée surchauffée et transformée affinée, et enfin en identifiant les constituants de la structure tertiaire issus du premier refroidissement. A noter que des éléments en faible teneur qui ne sont pas pris en compte par l'analyse principale, tels que O^2, S, P, Ti, Al, N^2, B, peuvent influencer la ténacité, soit directement, soit en favorisant la formation de certains constituants de la structure tertiaire. Rappelons que parmi ces constituants on distingue, la ferrite proeutectoïde ou primaire, les constituants lamellaires également désignés bainite granulaire ou ferrite avec alignements de phases M.A.C., et la ferrite aciculaire. Les deux premiers étant généralement associés aux mauvais comportements tandis que le troisième associé aux meilleurs.

Nuance		C	Mn	Si	Ni	Mo	Al	Ti	V	S	P	N2	O2
C-Mn	F1	0,08	1,33	0,32			<0,005	0,003		0,010	0,018	0,007	360
C-Mn-Mo	F3	0,08	1,35	0,38		0,50	0,009	0,005		0,013	0,023	0,008	292
C-Mn-Ni	E3	0,05	1,20	0,29	2,0		0,006		0,0029	0,012	0,015	0,009	

Tableau 1 - Composition chimique des zones fondues.

Nous avons choisi pour la présente description, trois produits de différentes nuances présentant des structures tertiaires et des comportements au traitement thermique typiquement différents, à savoir :
- Pour la nuance C-Mn, une combinaison fil-flux en poudre à très faible teneur résiduelle en Ti et Al, dont la structure tertiaire comporte une forte proportion de constituants lamellaires (repère F1).
- Pour la nuance C-Mn-Mo, une combinaison fil-flux en poudre produisant une structure tertiaire constituée par un fin liseré de ferrite primaire et par une matrice de ferrite aciculaire très fine (repère F3).
- Pour la nuance C-Mn-Ni, une électrode à 2 % de Nickel dont la structure tertiaire est composée outre la ferrite primaire, d'un mélange de constituants lamellaires et d'une forte proportion de ferrite aciculaire. Comme pour beaucoup d'électrodes basiques, le dépôt contient des traces importantes de Vanadium (0,029 %).
Les courbes de résilience déterminées pour ces trois assemblages, d'abord à l'état brut de soudage, puis après traitement thermique, sont données sur les diagrammes des figures 2, 3 et 4. On a distingué pour les assemblages sous flux F1 et F3 deux réglages A et B pour mettre en jeu des structures différentes. Au réglage A est associé une position de l'entaille de l'éprouvette de résilience dans la zone médiane où la structure impliquée est

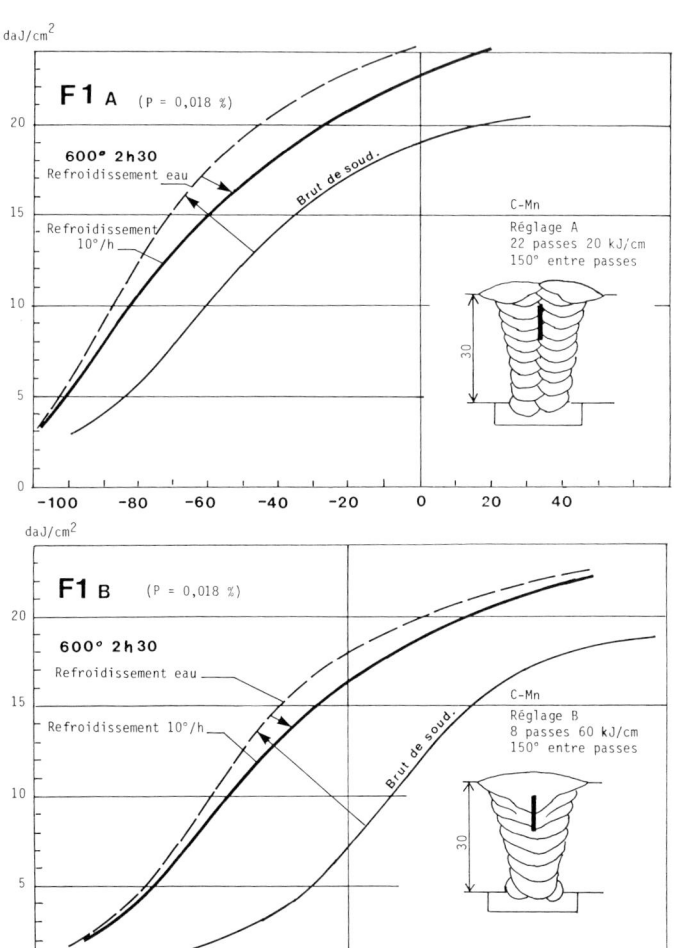

Figure 2 : Evolution des courbes de résilience de la nuance C-Mn sous l'effet d'un traitement thermique à 600°.

transformée, affinée à plus de 80 %. Au réglage B est associé une position de l'éprouvette qui laisse subsister plus de 50 % de structure basaltique non transformée. Cette distinction est indispensable pour la nuance C-Mn car l'écart entre les températures au niveau 5 daJ/cm (TK5) est de 72°. Elle met en évidence la médiocre ténacité des constituants lamellaires à l'état brut de soudage. Pour la nuance C-Mn-Mo, l'écart entre les courbes est bien moins considérable, c'est une conséquence de la très forte proportion de ferrite aciculaire qui caractérise sa structure tertiaire.

Quand on considère les déplacements des courbes de résilience produits par le traitement thermique, on remarque des évolutions d'amplitude et de sens variables selon les nuances. Elles sont favorables à la ténacité pour les nuances F1 et F3, de grande amplitude pour F1 et de faible amplitude pour F3. Par contre, le déplacement est défavorable (vers les températures positives) pour la nuance E3.

En distinguant les déplacements constatés en fin de palier de ceux observés en fin de refroidissement, on décèle pour les nuances F1 et F3 que la montée en température et le palier provoquent un déplacement "aller" vers les

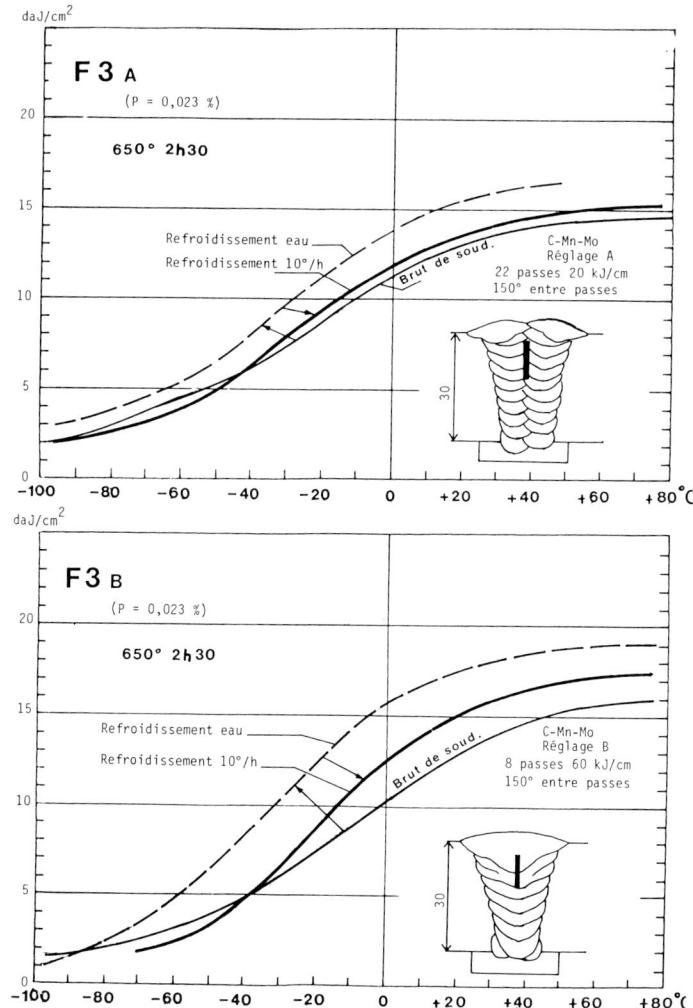

Figure 3 : Evolution des courbes de résilience de la nuance C-Mn-Mo sous l'effet d'un traitement thermique à 650°.

basses températures, alors que le refroidissement produit un déplacement "retour" en sens inverse laissant un solde largement bénéfique pour F1 et pratiquement nul pour F3. On remarquera que les structures fortement basaltiques donnent lieu à des déplacements de plus grande amplitude que les zones affinées. Pour l'électrode E3 le glissement de la courbe se produit continuellement vers la droite au cours du traitement.

Ainsi n'y a t-il pas un comportement unique des zones fondues mais des comportements multiples, propres à la nuance et à la structure qui lui est associée. On remarquera que c'est la soudure handicapée par la plus faible ténacité à l'état brut de soudage qui bénéficie le plus du traitement thermique. De ce fait, la structure de constituants lamellaires n'est plus rédhibitoire quand un traitement thermique après soudage est prévu. Le glissement continuellement défavorable observé pour l'électrode E3 parait être une constante des nuances faiblement alliées au Nickel, toutefois la présence de Vanadium a pu favoriser ce comportement en donnant lieu à des précipités durcissants.

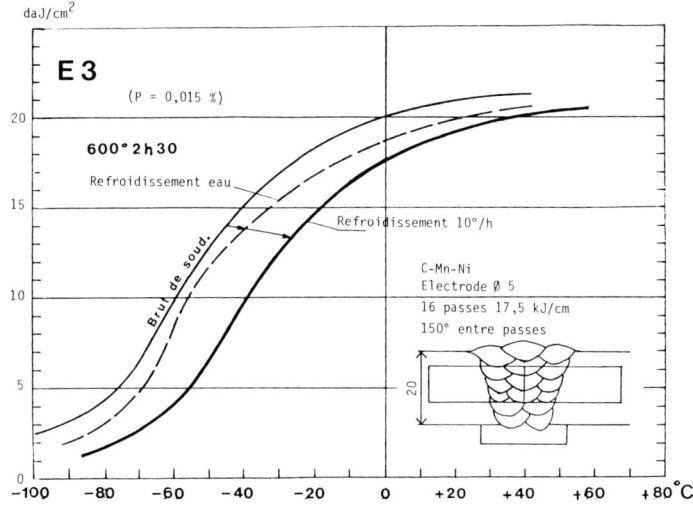

Figure 4 : Evolution des courbes de résilience de la nuance C-Mn-Ni sous l'effet d'un traitement thermique à 600°.

COMPORTEMENT EN FONCTION DE LA TEMPERATURE DU TRAITEMENT ET DE LA DUREE DU PALIER

Chacun a pu remarquer que les plages de température offertes par les codes de construction avaient été quelque peu abaissées ces dernières années. Cette évolution est justifiée par l'efficacité maintenant reconnue de la relaxation des contraintes dès 550° pour les nuances présentement concernées. Ces plages s'inscrivent dans les domaines suivants :
- 530° à 600° pour les nuances C-Mn ;
- 580° à 640° pour les nuances C-Mn-Mo.
Le choix de la température de traitement à l'intérieur des plages autorisées devrait être fait à bon escient en pleine connaissance des effets secondaires sur les propriétés mécaniques. Pour ce qui est des caractéristiques de traction, on peut être certain que les températures plus élevées produiront un adoucissement plus important. Pour la ténacité par contre, des expérimentations entreprises sur les présentes nuances montrent à l'exemple des résultats présentés par la figure 5 au moyen de la température TK5 qu'une même variation de température ne produit pas toujours le même effet quelle soit la nuance. On retiendra en règle générale que lorsque le traitement est globalement favorable, l'élévation de la température entraîne une amélioration de la ténacité, tandis que quand le traitement est globalement défavorable, la diminution de la température contribue à réduire la perte de ténacité.
Bien que le prolongement de la durée du palier au delà du temps nécessaire à l'homogénéité des températures dans l'épaisseur se justifie rarement, il importe de connaître les conséquences d'un prolongement anormal qui peut être, accidentel ou plus simplement imposé aux éléments minces d'un appareil qui comporte par ailleurs des éléments très épais. On trouvera à la figure 6 les résultats de trois expérimentations menées pour comparer les évolutions des ténacités lors de traitements courts et de traitements longs. Le prolongement important de la durée du palier de 2 h 30 à 24 h ne donne pas lieu à des modifications considérables du déplacement de la courbe de résilience.

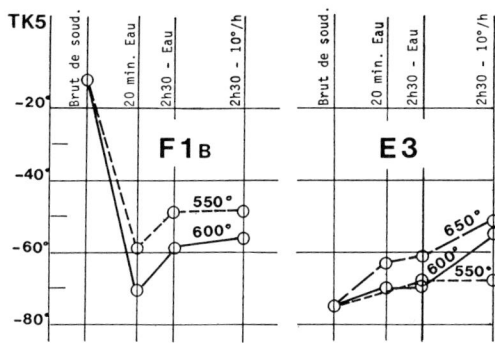

Figure 5 : Influence de la température du palier sur la ténacité.

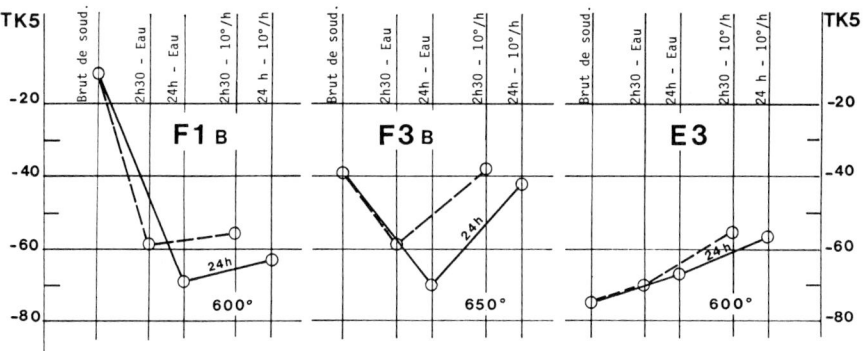

Figure 6 : Influence du prolongement de la durée du palier sur la ténacité.

On remarquera qu'elles sont même favorables pour les nuances F1 et F3 dans le cas du réglage B fortement basaltique. D'autres résultats relatifs au réglage A ont montré par ailleurs une évolution plutôt défavorable pour les zones affinées. On retiendra que contrairement à une idée reçue, le prolongement du palier ne constitue pas un inconvénient réellement dommageable à la ténacité de la zone fondue.

COMPORTEMENT EN FONCTION DE LA TENEUR EN PHOSPHORE ET DE LA VITESSE DE REFROIDISSEMENT

L'effet fragilisant au Phosphore apparait nettement dès l'état brut de soudage comme le montre le diagramme de la figure 7. L'intérêt des très faibles teneurs, inférieures à 0,015 %, est évident et on remarque une saturation de l'effet au delà de 0,030 %. Il est en grande partie responsable de la fragilité apparaissant au cours du refroidissement décrite plus avant, dont l'ampleur est accentuée par les faibles vitesses de refroidissement. Cette fragilisation est bien du type fragilité de revenu réversible semblable à celle qui est abondamment décrite pour les structures trempées faiblement alliées au Cr, Ni, Mo, car elle est en grande partie annulée par un traitement de remise en solution des précipités à 600° ou 650° suivi d'un refroidissement rapide par trempe à l'eau. Ce comportement est illustré par les exemples de la figure 8, qui montrent pour trois structures tertiaires l'évolution de la température TK5 produite par un refroidissement étagé fortement fragilisant de type step cooling et l'efficacité de la réversion produite par la remise en solution. A noter que cette fragilisation se distingue du phénomène classique car elle ne donne pas lieu lors des essais de résilience à une rupture typiquement intergranulaire. Il y a donc pour ces structures de zone fondue un processus particulier qui reste à expliquer.

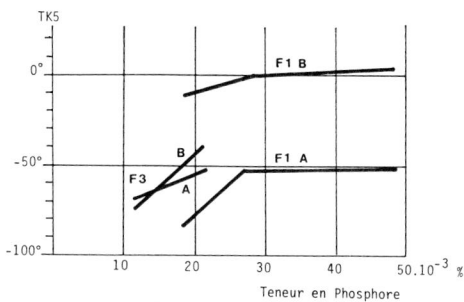

Figure 7 : Influence de la teneur en Phosphore sur la ténacité à l'état brut de soudage.

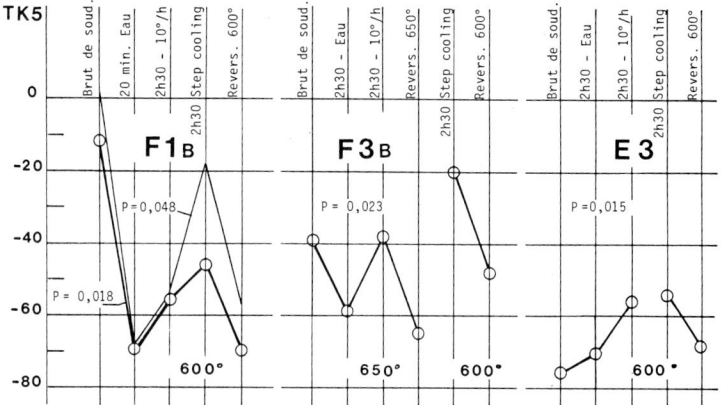

Figure 8 : Influence de la vitesse de refroidissement et vérification de l'efficacité de la réversibilité de la fragilisation.

L'efficacité de la réversibilité de la fragilisation au refroidissement a pu être complètement vérifiée par des expérimentations sur les trois nuances au moyen d'une succession de huit traitements thermiques de 2 h 30 de palier chacun, avec refroidissement intermédiaire jusqu'à 200° à la vitesse de 50° ou de 10°/h. A l'issue d'un tel traitement totalisant vingt heures de palier, on a pu constater que les températures TK5 étaient pratiquement les mêmes que celles déterminées après un traitement continu de 24 heures. Ce qui prouve que chaque mise en température a annulé la fragilité produite par le refroidissement précédent. Il semble qu'il y ait eu par contre cumul des effets produits durant les paliers.

En raison du glissement systématiquement défavorable de la courbe de résilience durant le refroidissement, il est souhaitable de maîtriser le mieux possible la descente en température des traitements thermiques et de rechercher les vitesses les plus rapides, compatibles avec une homogénéité suffisantes des températures dans la construction traitée.

RESTRUCTURATION DE LA TENACITE DEGRADEE PAR VIEILLISSEMENT TENSO-THERMIQUE

Il est un fait que les soudures multipasses de forte épaisseur sur préparation des bords en V ou en X dissymétriques engendrent des déformations locales permanentes notamment au voisinage des racines. Ces déformations peuvent être amplifiées par la rotation angulaire des éléments quand elle n'est pas entravée. Elles donnent lieu au processus de fragilisation par écrouissage à chaud impliquant les éléments interstitiels N^2 et C. Comme la déformation est produite dans la plage de températures de 150° à 400°, le processus est fortement activé. Il se traduit par une différence considérable de la résilience entre la racine et la zone extérieure supérieure. Une illustration de cette différence est donnée à la figure 9 par les courbes de résilience déterminées sur une soudure C-Mn en V, de 40 mm. L'écart de 43° entre les courbes est considérable.

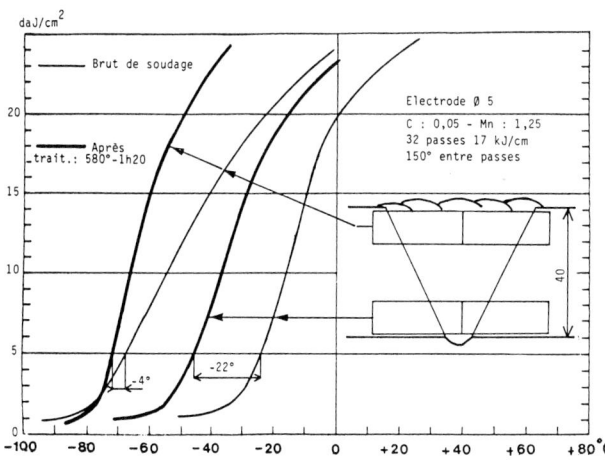

Figure 9 : Restauration de l'état vieilli par déformation à chaud des soudures multipasses par le traitement de relaxation.

A noter que des précautions de rechargement préalable des bords ont été prises pour éviter une modification de la composition chimique par dilution. Comme le traitement thermique appliqué après soudage est capable de restaurer en grande partie l'état vieilli, on observe un déplacement favorable des courbes bien plus considérable pour la racine (- 22°) que pour la peau extérieure (- 4°), alors que la structure est pratiquement la même.

CONCLUSIONS

Le traitement de relaxation peut améliorer ou diminuer la ténacité de la zone fondue des soudures C-Mn faiblement alliées ou non au Mo ou au Ni, selon la structure macrographique, la nature des constituants de la structure tertiaire et la présence d'éléments fragilisants. Pour décider en connaissance de cause de l'opportunité du traitement, il convient donc de connaître ces facteurs et d'être averti du comportement de la nuance. On retiendra particulièrement l'influence fortement bénéfique pour les structures comportant des constituants lamellaires et pour les états vieillis par processus tenso-thermique, de même que le comportement généralement défavorable des nuances faiblement alliées au Ni.

Il faut être conscient que le prolongement du palier ne produit pas de modification considérable de l'évolution, de même que la répétition des traitements n'est pas forcément pénalisante du fait de l'efficacité des processus de réversion.

Par contre, pour minimiser la dégradation systématique au refroidissement, il importe de rechercher les vitesses de refroidissement les plus rapides.

REFERENCES

DEBIEZ, S. et D. CARTAUD, J. TANGUY - Etudes des facteurs influant sur la résilience des soudures multipasses en acier faiblement allié. Rapport I.S. n° 13-470.

DEBIEZ, S. - Synthèse de résultats d'études sur l'influence de différents facteurs affectant la ténacité des soudures multipasses en acier non allié. Soudage et Techniques Connexes - Novembre - Décembre 1982.

DEBIEZ, S. et D. CARTAUD - Etude de l'influence de la durée du palier du traitement thermique de relaxation sur l'évolution de la résilience de soudures multipasses. Rapport I.S. n° 16-662.

DEBIEZ, S. et D. CARTAUD - Etude de l'influence de la multiplication des traitements thermiques de relaxation successifs. Rapport I.S. n° 19-139.

I.6

Influence of Post Weld Treatment on the Fracture Toughness in Welding Procedure Qualification- and Component Welds

J. G. Blauel and W. Burget

Fraunhofer-Institut für Werkstoffmechanik, Freiburg, FRG

ABSTRACT

Standard fracture toughness tests were done using full thickness (60 mm) single edge notched bend and compact specimens extracted from identically fabricated SA-tandem plate- and tube-weldments on a C-Mn steel. Critical values of CTOD were determined at -10°C for the weld metal of double V- and K-joints and for the HAZ of K-joints both in the as welded and post weld heat treated condition. To enable valid fracture toughness testing of specimens in the as welded condition mechanical precompression treatment was applied. Fracture toughness results obtained for weld metal and HAZ of the mechanically- and thermally-stress relieved qualification- and component-welds are compared. Whereas in the aw-condition fracture toughness results from WPQ-tests can be directly applied for tube welds there is some risk of overestimating the tube weld metal fracture toughness from WPQ-tests after p w h t.

KEYWORDS

Fracture mechanics, CTOD-testing, welding procedure qualification test, transferability to components, weld metal, heat affected zone, stress relief treatment

INTRODUCTION

The CTOD-procedure is widely used not only as a screening test to qualify the fracture toughness of different materials and welding procedures but also as the basis of a quantitative assessment of the structural integrity of load bearing members and components in which crack like defects have been found or have to be assumed to exist from non destructive examination arguments. In connection with the rapid development of the offshore technology and the introduction of higher strength steels in this field these concepts have proven successful and are now included in design rules and material specifications.

Since the CTOD is a local crack tip parameter it will depend on the material and mechanical inhomogeneity and anisotropy of the multiple compound that forms a weld-joint and which is a consequence of the materials involved and the welding conditions applied. This requires increased efforts and more detailed analyses in fracture toughness testing than for just plain material (Harrison and Anderson, 1985; Blauel and Burget, 1985). In addition, welding residual stresses depending on the restraints during fabrication and on post weld treatment have to be taken into account when applying fracture mechanics concepts.

In this context the paper describes investigations on weld material and heat affected zone of plate and tube welds and compares the results with respect to transferability of test results to the assessment of welded structures.

MATERIALS AND WELD FABRICATION

The experimental investigations were conducted on multipass submerged arc (SA) welds. The base material used was an offshore-modified TT St E 36 steel with an original plate thickness of 60 mm. The chemical composition and mechanical properties are given in Table 1. Standard welding procedure quali-

TABLE 1 Chemical Composition and Mechanical Properties of Test Material

C	Si	Mn	P	S	Al	Cr	Cu	Mo	N
0.11	0.34	1.52	0.018	0.002	0.026	0.08	0.19	0.02	-

Ni	V	Nb	Nb+V	CE
0.13	0.01	0.025	0.035	0.41

Material	Test Temp. (L - orientation)		UTS	YS	EL	Charpy Energy ($-40^\circ C$)
	$^\circ C$		MPa	MPa	%	J
TT St E 36	+20		516	372	32	246

fication (WPQ) welds on plates and longitudinal (C) welds on construction tubes of the same wall thickness were produced - see Fig. 1. Double V- and K-joint preparations were chosen in both cases to faciliate specimen extraction for weld metal (WM) and heat affected zone (HAZ) testing, respectively. During welding the plates were restrained by strong-backs to approximate the stiffness of the tube and to minimize angular distortion. For all welds the

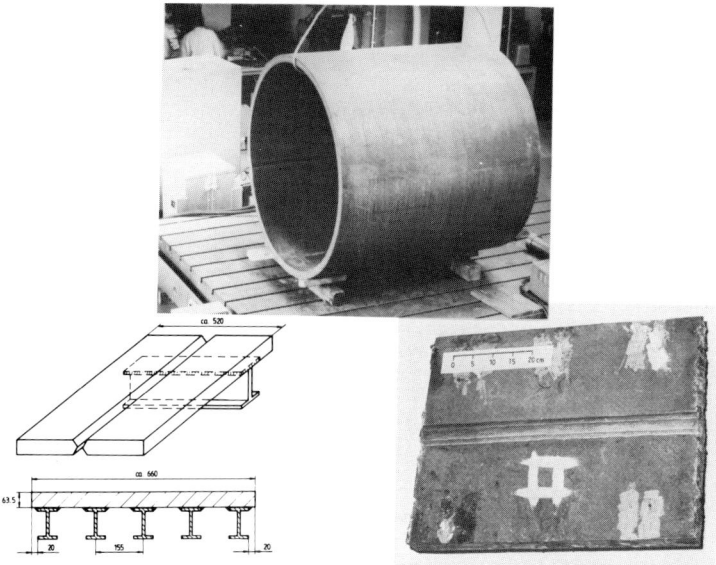

Fig. 1. Procedure qualification weld in a plate and section of a component weld in a construction tube as used in this program

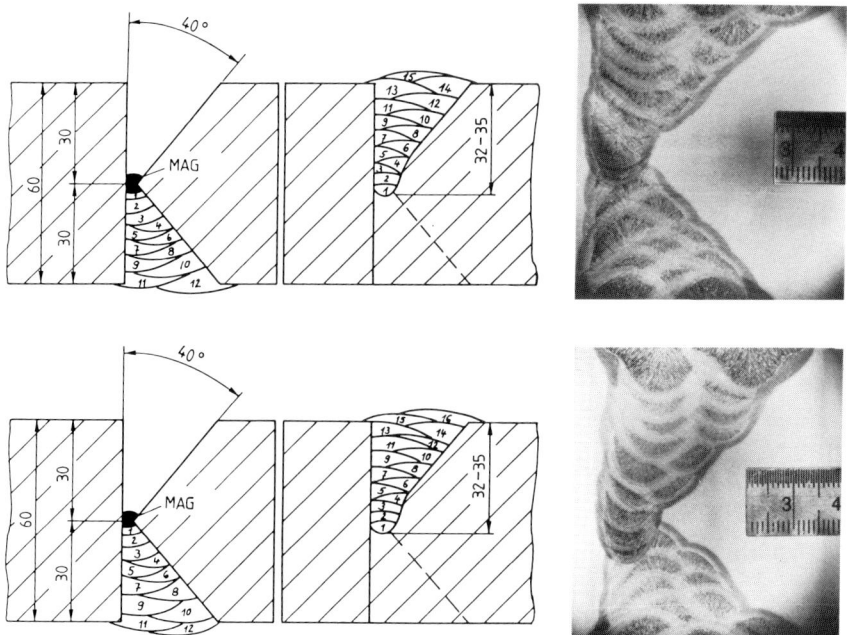

Fig. 2. Welding sequences and macrosections of K-joints in a plate (top) and in a tube (bottom)

wire/flux combination SD 3 / OP 121TT and a heat input of 30 kJ/cm were used. Figure 2 as an example shows the documentation of the welding sequences and the resulting macrosections of two K-joints in a plate and a tube. The tubes were mechanically rounded after welding. One half of each weld was subjected to a post weld heat treatment (pwht) of $570^\circ C/2.5$ h/air before testing.

TEST PROGRAM

Besides conventional material testing using tensile and Charpy impact specimens full thickness square shaped 3-point bend (SENB-) specimens and/or full thickness compact (CT-) specimens were extracted from the plate- and the tube-welds for fracture mechanics testing. The rules of BS 5762 (1979) and ASTM (1984) together with additions specific for the situation of a weld (see later) were followed to evaluate critical values of the crack tip opening displacement (CTOD).

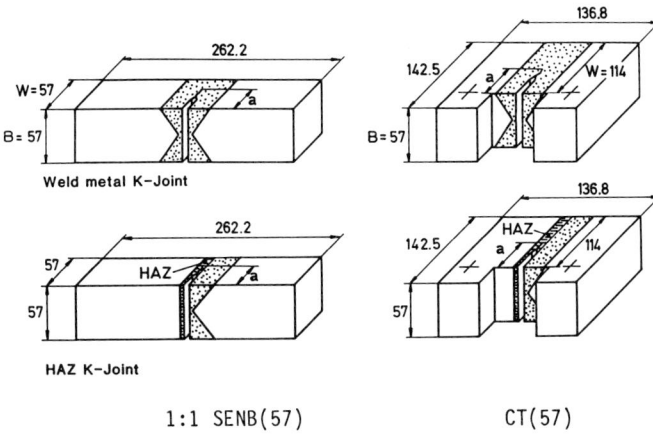

Fig. 3. Fracture mechanics specimens for WM- and HAZ-testing

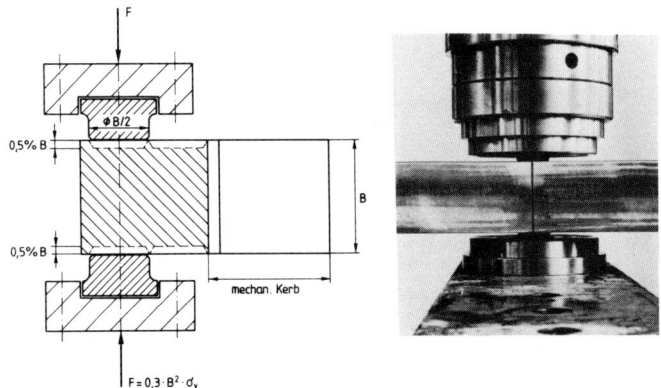

Fig. 4. Stress relief treatment by local plastic precompression

In Fig. 3 the dimensions of the fracture mechanics specimens and the notch positions in the different joints are shown for testing weld metal or heat affected zone. After the extraction the specimens were through thickness notched under metallographic control and were then fatigue precracked to a/W 0.3 or 0.64, respectively. For the specimens to be tested in the as welded (aw) condition a precompression treatment as proposed by Towers and Dawes (1985) - see Fig. 4 - was applied prior to fatigueing to diminish the detrimental influence of welding residual stresses on the development of straight crack fronts.

All fracture mechanics testing was done under displacement control at $-10^{\circ}C$. Computer based data acquisition and storage during the tests included force F, crack mouth opening displacement V, test temperature T, and AC-potential difference $\Delta \varphi$ for crack initiation detection and crack growth measurement. From the stored data critical values of CTOD (δ_{crit}) were determined using the formula and the definitions of BS 5762 (1979):

$$\delta_{crit} = \frac{K^2 (1 - \nu^2)}{2 \sigma_y E} + \frac{0.4 (W-a) V_p}{0.4 W + 0.6 a + z} \qquad (1)$$

where K = stress intensity factor, ν = Poisson's ratio, σ_y = yield stress, E = Young's modulus, V_p = plastic component of crack mouth opening displacement, a = crack length, z = distance of clip gage from test piece surface, W = test piece width,
Index crit = i for onset of stable crack growth
= c for crack instability without prior stable growth
= u for crack instability with prior stable growth
= m for first attainment of maximum load plateau

All HAZ-specimens were subjected to a detailed metallographic examination after the COD-test to control the correct positioning of the fatigue crack front. For a test result to be valid it was required according to actual offshore specifications that either the initiation point of unstable fracture (where it could clearly be detected on the fracture surface) or an adequate length of the fatigue crack front in the central 50 % of the specimen thickness was positioned within the critical coarse grained microstructure of the HAZ.

X-ray residual stress measurements were performed on the surfaces of the different weld joints in plates and tubes before and after sectioning into test pieces.

RESULTS: INFLUENCE OF THERMAL STRESS RELIEVING

Conventional Testing

Tensile properties of transverse specimen from the weld metal tested in the as welded (aw) condition and after post weld heat treatment are given in Table 2. Considering the mean values there is little difference between the plate and tube results whether they are from X- or K-joints.

The mean values of YS and UTS are somewhat lower and the deformation values are somewhat improved for the pwht condition. Comparing the individual values of YS obtained for sub-surface and root test specimens the degree of mechanical heterogeneity of the joint thickness is evident in the case of the X-joints and is less pronounced for the K-joints - the highest individual values are always from the root specimens.

TABLE 2 Tensile properties (transverse) for weld metal in aw and pwht condition at -10°C

Test piece Joint-preparation	Specimen Location	YS MPa aw	YS MPa pwht	UTS MPa aw	UTS MPa pwht	El. % aw	El. % pwht	RA % aw	RA % pwht
plate X	1. side	417	348	539	479	20	31	79	78
	2. side	-	461	-	570	-	30	-	74
	root	578	511	646	612	12	-	73	73
tube X	1. side	462	472	559	587	33	30	80	75
	2. side	438	470	549	564	33	28	79	77
	root	519	548	623	641	18	-	76	73
plate K	1. side	464	496	566	587	31	30	77	75
	2. side	435	496	562	587	36	29	81	75
	root	591	501	689	685	16	-	62	70
tube K	1. side	512	497	618	584	36	30	77	77
	2. side	503	502	618	587	27	27	77	77
	root	593	519	722	667	-	-	79	70

2. side

root

1. side

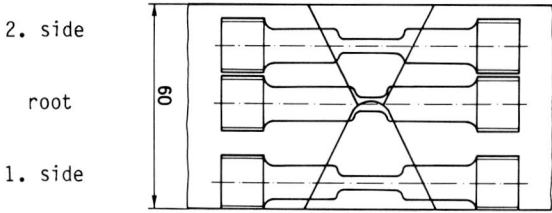

Charpy impact tests at -10°C using transverse specimens delivered values of 200 ± 20 J for the weld metal and the heat affected of X- and K-joints in the plate and in the tube; no influence of the thermal stress relief treatment was found; individual values were lowest in the respective root areas (Burget and Blauel, 1987).

Weld Metal Fracture Toughness

In Fig. 5 the results of weld metal CTOD-tests on X-joints in a plate using different specimen geometries are summarized. The fracture toughness is clearly improved by a post weld stress relief heat treatment - and this is evident for both specimen geometries. For both material conditions investigated subsidiary bend specimen tests have lead to higher fracture toughness results compared to compact specimens of the same ligament size. Crack

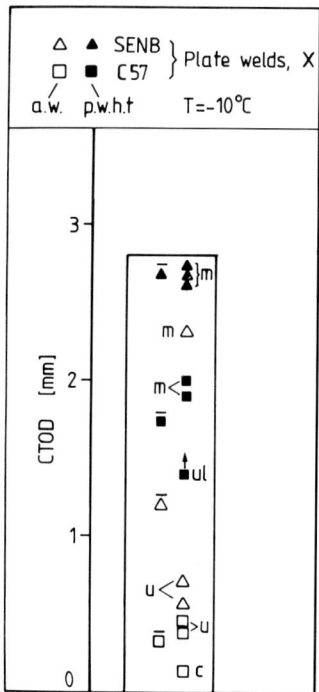

Fig. 5. Fracture toughness of weld metal in a plate X-joint comparing SENB- and C-specimens (symbols with a bar indicate mean values)

length measurements on fracture surfaces of the broken SENB-specimens revealed smaller increments of stable crack growth for maximum load (m) as compared to the CT-specimens. The above observations indicate a different crack growth behaviour with a steeper crack resistance curve due to diminished constraint for the bend specimens. This effect is also responsible for the higher toughness values obtained for bend specimens at instability (u).

In Fig. 6 (left) the CT-specimen results of the plate X-joint are compared to corresponding results obtained for the tube weld. Again the improvement of toughness by pwht is obvious. In the as welded condition almost identical results were obtained for the plate- and the tube-weld. In constrast weld metal toughness in the pwht condition is a factor of 2 higher for the WPQ-joint as compared to the component weld.

An explanation for this result may be found from differences in the effect of thermal stress relief treatment as a consequence of different original residual stress distributions in the WPQ- and tube-weld. Since the degree of restraint during fabrication is higher for the tube higher residual stresses (and stress gradients) are expected in the tube joint. In the course of thermal stress relief treatment residual stresses are diminished through the occurrence of plastic deformation which is mainly confined to the microstructural constituent with the lowest yield strength at stress relief temperature. In the case of C-Mn weld metal this will be the grain boundary ferrite. As a consequence of the higher residual stresses more plastic deformation will take place on the grain boundary ferrite during pwht in the tube weld as compared to the plate weld. Following the model of Tweed and Knott (1982) this will enhance low toughness cleavage type failure in the CTOD tests for the tube.

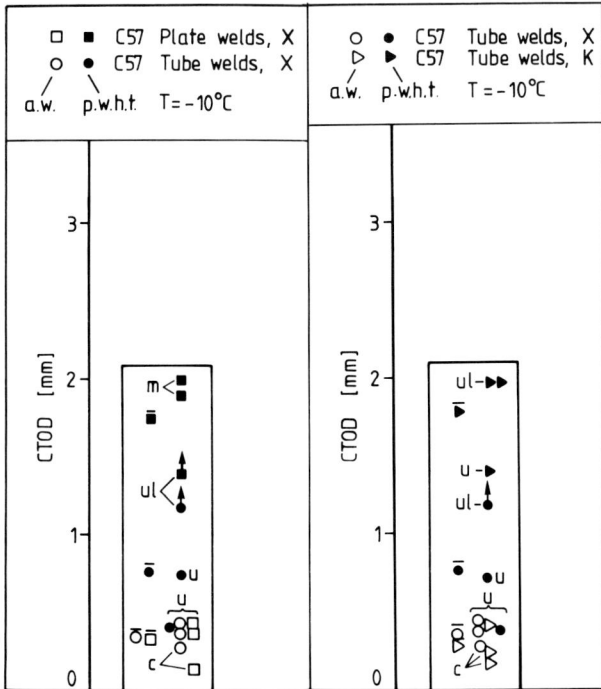

Fig. 6. Fracture toughness of weld metal comparing X-joints in plate and tube (left) as well as X- and K-joints in tube (right)

Finally in Fig. 6 (right) results of tube-welds are plotted to show the influence of joint geometry on the weld metal toughness. In the aw-condition both tube-welds gave almost the same fracture toughness results. In the pwht-condition the results for the K-joints are clearly higher than for the X-joints and the difference is of the same magnitude as that found in the left part of the figure comparing plate and tube-welds. Again it can be argued that this result is a consequence of higher restraint in the X-joint during welding compared to the K-joint leading to higher residual stresses, more local plastic deformation during pwht, and therefore increased probability for low toughness fracturing. In addition constraint may be lower in the case of K-joint as compared to X-joint specimens due to the near presence of lower yield strength base material over the whole specimen thickness at one side of the crack plane.

The "mechanical stress relieving treatment" of as welded specimens unlike the thermal stress relief treatment obviously does not have a comparable influence on the weld metal toughness.

During the fracture experiments weld metal toughness values have also been determined for the initiation of stable crack extension using an alternating current potential drop method. Initiation toughness values determined by evaluating the first distinct change in slope of the potential-displacement curves after the potential minimum are shown in Fig. 7. This method of

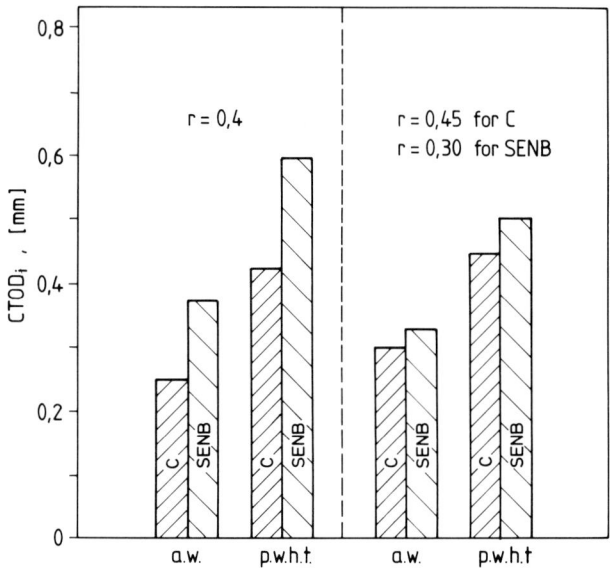

Fig. 7. CTOD at initiation of slow crack growth evaluated using a constant rotational factor r = 0.4 and variable rotational factors for the different specimen types

crack initiation detection was confirmed by early unloading some of the specimens and evaluating the fractographic observations on the fracture surfaces. The evaluation of the bend specimen results over those of the C-specimens may again be a consequence of their lower constraint. Taking this into account by using specimen geometry dependent factors of rotation - r = 0,45 for the CT- and r = 0,3 for the SENB-specimens (Dawes, 1978) instead of r = const. = 0,4 as in Eqn.(1) - the δ_i-values for both specimen types come close.

HAZ Fracture Toughness

HAZ fracture toughness values obtained for the WPQ-plate weld from SENB- and CT-specimens are summarized in Fig. 8. In the as welded condition two valid HAZ-CTOD (δ_u) results were obtained on SENB-specimens. For the third bend specimen metallographic investigation showed that unstable fracture (δ_c) had initiated in the weld metal. The highest toughness value out of the series of three CT-specimens is due to the fatigue crack front being located in the fine grained HAZ and in the parent plate material. The lowest δ_c-value is an invalid HAZ-result because fracture initiated in the weld metal.

In the post weld heat treated condition only one valid HAZ test result (δ_u) was obtained for each specimen type. The CT-specimen with the lowest δ_u result sampled fusion boundary material in the central part of the fatigue crack front. The results in Fig. 8 show the same trend as in Fig. 5 but because of the very limited number of valid results no final conclusion can be drawn concerning the influence of specimen geometry on HAZ-fracture toughness. But it is obvious that metallographic post evaluation of each broken specimen is essential to avoid misleading results.

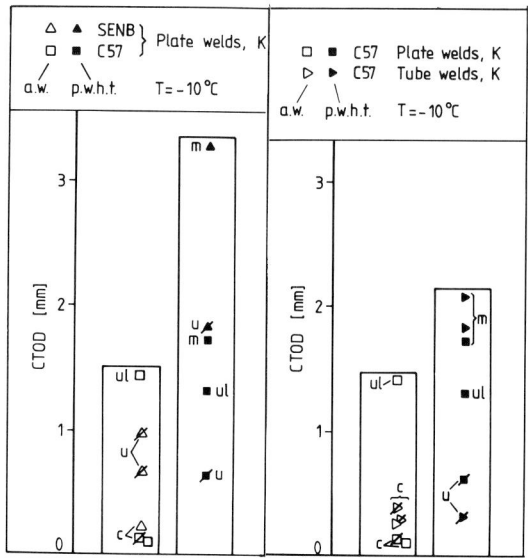

Fig. 8. HAZ fracture toughness
left: K-joint comparing SENB- and CT-specimens
right: CT-specimens comparing K-joint in plate and tube
(only symbols with bar indicate valid HAZ results)

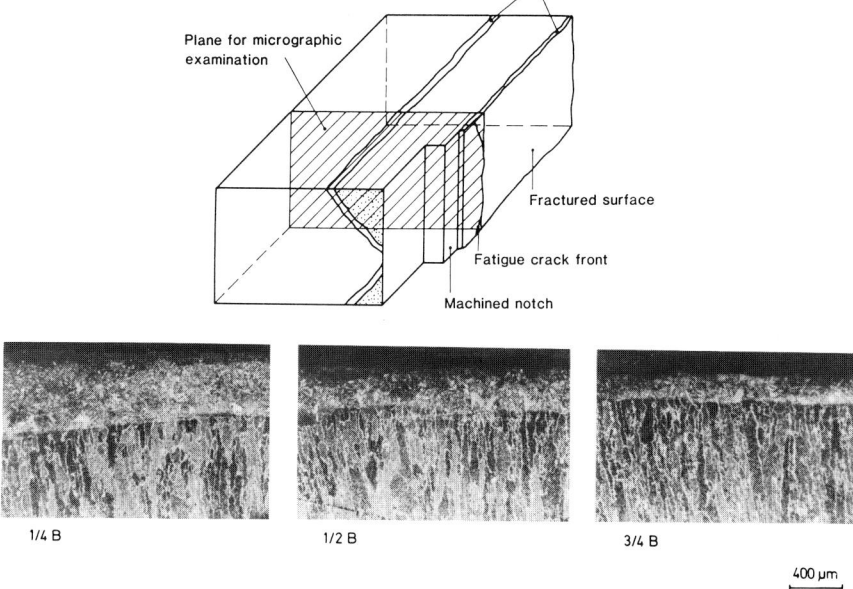

Fig. 9. Micrographic examination of correct crack front position in HAZ specimens

In Fig. 8 (right) results from WPQ- and component-welds are compared. All valid HAZ toughness values for the aw material condition are in the same scatter band; they failed by cleavage fracture without any preceeding stable crack extension. For the HAZ tests in the pwht-condition two valid test results were obtained which are of the same order of magnitude as for the aw HAZ material. As an example Fig. 9 shows the micrographic examination of the compact specimen extracted from the pwht tube weld giving a result δ_c = 0,34 mm.

CONCLUSIONS

1) In the as welded condition
   fracture toughness values determined for welding qualification (plate)- and component (tube) welds are almost the same. This was found for double V- and K-joints and using mechanical stress relieving by precompression in all tests. The mean values of CTOD from these tests are all higher than 0,25 mm, but a number of pop-in events have been recorded; the lowest individual result was obtained for the HAZ giving a CTOD value of 0,15 mm.

2) Post weld heat treatment
   improves the fracture toughness of the weld metal but the degree of improvement depends on the original residual stress distribution in the respective joint which in turn is determined by the groove geometry and the restraint conditions during welding.

   Specifically for the weld metal investigated a larger fracture toughness was found for the WPQ-plate as compared to the tube and for the K-joint as compared to the double-V-joint.

   Whereas all WPQ-plate results after pwht showed maximum load behaviour for the weld metal with δ_m-values above 1,3 mm, two tube specimens fractured after only little stable crack extension had occured at a δ_u-value of 0.42 mm.

   Little or no improvement of toughness by a post heat treatment could be derived from the limited number of valid (due to micrographic examination) HAZ test results.

3) Fracture toughness values for instability and maximum load determined using single edge notched subsidiary bend specimens are higher than those obtained from compact specimens with identical ligament dimensions; the difference is caused by lower constraint in bend specimens leading to different crack growth resistance behaviour; true initiation values are nearly independent of specimen geometry.

In summary these results show: Whereas in the aw-condition fracture toughness results from WPQ-tests can be directly applied for tube welds there is some risk of over-estimating the tube weld metal fracture toughness from WPQ-tests after pwht.

REFERENCES

Anderson, T.L., H.I. Mc Henry and M.G. Dawes (1985). Elastic-plastic fracture toughness tests with single edge notched bend specimens. In E.T. Wessel and F.J. Loss (Eds.), Elastic-Plastic Fracture Test Methods, ASTM STP 856, Amer. Soc. Testing Mat., Philadelphia, USA, pp. 210-229.

ASTM (1984). Draft test method for crack tip opening displacement (CTOD) testing. Working Document of ASTM Com. E24, Feb. 84, unpublished.

Blauel, J.G. and W. Burget (1985). CTOD-testing of welds. In K.H. Schwalbe (Ed.), The Crack Tip Opening Displacement in Elastic-Plastic Fracture Mechanics, Workshop on CTOD Methodology, Geesthacht, Germany, April 23-25, 1985, Springer-Verlag, Berlin/Heidelberg, 1986, pp. 225-250.

BS 5762 (1979). Methods for crack opening displacement (COD) testing, British Standard 5762:1979, British Standards Institution, London UK.

Burget, W. and J.G. Blauel (1987). Fracture toughness of welding procedure qualification- and component weld tested in SENB- and CT-specimens. In J.G. Blauel and K.H. Schwalbe (Eds.), Welding Fracture Mechanics, to appear as an EGF publication, Materials Engineering Publishers.

Dawes, M.G. (1978). A re-assessment of J-estimation procedures. Int. J. Pres. Ves. and Piping, 6, pp. 165-176.

Harrison, J.D. and T.T. Anderson (1985). The application of fracture mechanics to weldel construction. 18th ASTM Nat. Symp. on Fract. Mech., Boulder Co. USA, June 24.-26. 1985, to be published as ASTM STP.

Tweed, J.H. and J.F. Knott (1982). Microstructure-toughness relationships in C-Mn weld metal. In K.L. Maurer and F.E. Matzer (Eds.), Fracture and the Role of Microstructure, Proc. 4th European Conf. on Fracture, Leoben, Austria, Vol. 1, Eng. Mat. Advis. Serv. Cradley Heath UK, pp. 127-133.

Towers, O.L. and M.G. Dawes (1985). Welding Institute research on the fatigue precracking of fracture toughness specimens. In E.T. Wesseland and F.J. Loss (Eds.), Elastic-Plastic Fracture Test Methods, ASTM STP 856, Am. Soc. Testing of Materials, Philadelphia USA, pp. 23-46.

Effects of Heat Input and Stress-Relief Annealing on the Toughness of the HAZ in the Welding of the Boiler Steels Type HI, HII, 17 Mn^4 and 19 Mn5

S. Anik*, K. Tülbentçi** and A. Dikicioğlu*

*Technical University of Istanbul, Turkey
**Yildiz University, Kocaeli Engineering Faculty, Izmit, Turkey

ABSTRACT

In this work the effects of the welding heat input and stress relief annealing on the toughness of the HAZ in the welding of the steels type HI, HII, 17 Mn4 and 19 Mn5 which are used in the fabrication of the boilers are investigated. The toughness is evaluated with the help of Charpy impact test and hardness.

KEYWORDS

Welding heat input, HAZ toughness, stress relief annealing, boiler steels type HI, HII, 17 Mn4, 19 Mn5.

INTRODUCTION

During the last years by the means of newly developped methods, welding speed, melting rate and the thickness of the material which is joint by the welding are increased and as a result of these effects, investigations on the weldability became more important.

Most of the welding procedures, with exception of a few, involve local heating of the metal above the solidus; this heating up followed by rapid cooling, at the proximity of the weld forms a zone which is so called heat affected zone (HAZ). In this zone the microstructure and all the properties obtained by heat treatment or by cold forming are deteriorated. The most interesting and important zone in the welding is HAZ; it is the zone which is determining the quality of the welded joints.

One of the most important problem of the welding is the prediction on the extent of the deterioration of the mechanical properties and the selection of the welding procedure, welding data and the work piece geometry to minimize this deterioration.

It is clear that this deterioration is arised by the heat applied to the joint during the welding; this heating up over the solidus temperature followed by rapid cooling of the welding region, not only provokes several micro structural changes in the HAZ, is the cause of the residual stresses which are the mean cause of the several defects in the weld zone. After the welding, in the regions of the HAZ which were heated up to the austenitization temperature, a martenzitic structure occurs, if these regions excess the critical cooling rate, during the cooling period. Martensite usually lies in a narrow band along the fusion line and the existance of martensite provokes the formation and the developpement of the cracks. The possibility of the occurance of martensite in the HAZ depends on the composition of the base metal and the cooling rate of the welded joint; for this reason, many countries bring limitations to the carbon and manganese content of the steels which has to be used in welded constructions, by several regulations and standards.

It is well known that sudden fractures are caused by the rapid progress of the already existing cracks and these cracks are developped in hard and brittle materials easily. One can not imagine a material or a welded joint absolutely without any defect; there are allways some defects in the material itself and in the welded joint; these defects are so small that they are mostly not detectable by the conventional test methods.

Weldability Comission of the International Institute of Welding (IIW-Com. IX) recommend the hardness at the HAZ not to exceed 350 HV (kp/mm^2) as a precaution to cracking. The well known conventional method which is used to reduce the hardness of the HAZ, is to apply a preheat to the pieces before welding and keep this temperature during the process. The temperature of the preheat is determined by the carbon equivalent of the material. The developpement of an already existing crack into a brittle fracture is closely related with the fracture toughness of the material. The HAZ in which the materials properties are deteriorated by the welding heat, is lying as a very narrow band along the fusion line, therefore it is very difficult to determine the fracture toughness or NDT temperature with known fracture mechanics tests. Due to the close relation between NDT temperature and the Charpy V Notch Impact toughness (Pellini and co-workers, 1962; Ewald and co-workers, 1978) of the material, several autors, for the steels used in welded constructions, recommend the minimum Charpy V notch impact energy to be 47 J at 0 °C as an important precaution to prevent the brittle fracture; some other authors recommend to use as a criterium for the design against the brittle fracture the Charpy energy level (20-30 J) at the lowest service temperature.

EXPERİMENTAL WORK

In this paper, the influences of the stress relief annealing and the rate of energy input during the welding to the properties of the HAZ in the steels HI, HII, 17 Mn4 and 19 Mn5 are investigated. These types of steel are used in manufacturing of boilers and other kind of pressure vessels and their composition, according to the carbon equivalent method, enables welding without preheating. Technical Rules for Steam Boilers, TRD 201, is not taking compulsory the stress relief heat treatment for the boilers made out from these steels when the thickness is not exceeding 30 mm.

Three groups of specimen were prepared from this four types of steel by using metal-arc welding with coated electrod, CO_2 shielded arc welding

(MAG) and submerged arc welding methodes (fig. 1). The first group of specimens was keept as in welded condition, second and third group was subjected for two hours to a stress relief annealing heat treatment at 550 °C and 650 °C. Vickers hardness and Charpy ISO V impact test were carried out to the specimens take in from the HAZ of the test weldements. It is observed that the hardness and Charpy ISO V results differ according to the welding method and to the applied heat treatment temperature.

TABLE 1. Composition and carbon equivalent of experimental steels

Steel	t mm	Chemical Composition (%)							According IIW
		C	Mn	Si	Al	S	P	Cu	C'%
HI	15	0.145	0.510	0.070	--	0.035	0.011	0.130	0.239
HII	25	0.170	0.540	0.145	0.020	0.030	0.011	--	0.260
17Mn4	28	0.165	1.030	0.315	0.055	0.035	0.015	--	0.337
19Mn5	20	0.200	1.250	0.380	0.021	0.035	0 015	0.10	0.415

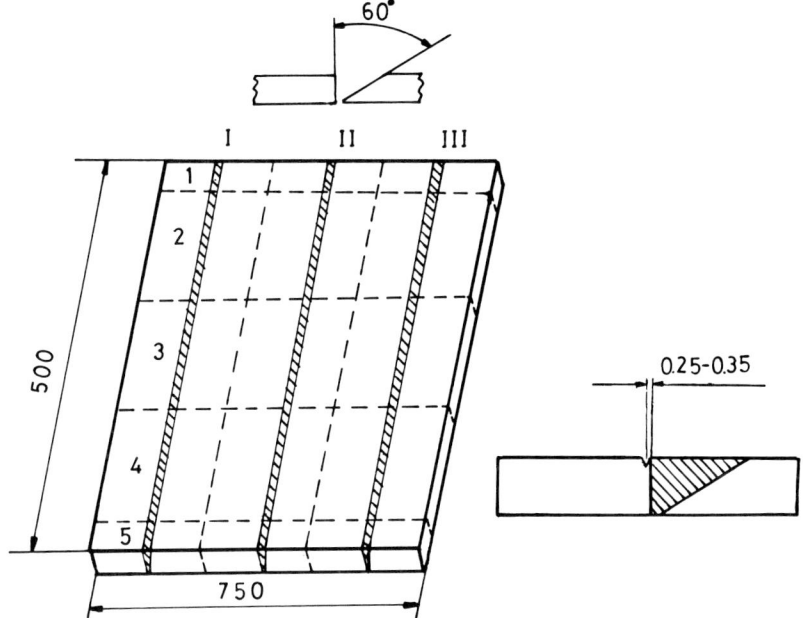

Fig. 1. Schematic view of the test plates and the position of the notch for the Charpy ISO V specimen
I: Metal-arc W.   II: Submerged arc W.   III: MAG W.
2: as welded   3: 550 °C - 2 h   3: 650 °C - 2h

During the welding process, HAZ does not melt and does not mix with the deposited weld metal; different methods of welding cause only a change in the heat input hence the heat distribution at the HAZ. From the theoretical work of Rosenthal (1941) and experimental studies of Rykalin (1952),

Christensen, Davies, Gjermundsen (1965), Uwer, Degenkolbe (1975, 1976) and many others, today one can determine the heat distribution and variation in the weld zone if the welding datas and the workpiece and edges geometry are known. In case of steels, from the transformation point of view, the cooling time between 800 and 500 °C and the regions of the HAZ which are heated above 900 °C are important. For this reason, the complicated heat distribution equation of Rosenthal, can be simplified and for three dimensional heat distribution, the cooling time between 800 and 500 °C can be expressed:

$$\Delta t_{8/5} = \frac{1}{2 \pi k} \cdot \frac{q}{v} \left(\frac{1}{500-T_o} - \frac{1}{800-T_o} \right) \qquad (1)$$

The energy input of the different welding methods and the $\Delta t_{8/5}$ calculated according the equation (1), which is derived from the Rosenthal (1941) three dimensional heat distribution equation are given at the table 2.

TABLE 2. Energy input and cooling times of the experimental weldings

Steel	Metal-arc W.		MAG W.		Submerged arc W.	
	E. input J/mm	$\Delta t_{8/5}$ sec	E. input J/mm	$\Delta t_{8/5}$ sec	E. input J/mm	$\Delta t_{8/5}$ sec
HI	1398	5,26	1125	4,23	1886	7,10
HII	1963	7,39	1303	4,90	3009	11,33
17 Mn4	1960	7,38	1385	5,21	3308	12,45
19 Mn5	1752	6,60	1190	4,48	2919	10,99

The results of the Charpy ISO V notch impact test and hardness measurements are given on the fig. 2 - 6.

RESULTS and CONCLUSIONS

1.- An increase in the net heat input causes an increase in the toughness, an decrease in the hardness and lowers the transition temperature of the HAZ.

2.- Stress relief annealing causes an increase of toughness at all the test temperatures in all the specimens which are welded with various net heat input and increase in the stress relief annealing temperature intensifies this effect.

3.- Cleavage fracture zones in the fracture surface of Charpy notch impact specimens at lower temperatures decrease by the increase of the net heat input and the stress relief annealing temperature.

The best HAZ caracteristics for all the steels used in the experiments are seen in the specimens which were welded with the highest heat input and were subjected to 650 °C stress relief annealing. So to obtain the utmost safety in the welding of these steels, methods which acquire the highest heat input should be used and even though it is not pointed out in regulations, a 650 °C stress relief annealing should be applied.

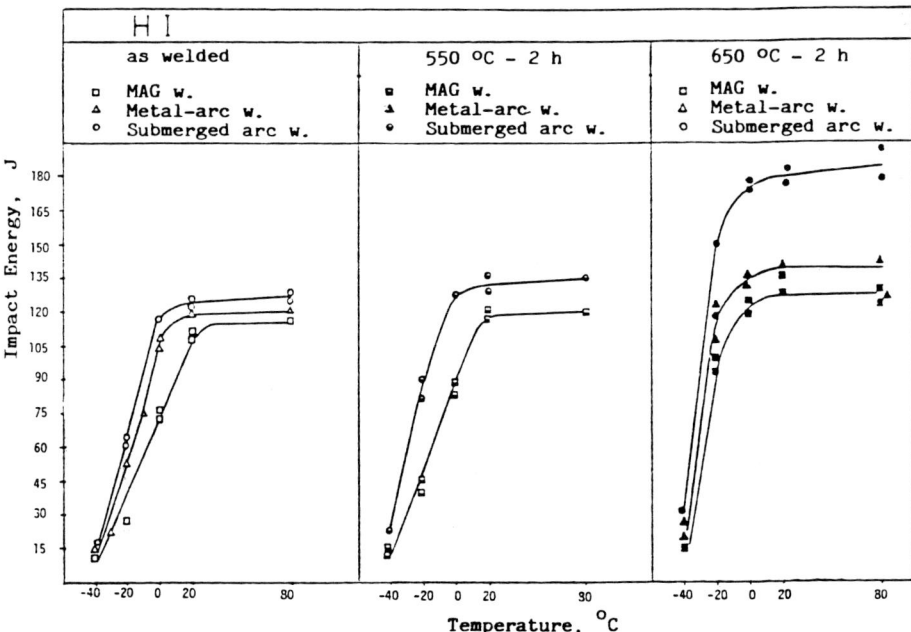

Fig. 2. Effect of the test temperature on Charpy ISO V impact energy (Material, HAZ of steel HI).

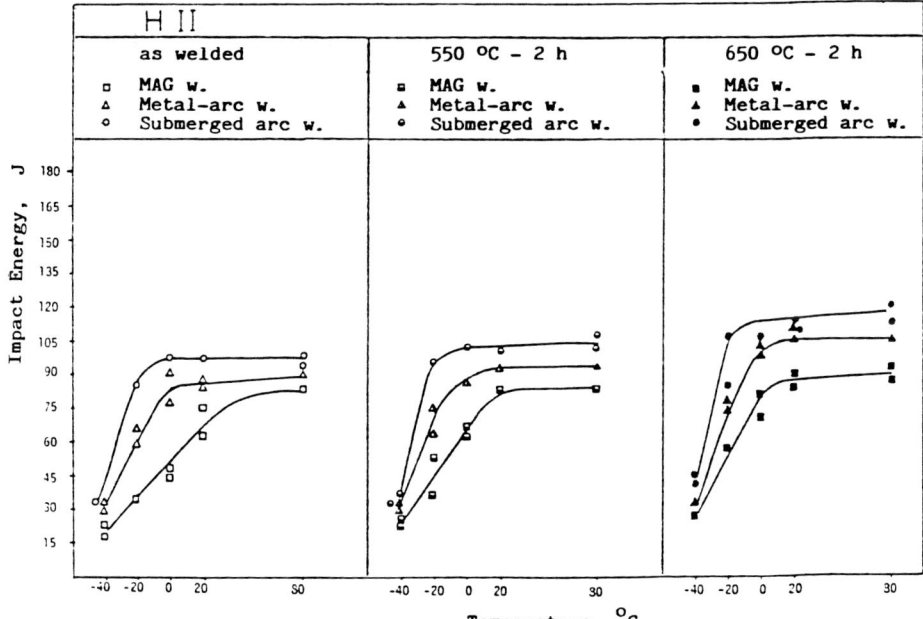

Fig. 3. Effect of the test temperature on Charpy ISO V impact energy (Material, HAZ of steel HII).

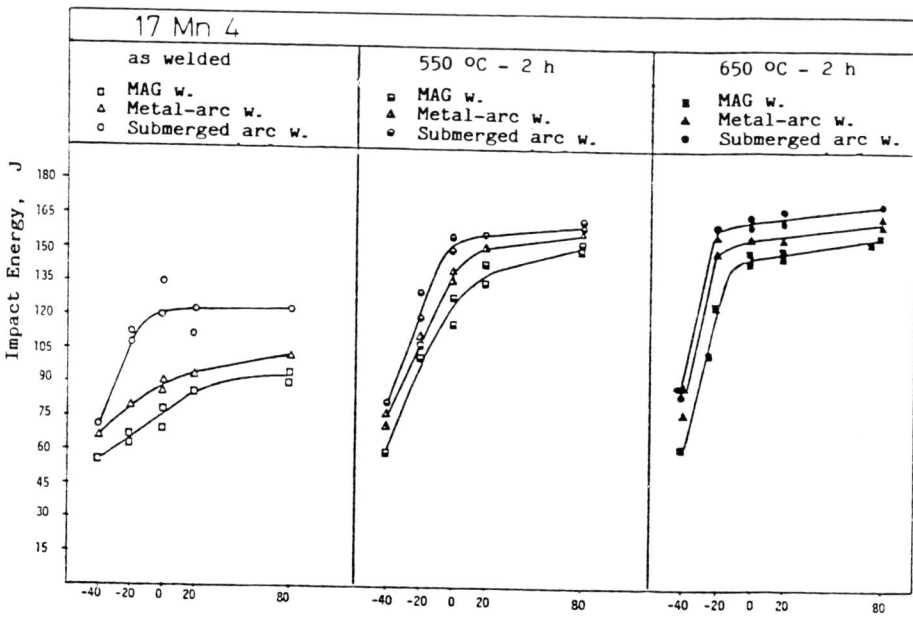

Fig. 4. Effect of the test temperature on Charpy ISO V impact energy (Material, HAZ of steel 17 Mn4).

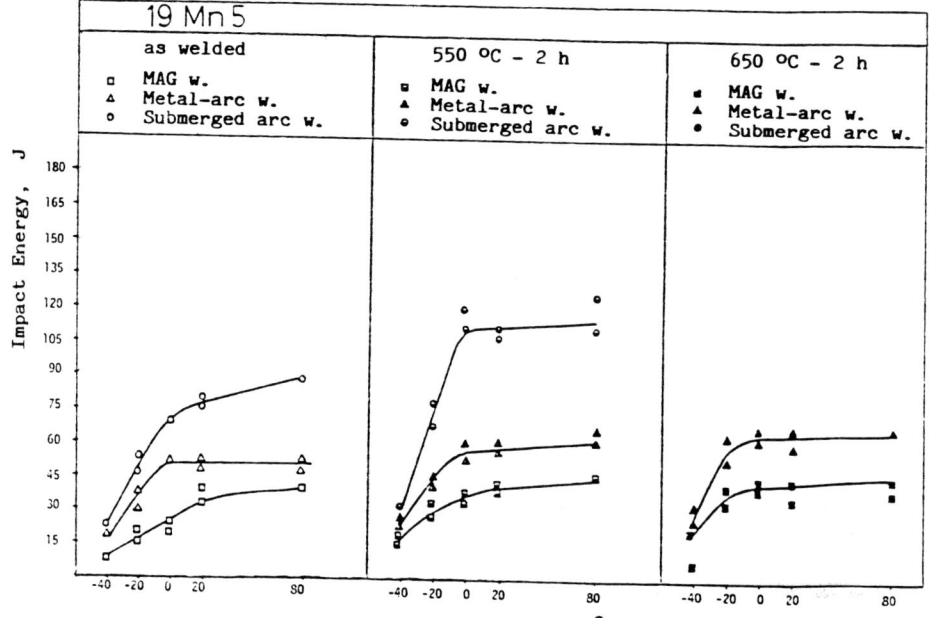

Fig. 5. Effect of the test temperature on Charpy ISO V impact energy (Material, HAZ of steel 19 Mn5).

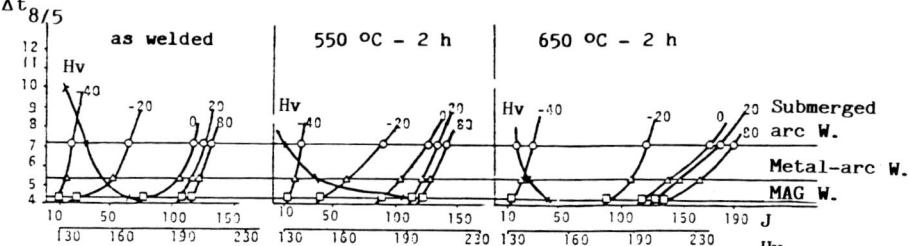
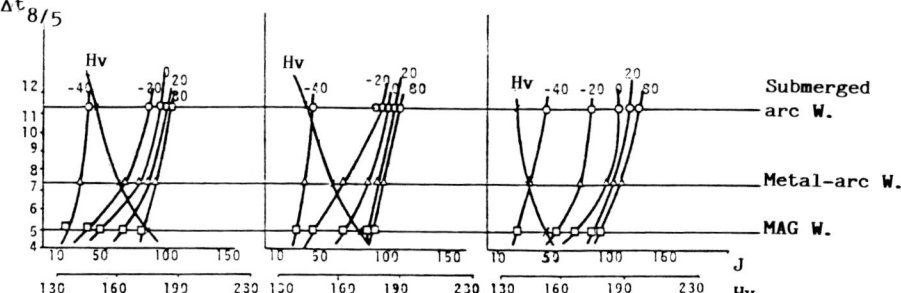
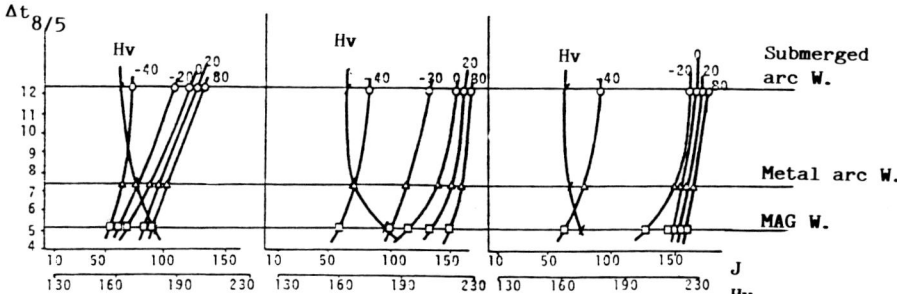
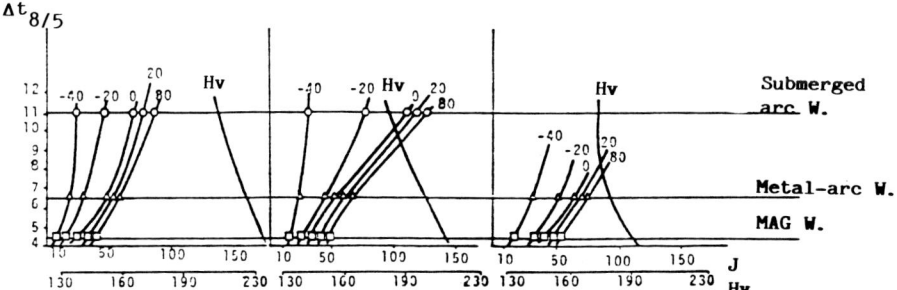

Fig. 6. Relations between $\Delta t_{8/5}$ cooling time, Charpy ISO V impact energy, hardness of the HAZ in as welded and heat treated conditions.

Fig. 7. Fracture surfaces of the Charpy ISO V specimens
(HAZ from the steel 17 Mn 4, Metal-arc welding
with coated elektrodes)

REFERENCES

Christensen, N., Davies, V., Gjermundsen, K. (1965). Distribution of temperatures in arc welding. British Weld. J. 1965/2.
Ewald, J., Schick, M., M.-Floera, B.L., Tülbentçi, K., (1978). Veranderung von Bruchzähigkeit und NDT Temperatur in der WEZ beim Spannungsarmgluhen. Schlussbericht, Forschungsvorhaben DVS 3069.
Pellini, W. S., Steele, L. E., Hawthorne, J. R. (1962). Analysis of engineering and basic research aspects of neutron embrittlement of steels. Welding Research Suppement, October 1962, 455-469.
Rykalin, N. N., (1952). Die warme grundlagen des schweissvorganges. Verlag Technik Berlin.
Rosenthal, D. (1941). Mathematical theory of heat distribution during welding and cutting. Weld. J. Research Supplement May 1941
The Welding Institute, (1971). Brittle Fracture of Welded Structures
Uwer, D., Degenkolbe, J., (1974) Einfluss der Schweissbedingungen auf the Kerbschlagzähigkeit in der WEZ von Schweissverbindungen hochfester Baustähle. Schweissen und Schneiden 1974/11
Uwer, D., Degenkolbe J. (1976) Temperaturzyklen beim Lichtbogenschweissen Einfluss von Schweissverfahren und Nahtart auf die Abkühlzeit. Schweissen und Schneiden 1976/4

Heat Treatment and Welded Structures Dimension Stability

V. M. Sagalevich and V. F. Savelyev

USSR

The Publisher regrets that the manuscript for this contribution was unavailable at the time of going to press and apologises for the inconvenience caused to readers.

I.9

The Improvement of Fatigue Resistance of Structure Welded Joints by Heat Treatmet

V. I. Trufiakov, P. P. Mikheev and Yu. F. Kudryavtsev

E. O. Paton Electric Welding Institute of the Ukrainian SSR Academy of Sciences, Kiev, USSR

Heat treatment is often recommended to improve the fatigue resistance of welded joints. However, such recommendations are not sufficiently grounded in all cases, since this kind of structure treatment can affect both positively and negatively the fatigue resistance of the welded joints /1...7 et al/. The expediency of heat treatment applied to improve the fatigue resistance of welded joints should be considered in terms of the conditions under which the negative effect of welding tensile residual stresses becomes apparent in fatigue processes.

The negative influence of tensile residual stresses on fatigue resistance of joints grows with the decrease in effective stresses, cycle asymmetry coefficient and stress concentration /1/. At the same time, the tensile residual stresses can exert their effect only in the presence of stress raisers /5, 1 et al/. In these cases the relieving of tensile residual stresses by heat treatment (by the high tempering conditions) should lead to the considerable increase in fatigue resistance of the welded joints.

The comparative fatigue tests of weldments showed that the general heat treatment raised the fatigue strength of butt joints at the symmetric loading cycle by 50...100%. Thus, in particular, the residual stress relieving resulted in the 65% rise of the fatigue strength of the low-alloy steel butt joint (σ_{yield} = 360 MPa, $\sigma_{ult.}$ = 523 MPa) (Fig. 1).

When going over from the butt joints to the higher stress concentration ones, the effectiveness of the welding residual stress relieving as the means of the fatigue strength improvement decreases. In the case of the rather sharp stress raisers high tempering can even cause the certain reduction in joint fatigue resistance (Fig. 2). Besides, the other conditions being equal, the efficiency of the heat treatment markedly decreases with the growth of the cycle mean stresses (Fig. 3).

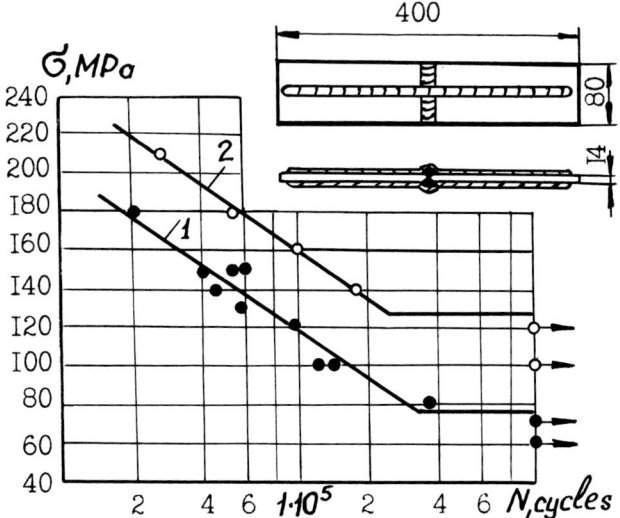

Fig. 1. Results of fatigue tests of low-alloy steel butt joints: 1-in as-welded condition; 2-after heat treatment.

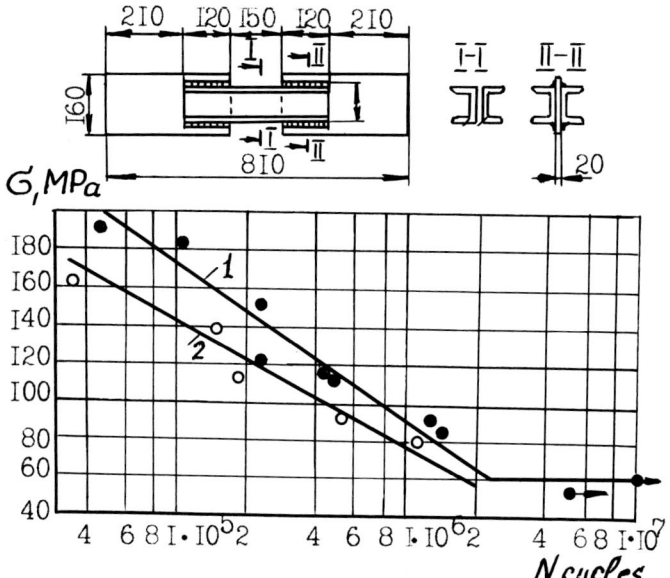

Fig. 2. Results of fatigue tests of lap joints with longitudinal fillet welds: 1-in as-welded condition; 2-after heat treatment.

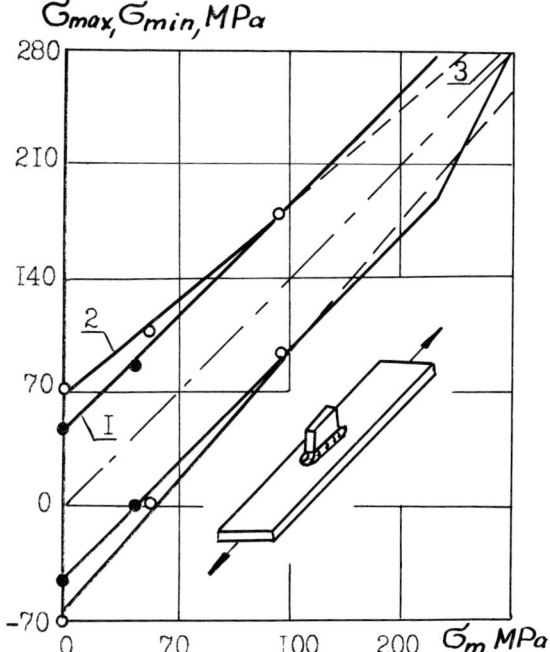

Fig. 3. Diagrams of limiting stresses of welded joints ($N = 2 \cdot 10^6$ cycles): 1-in as-welded condition; 2-after heat treatment; --- - estimates obtained from the relationships proposed; o,• - experimental data /8, 9/.

Such influence of the heat treatment on the fatigue strength value and the kind of welded joint limiting stress diagrams depends mainly on the fact that cyclic loading causes the changes in welding residual stresses of the as-welded joints. After the first loading cycles within the stress raiser regions of the as-welded joints the levels of residual stresses become stable (steady residual stresses) /10/. Since the steady residual stresses decrease with the growth of limiting maximum stresses under external loading /10/, the increase in nominal stresses causes the reduction of differences in fatigue strength between treated and as-welded joints. Starting from the certain level of external loading the general heat treatment of a weldment for the welding residual stress relieving will not result in its fatigue resistance increase.

If the amplitude of limiting stresses of a welded joint with the high tensile residual stresses σ_a^t is known, then the fatigue strength value σ_a^ℓ of such joint depending on the external loading cycle asymmetry coefficient R_σ can be determined from the relationship:

$$\sigma_R^\ell = \frac{2\sigma_a^\ell}{1-R_\sigma} \qquad (1)$$

In case when heat treatment completely relieves the welding residual stresses, the fatigue strength value σ_R^{treat} of such joint depending on the external loading cycle asymmetry coefficient can be determined from the equation:

$$\sigma_R^{treat} = \frac{\sigma_{ult}}{(1-R_\sigma)(\sigma_{ult}-\sigma_{yield}/\alpha_\sigma)/2\sigma_a^\ell + 1} \qquad (2)$$

where: σ_{yield}, σ_{ult} are the material yield and ultimate tensile strength, respectively; α_σ is the theoretical stress concentration factor of a welded joint.

Equation (2) is obtained on the basis of the analysis of stress-strain condition within the stress raiser regions by taking into account the residual stresses and their redistribution under the effect of cyclic loading and also of the welded joint limiting stress diagrams.

It follows from the (1) and (2) estimated relationships that with the growth of the asymmetry coefficient or the external loading cycle mean stresses the difference in the fatigue strength values between the as-welded and heat treated joints reduces. When $R_\sigma = R_{\sigma_1}$, where

$$R_{\sigma_1} = 1 - \frac{2\alpha_2 \cdot \sigma_a^\ell}{\sigma_{yield}},$$

the amplitudes of limiting stresses of welded joints with and without residual stresses coincide. This identity remains with the further growth of the external loading cycle asymmetry coefficient. The comparison of the estimation results about the effect of heat treatment for the residual stress relieving to improve the welded joint fatigue resistance with the experimental data obtained by using the various values of the external loading cycle asymmetry coefficient showed their good correlation (Fig. 3).

Estimated relationship (2) is valid not only for the case when the initial residual stresses reach the material yield strength value, but also for the cases when the said initial residual stresses are below σ_{yield} but above the certain limiting value of the initial residual stresses σ_{res}^1. Depending on the external loading cycle asymmetry at which it is necessary to estimate the effect of the residual stress relieving by heat treatment on the welded joint fatigue resistance, this value is determined from the following relationship

$$\sigma_{res}^\ell / \sigma_{yield} = 1 - \frac{2\alpha_\sigma \cdot \sigma_a^\ell}{(1-R_\sigma)\cdot \sigma_{yield}}$$

If the welded joint is subjected to heat treatment and the level of residual stresses σ_{res}^{treat} satisfies the condition $\sigma_{res}^{treat} \geqslant \sigma_{res}^1$, such redistribution of residual stresses will not lead to the joint fatigue resistance increase and the value of limiting stresses should be determined on the basis of equation (1). If $\sigma_{res}^{treat} < \sigma_{res}^1$, this redistribution of residual stresses will cause the welded joint limiting stress amplitude increase and the residual stress effect should be es-

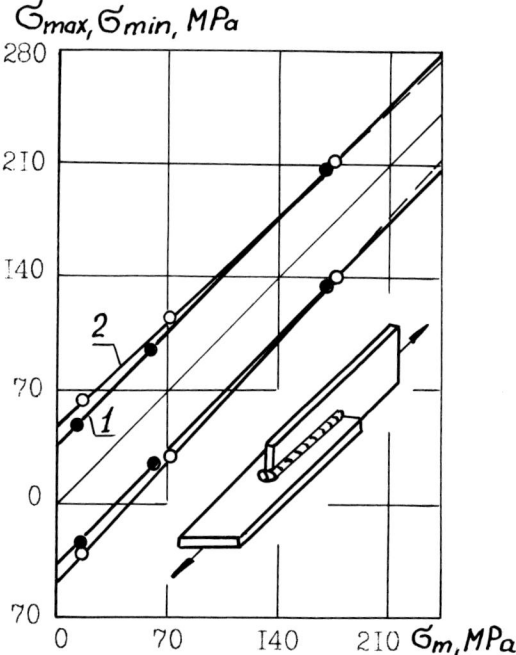

Fig. 3,b. Diagrams of limiting stresses of welded joints ($N=2 \cdot 10^6$ cycles): 1-in as-welded condition; 2-after heat treatment; --- - estimates obtained from the relationships proposed; o, ● - experimental data /8, 9/.

timated by using the equation

$$\sigma_R^{treat} = \frac{\sigma_{ult} - \sigma_{res}^{treat}/\alpha_\sigma}{(1-R\sigma)(\sigma_{ult} - \sigma_{yield}/\alpha_\sigma)/2\sigma_a^\ell + 1} \qquad (3)$$

In this case the treated joint limiting stress line will be parallel to the similar diagram for the welded joints without residual stresses (Fig. 4). Equation (3) allows to determine the fatigue strength values also for the cases when the welded joint is subjected to the local heat treatment /11/ which provides the favourable compression residual stresses being formed within the stress raiser regions. Here the residual stress values should be substituted into the said equation with regard to their sign.

When the residual stresses σ_{res}^{treat} are present within the stress raiser regions of the welded joint after local or general heat treatment, such joint fatigue strength values depending on the external loading cycle asymmetry coefficient at $R_\sigma < R_{\sigma_2}$, where

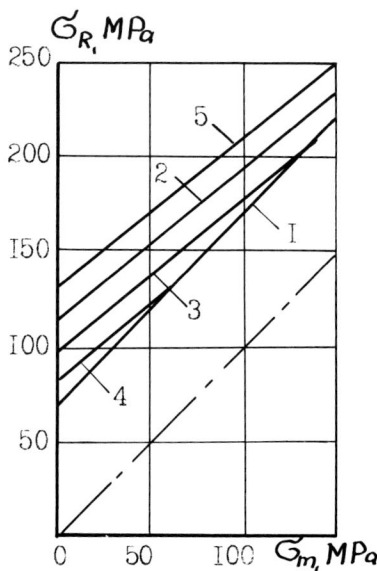

Fig. 4. Diagram of limiting stresses of a low-alloy steel butt joint (σ yield = 380 MPa, σ_{ult} = 540 MPa): 1 - σ_{res} $\geqslant \sigma_{res}^1$; 2 - σ_{res} = 0; 3, 4, 5 - σ_{res} = 100, 200 and -100MPa, respectively.

should be determined from equation (3). Starting from the value $R_\sigma \geqslant R_{\sigma_2}$, the welded joint limiting stress value is determined from equation (1) (Fig. 4).

Therefore, the redistribution of residual stresses by general or local heat treatment causes in a number of favours the significant improvement in fatigue resistance of structure welded joints. The effectiveness of heat treatment is reduced in transition from butt joints to the higher stress concentration ones and also with the growth of the external loading cycle asymmetry coefficient. The estimated relationships proposed allow with the accuracy sufficient for practical purposes to estimate the effect of heat treatment on the fatigue strength values of structure welded joints depending on the joint type, mechanical properties of the materials applied and the external loading parameters.

REFERENCES

Trufiakov V.I. (1973). Fatigue of welded joints. **Naukova Dumka,** 216.
Myunze V.Kh. (1968). Fatigue strength of steel welded structu-

res. Publ.House "Mashinostrojenije", 312.
Kudryavtsev I.V., Naumchenkov M.E. (1976). Fatigue of welded structures. Mashinostrojenije, 270.
Gokhberg M.M., and I.Tun Bao (1964). Effect of preliminary loading and high tempering on fatigue strength of welded joints. In book:Designs and machines. Proceedings of the Leningrad Polytechnical Institute, 236. Mashinostrojenije, 76-83.
Kudryavtsev I.V. (1951). Internal stresses as the safety factor in industrial engineering. Mashgiz, 278.
Navrotskii D.I., Zvyagintsev S.K., and N.I.Isaev (1954). Study of the effect of internal stresses on strength of weldments with the pronounced shape change inder vibrating loading.In book:Welding production. Proceedings of the Leningrad Polytechnical Institute, 4, Mashgiz, 74-86.
Gerney T.R. (1979). Fatigue of welded structures. Cambridge University Press. London, New York, Melburne, 456.
Gerney T.R. (1960). Influence of residual stresses on fatigue strength of plates with fillet welded attachments. British Welding Journal, 6, 415-431.
Overbeeke J.L., and J.Back (1983). The effect of residual stresses and R-value on the service life of welded connections subject to fatigue. Doc.IIW XIII-1095-83, 56.
Kudryavtsev Yu.F., and O.I.Gushcha (1986). Some regularities of changes in residual stresses under cyclic loading depending on their initial level and stress concentration. Problemy prochnosti, 11, 32-38.
Maksimovich V.N., Chabanenko A.A., Mikheev P.P., Gushcha O.I., and Yu.F.Kudryavtsev (1985). Determining the parameters of local heating to improve fatigue resistance of welded joints. Problemy prochnosti, 4, 32-36.

SESSION II

RESIDUAL STRESSES GENERATION-LOCAL RELIEVING HEAT TREATMENTS/ CREATION DES CONTRAINTES RESIDUELLES-TRAITEMENTS LOCAUX DE RELAXATION

II.1 Role of preliminary heat treatment on heat-affected zone behaviour as regards relieving of internal stresses
(K. Velkov, L. Vasileva — Bulgaria/Bulgarie)   109

II.2 Minimization of residual welding stresses by the superposition of thermal and transformation stresses
(P. Seyffarth, H. G. Gross, K. M. Gatovskij, S. P. Markov — GDR/RDA; USSR/URSS)   117

II.3 Local heat treatment by induction
(H. H. Müller — FRG/RFA)   125

II.4 Dans certain cas le traitement thermique par induction est non réussi. Quand et pourpuoi?
(S. Cundev — Yugoslavia/Yougoslavie)   131

II.1

Role of Preliminary Heat Treatment on Heat-affected Zone Behaviour as Regards Relieving of Internal Stresses

K. Velkov* and L. Vasileva*

*Higher Institute of Mechanical and Electrical Engineering, Sofia, Bulgaria

ABSTRACT

The paper considers the role of heat pretreatment on the residual local microstresses in HAZ of high strength steel welds. The criteria used are the weldability factors directly related to the above stresses. The positive effect of heat treatment processes providing for heterogeneous predecomposition austenite in welding is pointed out.

KEYWORDS

Heat pretreatment; local microstresses; heterogeneous predecomposition austenite; impact toughness; cold delayed cracking.

Stress relief annealing is the heat treatment process most widely applied in welding structures technology. Not minor is also known to be the role of the various heat treatment processes prior to and during welding, used to avoid undesirable structural and phase changes in welding process run (Granjon, 1978). The effect of heat pretreatment of the materials to be welded on final stressed state of the welds is comparatively less studied. And this effect becomes much stronger in welding of high strength steels. The presence of alloying elements in high strength steels makes them structurally sensitive at the non-equilibrium conditions of the process, affecting thus the efficiency and significance of stress relief annealing. On one hand, there is weakening or carbide segregation along the grain boundaries with the risk for reheat cracking. The presence of alloying elements, on the other, increases steel sensitivity to formation of microstructures in the heat affected zone (HAZ), related with high local microstresses. They are higher than the local stresses of geometric origin and exceed considerably the stresses of first order (Prochorov, 1967). As these stresses accumulate at critical conditions they could also provoke premature brittle fracture of the welded product. The sources of local microstresses, i.e. the phase transformations in the under bead area, point to the natural approach for their cutting down by control over structure formation during welding. Preheating and control over some parameters of the production cycle is most often applied for the above purpose. Yet insufficient is the use of the possibility for preliminary structure preparation that reduces or eliminates the need for preheating of the elements to be welded.

The present work attempts at demonstrating the role of heat pretreatment of high strength for cutting down the local microstresses in HAZ. The known experimental approaches for determining structural microstresses (in the main X-ray defraction) are not yet so reliable. That is why an indirect approach was chosen for their evaluation, using as criteria the typical factors related with the local structure stresses in HAZ such as the transient curve variation in impact toughness tests and resistance to cold, delayed cracking.

Standard Sharpy V test samples are used for determining the pattern of transient curve. To specify the resistance to delayed fracture use is made of notched samples of 3x10x60 mm, subjected, after a simulated welding cycle, to continuous loading with constant load at the conditions of four-point bending (Velkov, 1985). All samples are heat pretreated through a programmed thermal cycle on a SMITWELD apparatus for thermal cycle simulation.

The studies are made on high strength microalloyed steel of the following composition: 0,23%C; 1,58%M; 0,48%Si; 0,08%Cr; 0,04%Ni; 0,14%Cu; 0,11%V; 0,0156%N_2; 0,013%S and 0,025%P.

Fig.1 shows the schematic diagrams of the simulated heat pretreatment cycles followed by a simulated welding cycle, characteristic for the area around the joint, by manual electric arc welding (140°C/sec heating rate, T_{max}=1350°C and cooling rate, expressed in terms of the time period between 800 and 500°C (Δt8/5), 3 sec).

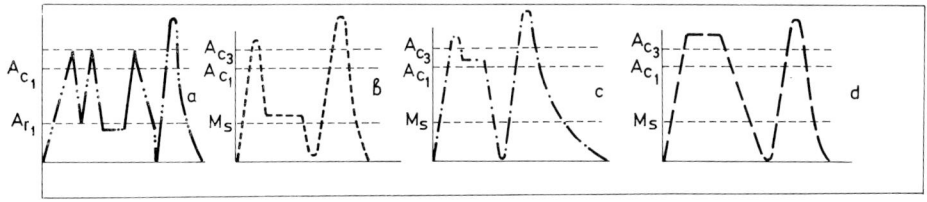

Fig.1 Schematic diagrams of heat pretreatment and welding cycle.

The heat pretreatment processes chosen, i.e. a) thermocyclic treatment; d) normalizing; c) short heating above A_3 and holding betwen A_1 and A_3 and b) bainitization, aimed to give variable predecay state of the austenite and hence different final structure in the area around the joint providing for different levels of local microstresses.

The thermocyclic pretreatment is a much efficient method for control over the predecay state of austenite. Admittedly, optimum thermocycling should be sought for every individual case. This heat treatment enables the matching of the competing processes of stress accumulation (particulary microstresses related to phase dilatation) and their relaxation. The quantity of stresses accumulated at the beginning of martensite transformation affects process run. The higher the microstresses the higher the temperatures whereat the martensite transformation begins, and this results in lower residual stresses (self tempering). Stress reduction during martensite transformation could also be related to the abnormal plasticity phenomenon, determining their intensive relaxation. Further, relieved are also the peak stresses responsible for the local brittle fracture of the welds. The thermocyclic treatment guarantees a finer grain structure of better plasticity. The latter favours the sharp raise in relaxation capability of the metal to local stresses on the account of local microplastic deformation. The thermocyclic treatment also improves the resistance to crack propagation, being closely related to welded structures reliability. According to data from literature the total residual stresses of first order and second order become lower as there is 4 to 8 times reduction of residual microstresses (AN USSR, 1984). Thermocycling is applicable for alloys containing phases with sharply distinguished thermal factors of linear expansion resulting in occurence of microstresses on phase interfaces. The microstresses are superimposed on the residual stresses of first order and in heating they provoke plastic deformation. In this way creep in different regions becomes more intensive, and contributes to the elimination of residual stresses (Novikov, 1978).

The choice of bainitization heat pretreatment is first of all determined by the high mechanical characteristics made available with this structure, and by the possibility for fine-grain inhomogeneous predecomposition austenite to be provided in the welding thermal cycle. Bainitization gives full fragmentation and desorientation of the phases, strong carbon inhomogeneity, because of the specific nature of the lower bainite process. Microstresses develop, the density of dislocations becomes higher, the mosaic blocks go finer, and well-built interface forms. At fast heating this structure both stimulates the occurence of austenite nuclei and suppresses their growth. Thus a fine-grain inhomogeneous predecomposition austenite is formed.

The above compeex heat treatment process that covers heating and short holding above A_3, and intermediate holding between A_1 and A_3 ("c" cycle of fig.1) provides for a ferrite-martensite initial structure to be obtained prior to welding. It guarantees marked carbon heterogeneity, and respective high degree of inhomogeneity of predecomposition austenite for the high heating

rates in the process of welding. Ferrite-martensite initial
structure will, on its part, provide for a mertensite-bainite
structure in the area under bead area, with mertensite present
in the lower bainite matrix, and this has to reduce the microstresses.

The normalized initial structure ("d" cycle) should provide for
the most homogeneous predecomposition austenite with most favourable conditions for accelerated diffusion during heating and
alloying of the solid solution. In the long run it should result
in lower temperatures of martensite transformation outset, and
in higher microstresses of second order.

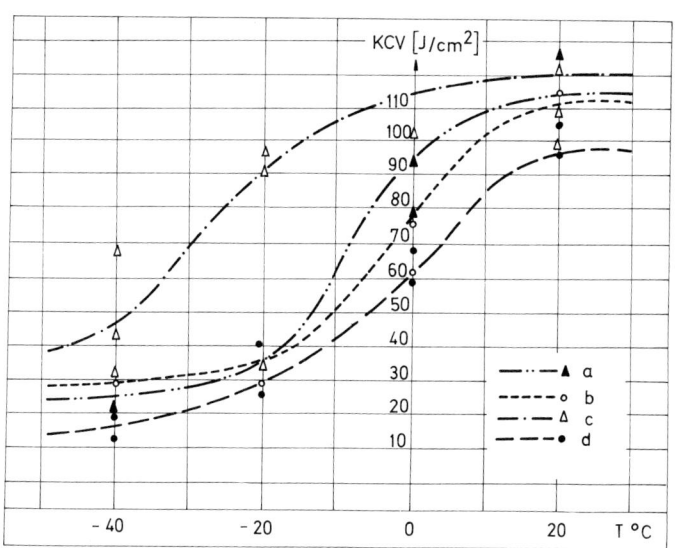

Fig.2 Effect of preheat cycles on the
variation in the transient curve.

The experimental results obtained were in support to what was
expected. Fig.2 shows the effect of the four heat treatment processes on brittleness transient temperatures. This effect is
quite evidently due to the favourable structural-stressed state
obtained after the welding cycle. "c" cycle occurs to be most
favourable for the specified conditions, and "d" cycle (normali-

zing) - most unfavourable one. Only these results suffice to confirm that it is the degree of homogeneity of the predecomposition austenite to be the cause for the differing final structural stressed state.

The tests for delayed fracture, being typical for the local microstresses, gave the same result (Fig.3). Lowest critical stress of delayed fracture (R_{cr}) is exhibited by the prenormalized structure, and the highest is for "c" cycle. Cold, delayed cracking is the most typical result from local microstresses caused by microscopic processes (structural transformations and hydrogen diffusion) localized intensively in microvolumes. That is why, kinetics of crack formation is determined on the whole by the size and distribution of stresses of second order.

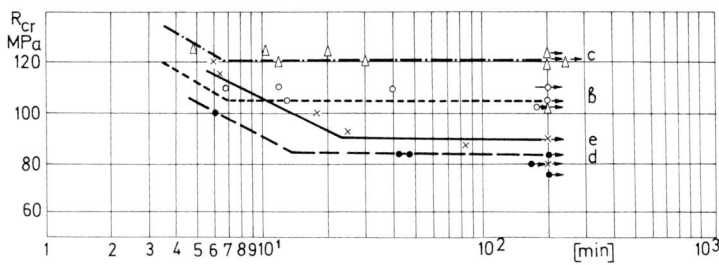

Fig.3 Effect of some of the tested heating cycles on the critical stress of delayed fracture.

The plot for austenite decomposition for continuous cooling in the conditions of welding thermal cycle, for three of the initial structures, explains best the variable levels of microstresses obtained by the heat pretreatment considered. For the case of most favourable structural-stressed state after welding there is a great shift to the left in the outset of bainite transformation and higher initial and final temperatures of martensite transformation, particular for the high cooling rates (the short time periods $\Delta t 8/5$) characteristics for the under bead area. The prestructures providing for heterogeneous predecomposition austenite provoke partial replacement of low-temperature martensite transformation with phase transformation of lower second

order stresses. A lower bainite structure is obtained, made up
of dispersion carbides allocated in a fine-grain ferrite matrix.
At the presence of a structural microinhomogeneity the different
regions differ sharply in properties as well. They have variable
resistance to metal plastic deformation and creep in the conditions of the constantly acting first order stresses in the welds,
and it is the reason why minimum local stresses are desirable.
Of much importance to the local stresses is also the size of the
real austenitic grain. The local microstresses, particularly
along the grain boundaries, sharply increase at the presence of
coarse-grain martensite. The heat pretreatment processes applied,
exhibited a favourable effect in this respect, too. For the normalized initial structure coarser grains are observed, and for
"c" cycle in HAZ the result is finegrain martensite-bainite
structure.

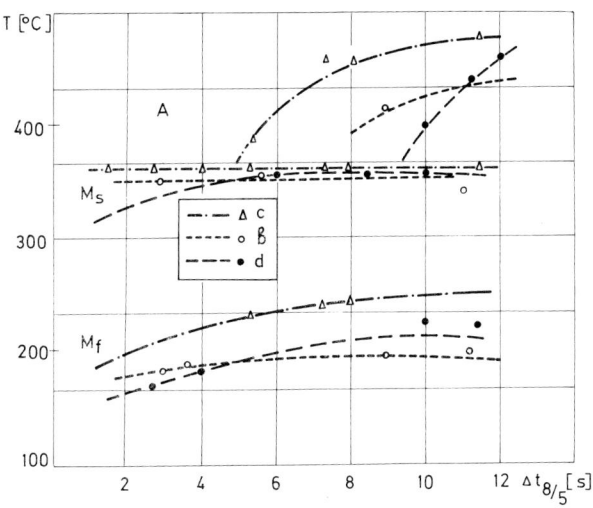

Fig. 4 Effect of the simulated heating cycles on the continuous cooling transformation diagram.

Proceeding from the results obtained it could be stated that
heat pretreatment of steels to be welded could be an efficient
approach for control over the predecomposition austenite state,
and hence for an influence on the local microstresses, that
from a considerable part of the total residual stresses in high

strength steel welding. The more heterogeneous is the predecomposition austenite, the more favourable is the respective heat pretreatment process.

REFERENCES

Granjon, H., Traitment thermique avant, pendant et après soudage. Conférence pléniaire. Journées d'Information. Soudage et traitments thermiques. Marseille. 1978.

Prohorov, N.N. and coautors, Thermodynamics of britle fracture of welded Constructions, Proceedings Reliability of welded Joints and Constructions, "Mashinostroenie" (in Russian), M, 1967.

Thermocycling of Steels, Alloys and Composite Materials. "Nauka" 1984, Academy of Sciences USSR (in Russian).

Velkov, K., Tz.Stoinov, Role of heterogeneous area in HAZ on formation of cold welding cracking, III Symposium "Cracks in Welded Joints. Proceeding Bratislava, 1985, v. 2 (in Russian)

Novikov I., Theory of Heat Treatment of Metals, "Metalurgia" M. 1978 (in Russian)

II.2

Minimization of Residual Welding Stresses by the Superposition of Thermal and Transformation Stresses

P. Seyffarth*, H.-G. Groβ*, K. M. Gatovskij** and S. P. Markov**

*Wilh.-Pieck-University, Faculty of Ship Technology, Welding Department, Rostock, GDR
**The Shipbuilding Institute, Welding Department, Leningrad, USSR

ABSTRACT

Residual welding stresses may be minimized by the choice of welding conditions. Using a MAG-surfacing bead with $E = 14,8$ kJ/cm as an example, four different types of heat input are compared: welding without preheating, welding with preheating at 340 °C, welding with preheating at 470 °C, and welding with short-time post-heating during cooling at 600 °C to 500 °C for the period of 60 s. The calculation model takes into account the superposition of thermo-mechanical and structural transformation stresses. The input data concerning the behaviour of the material in relation to cooling rate and temperature and in this way to the given structure were determined experimentally.
A favourable combination of minimum residual stresses, increased ductility values and lower costs for heat input is to be found with short-time post-heating.

KEYWORDS

Residual welding stress; preheating; short-time post-heating; transformation stress; phase transformation; TTT-diagram; FEM-calculation model.

INTRODUCTION

Residual welding stresses are additionally influenced by phase and structural transformation. In calculating the residual welding stresses, it is therefore necessary to take into account structural transformations as well as the behaviour of material in relation to temperature and structural composition in the heat-affected zone.
Transformation stresses are taken into account in recent publications (Denis and Sjöström, 1986; Heeschen and co-workers, 1986), but in calculation programmes for the determination of residual welding stresses only the thermo-mechanical problem has so far been solved and the influence of transformations has been neglected (Vinokurov and Ship, 1985). One paper only (Ueda and co-workers, 1985) was found in relation to this topic.

In the present paper material behaviour in the heat-affected zone is experimentally determined for different cooling rates depending on temperature, and the values so determined are applied as input data for the calculation of the residual stress distribution. For the calculation the finite element method is used.
In this way the determination of residual welding stresses can be performed much more exactly.
Since the redistribution of hydrogen during cooling is influenced by the transformation temperatures and the resulting transformation products as well as by the residual stress distribution, the thermal cycle has an influence on the local concentration of the diffusible hydrogen. Calculations of the residual stress distribution with respect to the transformation behaviour are thus an important precondition for the future calculation of the temperature- and time-dependent hydrogen concentration.

CHOICE OF WELDING CONDITIONS

To insure a welding connection free of hydrogen-induced cold cracks, the criteria *structure and mechanical properties, residual stresses,* and *diffusible hydrogen* should be taken into account when choosing suitable thermal welding conditions. In actual practice only the first criterion is applied in most cases, for the determination of residual stresses and of hydrogen is complicated and expensive.
The choice of the thermal welding conditions can be performed by means of weld ttt-diagrams, Fig. 1a. Depending on the chosen welding cycle, very different mechanical properties are obtained in the heat-affected zone, as can be seen in Fig. 1b. It is possible to avoid a hardened structure with a high proportion of martensite and with decreased ductility by increasing the heat input or preheating, Fig. 1. At the same time the resulting variation of the transformation mechanism and of the transformation temperature has an impact on residual stress.
For the chosen example of welding a surfacing bead on an HS 52-3 steel plate of 28 mm thickness the three cycles A, B, and C, Fig. 1a, have been chosen in order to investigate the influence of different transformation temperatures and varying material behaviour on the values of residual welding stresses. Keeping the heat input of cycle A constant at 14,8 kJ/cm, the thermal cycles B and C have been obtained by preheating to 340 °C and 470 °C respectively. With this approach the influence of the preheating temperature on the minimization of residual welding stresses can be investigated simultaneously.

DETERMINATION OF RESIDUAL WELDING STRESSES

To solve the thermo-mechanical problem, the method of finite elements has been applied. Algorithm and calculation programme are based on the solution of two-dimensional instationary heat conduction and thermoplasticity. Here the theory of non-isothermal yielding is applied to ideally elastic-plastic behaviour.
For each element of the temperature field the temperature-dependent yield point ist applied while Young's modulus is assumed to be temperature-indepedent.
In the programme the influence of structural transformations has been taken into account by a corresponding function of free volume dilatation and by the dependence of the yield point on the temperature and on the cooling rate in the transformation interval between 850 °C and 500 °C.

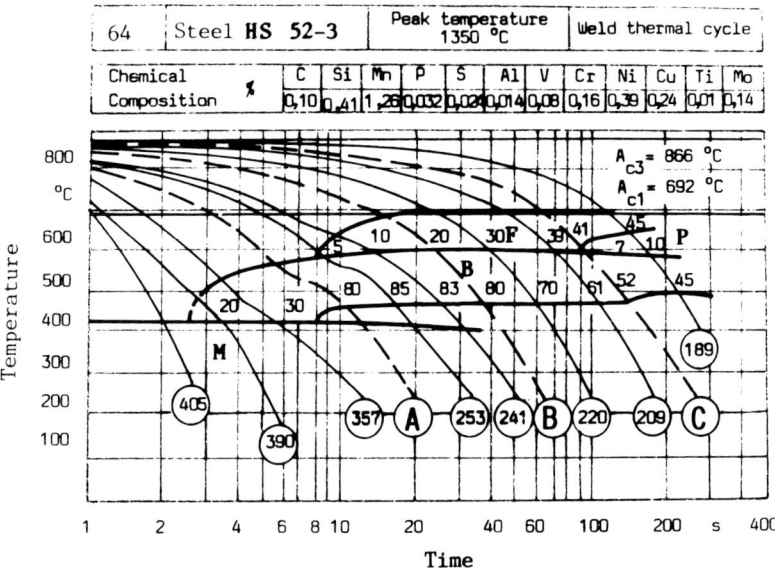

Fig. 1a. Time-temperature transformation diagram for steel HS 52-3

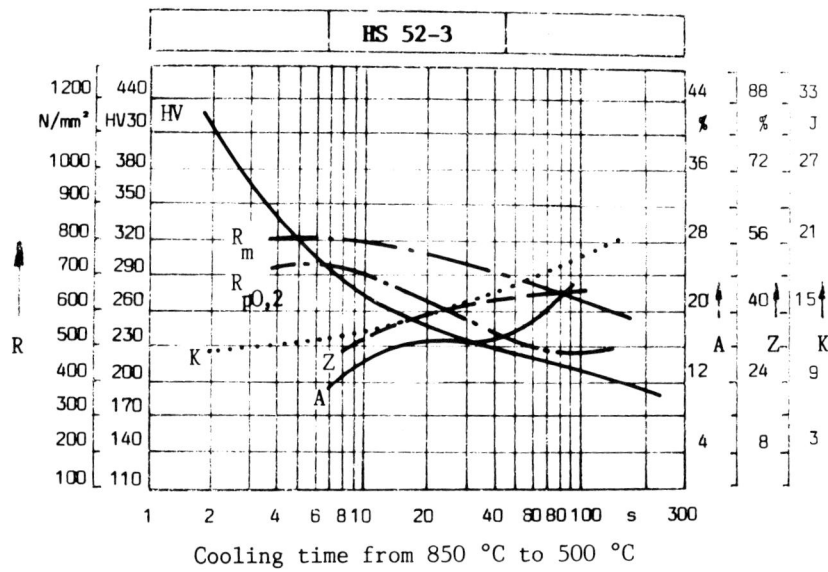

Fig. 1b. Mechanical properties of steel HS 52-3 as function of cooling time

In the range of the physical non-linearity caused by transformation, it is not possible to give a mathematical formulation of the opposite effects of contraction in volume due to transformation and temperature expansion (transformation being additionally dependent on the rate of temperature variation). Therefore the coefficient of linear thermal expansion α during transformation has been determined experimentally from dilatometric curves.

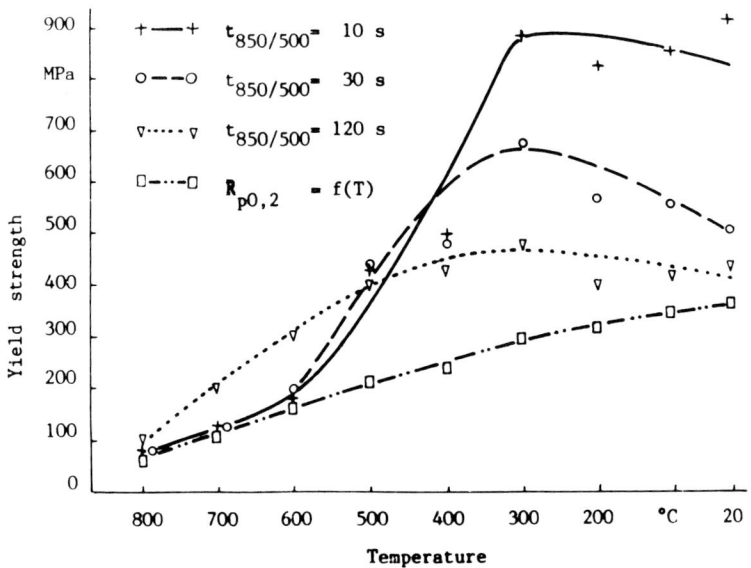

Fig. 2. Yield strength versus temperature for different structures as function of cooling time

EXPERIMENTAL INVESTIGATION OF MATERIAL BEHAVIOUR

The free volume dilatation during structural transformation has been determined by means of dilatometric curves for the three cycles A, B, and C at maximum temperatures of 1350 °C. The yield point has also been determined experimentally in the course of simulated thermal welding cycles at T_{max} = 1350 °C and $t_{850/500 \, °C}$ = 10, 30 and 120 s correspondingly to cycles A, B, and C. Figure 2 presents the resulting values used as input data for the yield point in the calculation.
In most previous calculation methods for the determination of stresses the tmperature-dependent yield strength of the ferritic-pearlitic base metal has been applied as input data. In the present calculation, however, this curve, which is exlusively valid for the initial state, is only applied to the heating process until the α-γ transformation has been reached. For the cooling process the curves from Fig. 2 referring to the cooling cycles A, B, and C are used, taking structural transformations into account. Whereas at 800 °C 100 % austenite is present in all cases and thus the same value of yield strength has been determined for all three cooling rates, according to Fig. 1a the decomposition of austenite is initiated even at 700 °C for the cycle C with $t_{850/500 \, °C}$ = 120 s. The increasing share of the ferritic-

pearlitic transformation product and the decreasing share of austenite in
the given structure increases the yield strength, whereas in cycles B and
C 100 % austenite is still to be found at 700 °C. Because of this fact only
the increase of the temperature-dependent yield strength must be taken
into consideration. Corresponding to the higher cooling rate of cycles A and
B, the temperature for the initiation of the decomposition of austenite decreases. For these cycles the yield strength increases with decreasing
shares of austenite during transformation, too.
Above approximately 450 °C the dependence on temperature predominates, but
during further cooling the material behaviour of the transformation product
and thus dependence on structure are much more important. Depending on
the cooling rates considered, the difference to the yield strength of the
non-influenced base metal can reach factor 3.
Because of the application of yield strengths depending on temperature
and cooling rates as input data referring to material behaviour, the calculation programme is superior not only from the qualitative but also from the
quantitative point of view.

RESULTS

In Fig. 3 the residual stress distributions obtained for the chosen cycles
with regard to the transformation behaviour are shown for the heat-affected
zone and adjacent regions. From the comparison of Fig. 3a (cycle A) and 3b
(cycle B) it is obvious that preheating in the practicable temperature range
approximately up to 300 °C does not cause a reduction of maximum stresses.
Preheating causes an increase of the maximum longitudinal residual stresses
from 450 MPa for cycle A to 550 MPa and the breadth of the plastified region
rises, too. In the immediate range of the fusion line the residual stresses
decreases. A decrease of the maximum stresses occurring outside the heat-affected zone beneath the level of the basis cycle A (without preheating) is
observed only after the application of much higher preheating temperatures,
which are not feasible in actual welding practice. In Fig. 3c the residual
stress distribution is shown for cycle C with 470 °C preheating. Here the
maximum residual stresses reach 350 MPa and are thus 100 MPa lower than in
cycle A.
The reason for the decrease of maximum stress lies in the reduction of
the degree of rigidity due to the high preheating temperatures. Because
volume dilatation nevertheless occurs during cooling in the course of transformation, the formation of compressions can be observed. Hence in the
region of the heat-affected zone compressions may occur both with very fast
cooling in connection with the formation of martensite but also with a very
high preheating and the slow cooling resulting from this. The "islands" of
maximum stresses occuring in Fig. 3 are numerically determined.

REDUCTION OF THE STRESS LEVEL BY SHORT-TIME POST-HEATING

In order to utilize the higher volume dilatations contributing to the reduction of the residual stresses at lower transformation temperatures and to
obtain improved mechanical properties in the heat-affected zone in spite of
this, the impact of a short-time post-heating on the residual stress distribution has been calculated. In Fig. 4 cycle D is compared to cycles A to C.
The short-time post-heating accompanying the cooling process makes control
of the desired transformation product possible by means of temperature and
the duration of post-heating, and resulting from this up to 95 % of the
energy required for preheating can be saved. In the example under consideration cycle D leads to a quasi-isothermal bainitic transformation. In spite

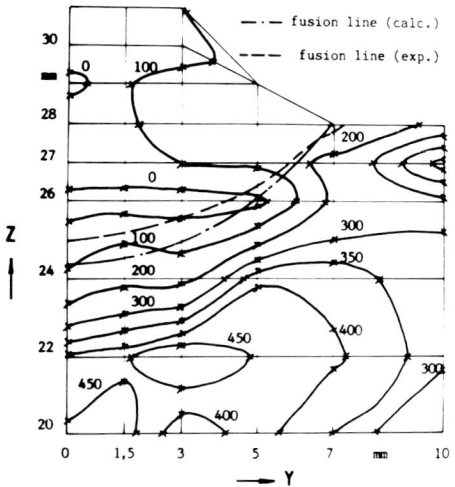

Fig. 3a. Longitudinal residual stresses R_x calculated for cycle A

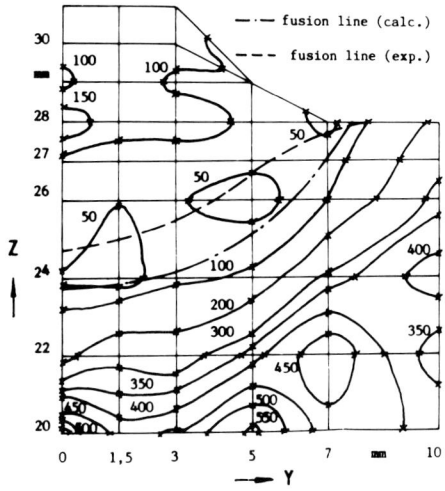

Fig. 3b. Longitudinal residual stresses R_x calculated for cycle B

of substantial energy reductions the ductility values obtained here are higher (Grobelin, 1987) than those shown in Fig. 1b with preheating.
Figure 3d showing stress distribution for the chosen most unfavourable case of short-time post-heating illustrates, that even that case leads to lower residual stresses (500 MPa) than preheating, Fig. 3b (550 MPa), although cooling time between 850 °C and 500 °C is nearly the same as in the preheat-cycles B and C. Minimized residual stresses can be observed especially with more decreasing post-heating temperatures due to the increasing volume

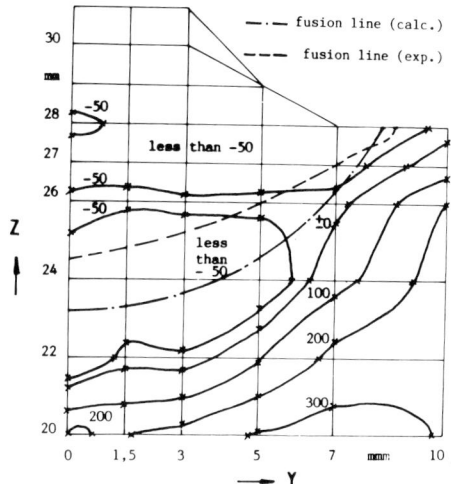

Fig. 3c. Longitudinal residual stresses R_x calculated for cycle C

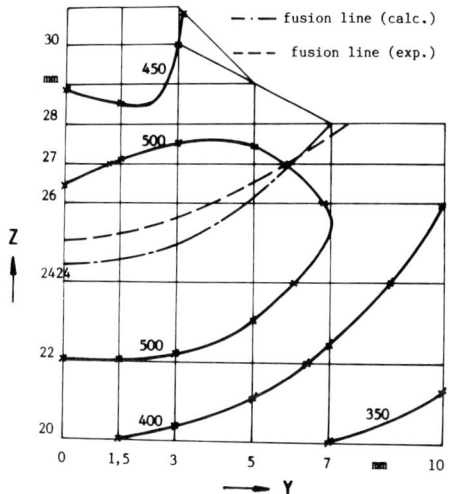

Fig. 3d. Longitudinal residual stresses R_x calculated for cycle D

dilatation effect caused by structural transformation.
Experimental checks (Groß 1986) using the trepanation method provided the same qualitative order.

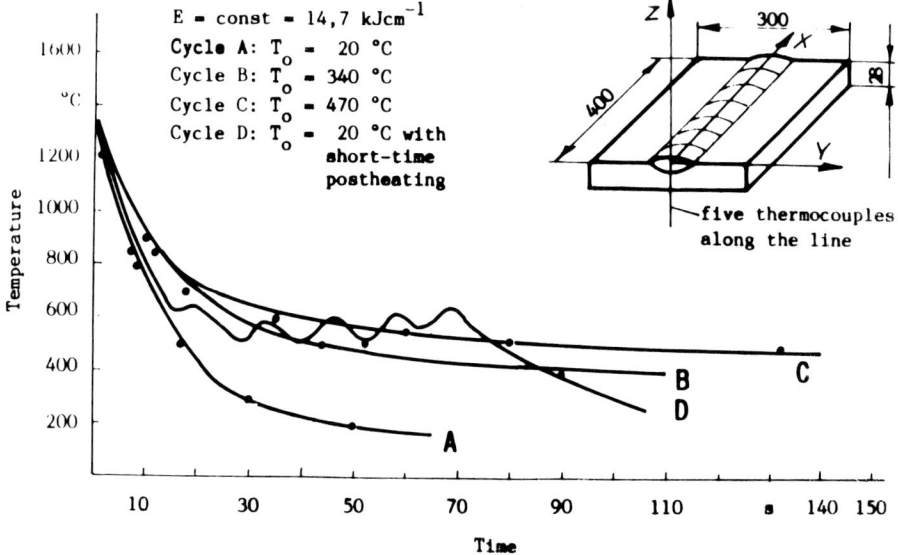

Fig. 4. Cooling cycles A to D

SUMMARY

If structural transformations are taken into consideration with the aim of minimizing residual welding stresses, short-time post-heating accompanying the welding process is to be recommended. In addition to this, short-time post-heating leads to an improvement of mechanical properties and the reduction of the energy required as compared to preheating.

REFERENCES

Denis, S., and S. Sjöström (1986). Coupled temperature, stress, phase transformation calculation model. Proc. Intern. Conf. on Residual Stresses, Garmisch-Partenkirchen, 15.-17. 10. 1986.

Grobelin, K. (1987). Kurzzeit-Nachwärmung beim Schweißen. Dr.-Ing.-Diss. Wilh.-Pieck-Univ. Rostock, 1987.

Groß, H.-G. (1986). Experimentelle Spannungsermittlung. Non-published report Wilh.-Pieck-Univ. Rostock, 1986.

Heeschen, J., T. Nitschke, and H. Wohlfahrt (1986). New results on the formation of residual stresses due to phase transformations in the welded structural steels St 52-3 and St E 690. Proc. Intern. Conf. on Residual Stresses, Garmisch-Partenkirchen, 15.-17. 10. 1986.

Ueda, Y., and co-workers (1985). Mathematical treatmentof phase transformation and analytical calculation method of restraint stress-strain. Trans. of JWRI, 14, 153-162.

Vinokurov, V. A., and V. V. Ship (1985). Abstracts list of papers on the application of numerical and mathematical methods for the analysis of strains, stresses and other effects arising at welding.IIW-Doc. X-1098-85.

II.3

Local Heat Treatment by Induction

H. H. Müller

AEG-ELOTHERM GmbH, Remscheid, FRG

ABSTRACT

For the local heat treatment prior to, during and after welding induction heating has been economically applied to a high degree during the last decades besides other methods. Mains frequency as well as medium frequencies ranging from 1 to 10 kHz are used for this process.

The process is applied in pipeline construction for petrochemical plants and power stations, in vessel- and tank fabrication, as well as for big components of nuclear power stations.
The advantages of this process are:

- precise dosing of power allowing a sensitive controlling of temperature,

- feed of energy is economically limited to the weld area to be heat treated,

- ecologically beneficial application of electroheat without encumbrance of the operating personnel by heat.

KEYWORDS

Local heat treatment of welds; induction heating; pipeline construction; vessel fabrication; nuclear components.

INTRODUCTION

When welding steel, mechanical stresses will develop within the vicinity of the welds as a consequence of the high energy infeed which is concentrated on a narrow area. The heated workpiece section cannot expand unhindered within the surrounding cold parts. Analogously, during the subsequent cooling process, shrinking is also hindered. Heating prior to and during welding will keep developing stresses on a low level. After the welding process tensile- and compressive stresses can be reduced by a heat treatment to values which are not longer harmful.

Heat treatment of the complete workpiece in a furnace is a perfect technical solution for the stress relieving of welded constructions. However, such workpieces are often too large to be introduced into an annealing furnace. Quite frequently, it is necessary to treat welds on components which form part of a large and extensive construction, e.g., pipelines in power stations or petrochemical plants. In such special cases partial heat treatment can be effected, provided that the competent supervisory bodies agree. If local heat treatment is practised, a strict observance of the applicable requirements and codes is made an essential condition (Müller, 1986).

Induction heat treatment fulfills these conditions and has been applied on welded constructions having diameters up to approx. 13 meters or having wall thicknesses of 800 mm and more with excellent succes for several decades. At this point it seems appropriate to emphasize that the heated workpieces can expand unhindered in any direction so that additional stresses cannot develop.

Principle of Induction Heating

The principle of induction heating may be compared to that of a transformer. A transformer has 3 main parts:

Primary winding
Secondary winding
Core.

If an alternating voltage is applied to the primary winding, a magnetic flux is induced in the core and, in turn, generates in the secondary winding, a voltage which is galvanically separated from that in the primary. Now consider the induction heating of a circumferential weld on a pipe joint. A flexible cable wound in several turns around the area of the weld constitutes the primary. The skin of the pipe in the area of the winding can be regarded both, as the transformer core and as a short-circuited single-turn secondary winding. When the annealing cable is connected to a voltage source of appropriate frequency, the workpiece is heated in the area of the winding, firstly by the continous magnetic reversals, and secondly by the induced eddy currents.

The choice of frequency of the voltage source affects the depth of penetration of the induced currents. However, since the heat treatment of steel is a long-term process which may last for several hours, the influence of thermal conductivity in the material is substantially greater than that of the depth of penetration.

Heating Equipment

High frequencies (in the region of 200 to 500 kHz) cannot be used for induction heat treatment of welds on sites and in factories because of the high inductor voltage and the considerable radio interferences that would probably result. On the other hand, mains frequency and medium frequency equipment have proved successful in practice. At first sight, mains frequency appears to offer the advantage of lower capital cost. This advantage is, however, offset by certain technical disadvantages which must be considered if the technique is to be economic. The principal disadvantages are:

- high currents due to the necessary low voltage

- consequently, rigid cables with substantial copper cross-sections

- single-phase operation, imposing a load on the mains which may not always be permissible

- low power factor of energy transfer.

In addition, the use of mains frequency precludes continuous preheating during welding, because the welding arc is affected by the high induction currents and the low frequency.

Medium frequency, on the other hand, offers a wealth of technical advantages, which must, however, be weighed against the price disadvantage of higher capital cost. The capital cost is higher because of the need to convert the usual 50 Hz mains frequency to a technically more suitable medium frequency in the range 1000 to 10000 Hz. Conversion is effected either by traditional motor-generators(three-phase motor/single-phase synchronous generator) or by thyristor-based static converters. The latter type has gained ground enormously in the last twenty years owing to its many technical advantages as well as those of cost. The chief advantages of the use of medium frequency are as follows:

- lower loading of annealing cables

- hence, flexible annealing cables

- uniform three-phase loading of mains

- better power factor

- continuous controllability

- welding no longer affected during preheating.

On account of these disadvantages and advantages, mains fequency equipments are used predominantly for lower powers, of up to about 25 kVA (in very few cases, also up to 85 kVA) and medium frequency equipments are used for powers between 20 and 400 kW. It should at this point be noted that MF-equipments subdivided into two units have proved very successful. Motor generators or converters - i. e., the heavy parts - are accommodated in a separate housing which is set up close to the mains supply. The other unit, containing the capacitor bank and control panel, is smaller and easily transportable, and is set up close to the working position. The two units are connected by control- and power cables.

Energy Transfer

As ready stated, highly flexible annealing cables are used with medium-frequency heating; their cross-sections can be minimized because water cooling is possible. These cables can be wound around the weld and surrounding parts which are to be heated true to the contours of the workpiece. The cables can be repositioned relative to each other and also to the workpiece during the heat treatment process, so that correction on the temperature characteristic is easily possible (Brühl, Müller, 1972).

Owing to the water cooling, the cables are exposed only to a slight heating from the current; this ensures long life (more than 200 annealing operations).

Thermal Insulation

The use of thermal insulating materials in induction heating of welds is necessary for various reasons. In partial heat treatment, there is normally a sharp decline in temperature between the heated and unheated parts, this temperature gradient can be substantially flattened by an appropriately fitted insulation. In this way the creation of additional stresses due to intensive temperature gradients is avoided (Müller, 1973). Furthermore, the annealing cable must be protected from the intense thermal radiation from the area of the weld. The use of an efficient thermal insulation is also desirable on economic grounds, to minimize convection and radiation losses. Glass fabric with no free crystalline silica is being used increasingly for insulation. Blue or white woven asbestos should on no account be used because of the health hazard (danger of asbestosis).

Temperature Measurement

Flexible thermocouples are normally used in practice. They are made of thermocouple wire, generally nickel-chrome and nickel. They are fixed to the workpiece by means of capacitor discharge units with a short welding pulse. When using temperature measuring devices and recorders, it is important so ensure a correct electrical match between the thermocouple, compensating cable and display or recording unit.

Examples of Economic Applications

Heat losses are minimized by controlled heating in the desired area of the workpiece and by the insulation used. The efficiency of energy transfer into the workpiece, referred to mains consumption, is probably higher than 60 %.

Since there is no flame as in the case of gas heating, thermal molestation of the welder during preheating is extremely small. Complex and expensive heat protection measures are unnecessary, and in addition welding times are reduced because of the improved working conditions.

Owing to the high flexibility of the annealing cables and simplified possibilities of connection, preparation- and dismantling times for the heat treatment are substantially shorter than with non-inductive techniques.

Heat treatment equipments having higher powers, can anneal several welds simultaneously. This saves time and labour costs.

Electronic two-point controllers and programmers permit automatic control of holding temperature or even of entire heat treatment cycles. This allows particularly an economic employment of personnel, because the preparations for the next preheating- and annealing cycle can take place while the proceeding program is running down automatically.

Avoiding Flexural Stresses

In the partial heat treatment of hollow cylindrical parts, there is a serious risk of flexural stresses arising if the heated area is too narrow. This was investigated in U.K. by Rose (1960) and Burdekin (1963). A few years later, the Netherlands technical supervisory organization "Dienst vor het Stoomwezen" turned its attention to the same problems. The technical and scientific department of this body drew up the Specification W0701. A minimum width of the heating zone is suggested to ensure the avoidance of flexural stresses in partial heat treatment. This width is given by 3,5 $\sqrt{D \cdot t}$, in which D is the vessel diameter to the middle of the wall and t is the wall thickness.

Before the partial heat treatment of nuclear reactor components - which are of a very high value - are actually carried out, the heat treatment cycle is simulated by computer programs which were previously developed. Practical experience of the temperature distribution in low-stress induction annealing of thick-walled vessel welds and theoretical calculations of the expected reference stresses in the area of the weld formed the basis for the first specifications of a temperature pattern along the longitudinal axis of the vessel, i. e., perpendicular to the weld (Barkmann, Geisel, Müller, 1978).

Conclusion

These economic advantages, as well as its universal applicability, have assured the induction heat treatment technique of a wide and still increasing degree of acceptance over the last four decades. In this connection it should be mentioned that, during this period several hundred thousands of heat treatments have been succesfully carried out with annealing equipments supplied by only one manufacturer.

REFERENCES

Barkmann, C. G., H. Geisel and H. H. Müller (1978). Messen und Berechnen des örtlichen und zeitlichen Temperaturverlaufes bei induktiver Wärmebehandlung von Kernreaktor-Komponenten. DVS-Berichte, Band 52, 44-47.
Brühl, F. and H. H. Müller (1972). Beeinflussung der Temperaturverteilung bei der induktiven Wärmebehandlung von Hochdruckrohrleitungen aus warmfesten Stählen. Schweißen und Schneiden, 24/1, 9-12.
Burdekin, F.M. (1963). Local stress relief of circumferential butt welds in cylinders. British Welding Journal, 10(9), 483-490.
Müller, H. H. (1973). Induktive Wärmebehandlung im Kernreaktorbau. Werkstatt und Betrieb, 106/6, 357-359.
Müller, H. H. (1986). Das internationale Regelwerk über die örtliche Wärmebehandlung von Schweißnähten. Fachbuchreihe Schweißtechnik DVS, Band 85.
Rose, R. T. (1960). Stresses in cylindrical vessels. British Welding Journal, 7(1), 19-21.

Dans Certain cas le Traitement Thermique par Induction est non Reussi. Quand et Pourquoi?

S. Čundev

Faculté d'électrotechnique, Skopje, Yougoslavie

RESUME

Malgré toutes les précautions prises, lors du traitement thermique par induction des constructions soudées, la température atteinte est parfois au-dessous de la grandeur admissible. La cause de ce phénomène consiste sur: l´inconcordance entre la fréquence et l´épaisseur de la tole traitée, la position incorrecte de l´inducteur par rapport au cordon de la soudure, les paramètres inconvenables du régime etc.-en un mot, sur le fait que le flux thermique est insuffisant. Le texte proposé donne les explications de ces phénomenes et recommande les solutions qui garantissent un traitement thermique favorable.

MOTS CLEFS

Chauffage par induction; chauffage par proximité; inducteur; charge; rendement électrique; facteur de puissance; effet de bord.

INTRODUCTION

Le chauffage par induction donne la possibilité d´exercer le traitement thermique indépendamment de la phase du montage d´une construction dans laquelle on effectue le soudage. L´utilisation des cables souples pour former l´inducteur, donne des vastes possibilités mais impose une bonne connaissance des phénomènes se produisant comme conséquence de la propagation électromagnétique de l´énergie. Le texte qui suit est le résultat d´une étude théorique et pratique. Sans donner les déductions des relations qui définissent le transport de l´énergie, on cite les conclusions et les recommandations pour les modèles fondamentaux sur lesquels se réduisent tous les cas de chauffage par induction.

Liste des symboles utilisés dans le texte

H - intensité du champ magnétique (At/m),
J - densité du courant de conduction (A/m^2),

ρ - résistivité électrique (Ωm),
γ = 1/ρ - conductivité électrique (1/Ωm),
μ_r - perméabilité magnétique relative,
cosφ - facteur de puissance,
I_1 - courant dans l'inducteur (A),
$\delta = 503\sqrt{\rho/\mu_r f}$ - profondeur de pénétration (m),
note: δ_1 - profondeur de pénétration dans l'inducteur;
δ_2 - profondeur de pénétration dans la charge,
$\alpha = \sqrt{\omega \mu_r \mu_0 \gamma/2}$ - coeficient d'atténuation (1/m).
Les autres symboles sont présentés dans les figures.

APERÇU DES MODELES DE CHAUFFAGE PAR INDUCTION UTILISES
LORS DU TRAITEMENT THERMIQUE DES CONSTRUCTIONS SOUDEES

Supposant que le lecteur connait le principe et les possibilités du chauffage par induction, nous donnons en bref les différents modèles utilisés pendant le traitement thérmique.

Groupe A: Chauffage a l'aide d'un inducteur qui enroule (une ou plusieurs fois) la charge conductrice

Modèle 1. : chauffage d'un cylindre plein (Fig. 1a).
Modèle 2. : chauffage d'un cylindre creux (Fig. 1b).

    Note: la figure 1b contient seulement les supplément par rapport à la figure 1a.

Modèle 3. : chauffage d'une plaque (Fig. 1c).

    Note: la longueur de la plaque l_2 est égale ou superieure à la longueur de l'inducteur l_1.

Groupe B: Chauffage par proximité

Modèle 4.: Inducteur linéaire posé d'une coté de la charge (Fig. 2a).

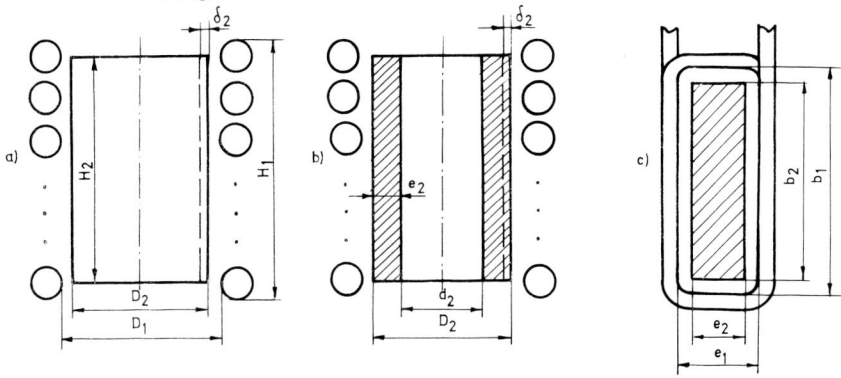

Fig.1. a) Chauffage d'un cylindre plein; b) Chauffage d'un cylindre creux; c) Chauffage d'une plaque

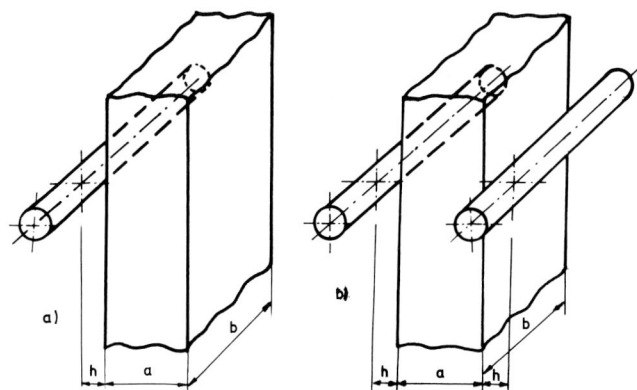

Fig.2. a) Inducteur linéaire posé d´une côté
       b) Inducteur linéaire posé des deux côtés

Modèle 5.: Inducteur linéaire posé des deux côtés de la charge (Fig.2b).

> Note: le modèle 5. peut être examiné comme cas particulier du modèle 3. quand le dernier est avec une seule spire, ou bien dans le cas quand la largeur de la plaque b_2 est très grande par rapport à l'épaisseur e_2.

Modèle 1.

Dans ce cas, la puissance active injectée est donnée par l'équation $W_1 = H_o^2 F/2\gamma\delta$ où la fonction F prend les valeurs correspondantes au rapport $2a/\delta$ selon la courbe présenté par la figure 3. On peut conclure facilement que W_1 augmentera avec l'augmentation de la fréquence utilisée, en tenant compte en même temps, de ne pas permettre que le rapport $2a/\delta$ soit inférieur à 5 ce qui dépend, de nouveau, de la fréquence.

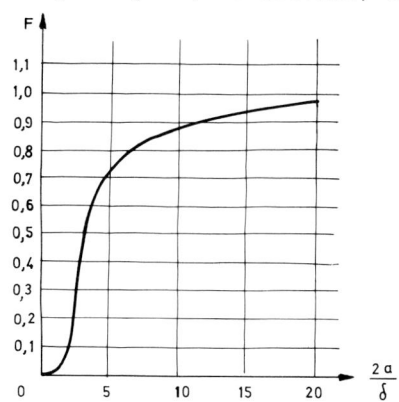

Fig.3. Courbe F en fonction de $2a/\delta$

Si on traite ce cas comme chauffage par proximité (ce qui est très fréquent lors du traitement thermique des constructions soudées), l'analise du transport de l'énergie (voir plus detaillés dans le texte qui décrit le modèle 4.) donne la possibilité de conclure que la distance "c" entre l'inducteur et la charge doit être minimale. Et la valeur du $\cos\varphi$ de l'instalation dépend de cette distance:

$$\cos\varphi = \left[\sqrt{1 + (1 + c\,2/\mu_r\delta)^2}\right]^{-1}$$

ou bien, $\cos\varphi = 356 \dfrac{\mu_r^{1/2}}{C} \rho/f$ pour $\mu_r\delta \ll C$.

ce qui recommande la diminution extrême de "c". Notons que, en analisant les formules pour $\cos\varphi$ on peut facilement obtenir que $\cos\varphi \equiv 1/f$, ce qui est en contradiction avec la constatation qui concerne la grandeur de la puissance injectée. Ce fait impose une analise serieuse du problème et une solution qui sera le compromis des deux éxigences contradictoires.

Modèle 2.

Le chauffage d'un cylindre creux peut être traité comme le cas du chauffage d'un cylindre plein. La différence concerne les limitations qui existent pour les fréquences. Au lieu de donner les courbes de la puissance specifique injéctée, nous nous limitons de donner les recommandations suivantes:

e_2/D_2	0,05	0,1	0,2	0,3	0,4
$(e_2/\delta)\min$	0,3	0,4	0,6	0,8	1,25
$(\delta/e_2)\max$	3,33	2,5	1,67	1,25	0,8

La relation $\delta = 503\sqrt{\rho/\mu_r f}$ donne la possibilité de définir la fréquence minimale: $f_{min} = 250 \cdot 10^3 \rho/\mu_r \delta^2$.
Et cette fois nous ne conseillons pas d'augmenter considérablement la fréquence pour n'abaisser pas le facteur de puissance $\cos\varphi \equiv 1/\sqrt{f}$.

Modèle 3.

Ce modèle peut etre analisé comme:
- modèle 1., quand la largeur de la plaque b_2 est comparable avec son épaisseur e_2 et comme
- modèle 5., quand $b_2 \gg e_2$.

Modèle 4.

Avec les symboles presentés sur la Fig.2a, dans le cas d'un inducteur linéaire ponctuelle sont valables les relations suivantes: $J_o = 2I_1 h/\pi\delta(h^2+z^2)$ (A/m² - densité du courant à la surface de la charge, x=0), $J_r = J_o e^{-\alpha x}$ (A/m² - densité du courant en fonction de x mais pour z=const), $W_1 = \rho_2 J_o^2 \delta/4 = 126,5 \rho_2^{3/2} J_o^2 \sqrt{1/\mu_{2r} f} =$
$= 0,1 \rho_2 h^2 I_1^2 \sqrt{\mu_{2r} f/\rho_2}/(h^2+z^2)^2$ (W/m² - flux thermique dû à la propagation du champ électromagnétique). Pour une longueur de l'inducteur "l" la puissance active injéctée dans la charge conductrice est: $W = \rho_2 \ell I_1^2/2\pi\delta h = \ell I_1^2 \cdot 10^{-3} \sqrt{\mu_{2r} \rho_2 f/\pi h}$ (W).

Les courbes de la Fig.4 donnent la possibilité de citer les

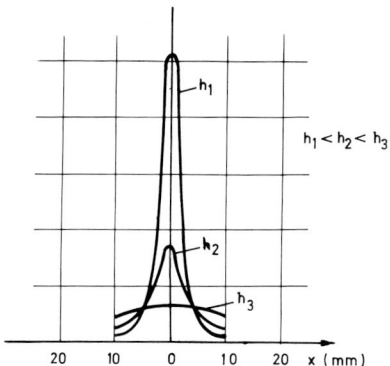

Fig.4. Courbes J_o en fonction de x

conslusions suivantes:
1. La densité du courant décroit sensiblement en s'eloignant (dans le sens des coordonnées z) de l'inducteur: $J_o \equiv 1/r^2$; $r^2 = h^2 + z^2$.
2. Quand on obtient des grand vitesses de l'augmentation de la température (chauffage en défilé), la largeur de la zone chauffée dépend de la largeur du conducteur qui représente l'inducteur et il faut en tenir compte pour que la largeur de cette zone soit suffisante. Si la largeur du conducteur (de l'inducteur) est plus petite que le cordon de la soudure, on sacrifie le rendement électrique (η) en augmentant la distance h entre l'inducteur et la charge. Au contraire, si le temps de chauffage est long (le cas le plus souvent, quand on utilise des cables souples comme inducteurs) la conduction thermique donne la possibilité d'obtenir une zone assez large même en diminuant la distance h pratiquement jusqu'à zéro.
3. La densité maximale du courant (pour z=o et x=o) est:
$J_{omax} = 1,265.10^{-3} I_1 \sqrt{\mu_{2r} f/\rho_2}/h$ (A/m^2)
4. Le flux maximal de la source thermique (au-dessous de l'inducteur: z=o; x=o) est: $W_{1max} = 0,1 \rho_2^{1/2} I_1^2 \sqrt{\mu_{2r} f}/h^2$ (W/m^2).
Donc, J_{omax} et W_{1max} dépendent directement du $\sqrt{f}$ se qui veut dire que, pour obtenir le même effet thermique pour deux distances différentes h_1 et h_2 il faut respecter la rélation suivante:
$f_2 = f_1 (h_2/h_1)^2$.
Modèle 5.
Selon le sens du courant on peut distinguer deux variantes différentes:
a) le courant est de sens opposé des deux côtés de la charge, ce qui est le cas le plus fréquent
La Fig.5a représente la distribution des deux composantes du vecteur H (H_2' comme résultat du courant dans les conducteurs du côté gauche et H_2'' produit par ceux du côte droite) et le

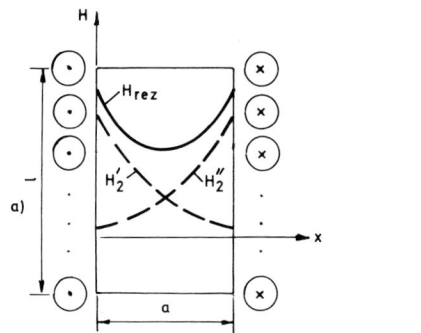 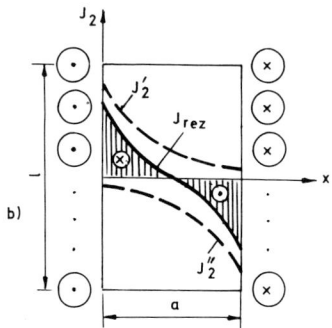

Fig.5. Répartitions de H et de J_2 pour modèle 5. variante a)

champ magnétique resultant: H_{rez}.

$H'_2 = H_o e^{-\alpha x}$ $\qquad H'_{2a} = H_o e^{-\alpha a}$ $\qquad H''_2 = H'_{2a} e^{\alpha x}$

$H''_2 = H_o e^{-\alpha a} e^{\alpha x}$ $\qquad H_{rez}=H'_2+H''_2=H_o(e^{-\alpha x}+e^{-\alpha a} e^{\alpha x})$ $\qquad 0 \leqslant x \leqslant a$

La Fig.5b. montre la répartition de la densité du courant resultant J_{rez} et de ses deux composantes J'_2 et J''_2 provoquées par les conducteurs correspondants: $J_{rez}=J_o(e^{-\alpha x}-e^{-\alpha a}e^{\alpha x})$.

La puissance thermique injectée dans la charge est:

$$W_a = \int_{x=0}^{x=a} \rho \frac{b}{\ell dx}(\frac{J_{rez}}{\sqrt{2}})^2 (\ell dx)^2 = \rho \frac{\ell b}{2} J_o^2 \frac{1-e^{-2\alpha a}-2\alpha a\ e^{-\alpha a}}{\alpha}$$

ce qui porte à la conclusion générale que l'épaisseur de la plaque "a" doit être au moins deux fois plus grande que la profondeur de pénétration $\delta (a>2\delta)$ pour que les composantes J'_2 et J''_2 n'influencent pas l'une l'autre. En comparant les figures 5a. et 5b. on peut verifier la loi $\bar{J} = rot\bar{H}$ qui se transforme en $J = dH/dx$ dans le modèle traité.

b) le courant des deux côtés de la charge est de même sens

Pratiquement, ce modèle peut être realisé de la façon suivante (Fig.6.): la plaque épaisse est sous l'influence d'un inducteur ordinaire, tandis que la plaque mince, qui a une épaisseur $a<2\delta$, est sous l'influence de deux conducteurs, dont le sens du courant est le même. Au point de vue énergétique, on peut réaliser le même flux thermique specifique dans les deux plaques, parce que le nombre des spires autour de la plaque épaisse est plus grand par rapport à celui autour de la plaque mince (pour $n_{mince}= 1$, $n_{épaisse} \simeq 2$).

Cette fois, les répartitions de l'intensité du champ magnétique et de la densité du courant sont comme à la Fig.7. Les rélations correspondates sont:

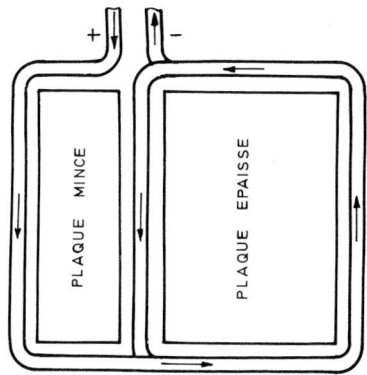

Fig.6. Chauffage simultané d'une plaque mince et d'une plaque épaisse

$$H'_2 = H_o e^{-\alpha x} \qquad H''_2 = H_o e^{-\alpha a} e^{\alpha x}$$

$$H_{rez} = H'_2 - H''_2 = H_o(e^{-\alpha x} - e^{-\alpha a} e^{\alpha x})$$

$$J_{rez} = J_o(e^{-\alpha x} + e^{-\alpha a} e^{\alpha x}) \qquad 0 \leqslant x \leqslant a$$

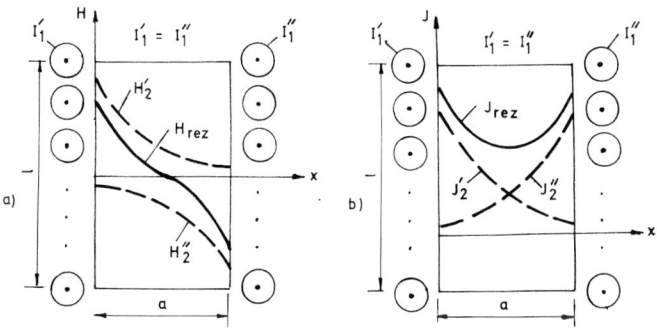

Fig.7. Répartition de H et de J_2 pour modèle 5. variante b)

$$W_B = \rho \frac{\ell b}{2} J_o^2 \frac{1 - e^{-2\alpha a} + 2\alpha a e^{-\alpha a}}{\alpha}$$

C'est évident que pour la plaque mince, surtout si la fréquence utilisée est assez basse, ce mode de montage de l'inducteur est favorable et donne la possibilité d'exercer le traitement thermique d'une plaque de n'importe quelle épaisseur avec les fréquences de quelques centaines de Hertz.
Le tableau I donne le rapport $k = W_B/W_A$ pour 5 fréquences différentes en fonction de l'épaisseur de la plaque dans le cas du chauffage par induction d'un corps en acier à une température

de 600 °C ($\mu_r=15$; $\rho=0,5.10^{-6}$ Ωm).

TABLEAU I

f|Hz|	50	600	2000	3000	10000
a|mm|					
1	2000,4	167,1	50,4	33,74	10,42
4	125,4	10,8	3,58	2,56	1,25
7	41,2	3,8	1,57	1,29	1,01
10	20,4	2,2	1,16	1,06	1
13	12,3	1,5	1,04	1,01	1
16	8,2	1,3	1,01	1	1
19	6,0	1,1	1	1	1
22	4,6	1	1	1	1
25	3,6	1	1	1	1

Le tableau II correspond au meme acier à une température de 900 °C ($\mu_r=1$; $\rho=1.10^{-6}$ Ωm).

TABLEAU II

f|Hz|	50	600	2000	3000	10000
a|mm|					
1	60000	5000	1500	1000	300
4	3750	313	94	63	19,2
7	1224	102	31	20,8	6,6
10	600	50	15,4	10,4	3,5
13	355	30	9,3	6,4	2,3
16	234	20	6,3	4,4	1,7
19	166	14	4,6	3,2	1,4
22	124	11	3,6	2,6	1,2
25	96	8	2,9	2,1	1,1

La conclusion est bien évidente: en utilisant le système de montage de l'inducteur autour d'une plaque mince d'après la Fig.6. (il n'est pas obligatoire de chauffer en même temps une plaque épaisse) la puissance thermique injectée sera beaucoup plus importante que celle dans le cas du chauffage ordinaire et ne sera pas, pratiquement, limitée par l'inconcordance entre la fréquence et l'épaisseur de la plaque.

Note (commune pour les deux variantes du modèle 5.):
D'après l'équation $J=J_{Om}e^{-\alpha x}\cos(\omega t-\alpha x)$, la densité du courant change de sens après une profondeur $\alpha x=\pi$ donc $x=\frac{\pi}{\alpha}=\pi\delta$. Cela veut dire que la demi-onde du courant avec sens inverse par rapport à celui dans la couche avec une épaisseur égale à 3δ n'influencera pas le courant résultant de l'inducteur qui se trouve de l'autre côté de la charge. Cette constatation est très importante parce que dans le cas contraire les diagrammes présentés par les figures 5. et 6. ne seront pas valables.

EFFET DE BORD

L'effet de bord peut être très prononcé. Les expériences realisées prouvent que, parfois, la répartition non uniforme du champ thermique dû a l'effet de bord peut être remarquable. Cet effet existe régulièrement quand l'inducteur s'éloigne de la charge métallique. Les illustrations données a la Fig.8. présentent les cas les plus fréquents et les mésures qu'on doit entreprendre pour diminuer les conséquences. Puisque le but de cet ouvrage

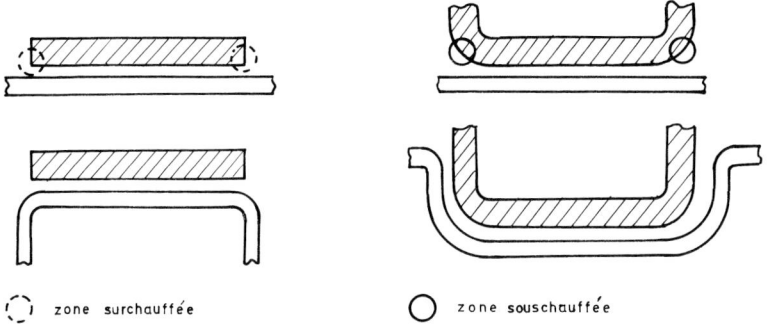

◌ zone surchauffée   ○ zone souschauffée

Fig.8. Effet de bord et son élimination

n'est pas de resoudre tous les problèmes qui paraissent pendant le traitement thermique des constructions soudées, nous recommandons aux utilisateurs du chauffage par induction d'acquerir l'expérience necessaire en exécutant plusieurs essais en changeant l'allure de l'inducteur surtout au voisinage des zones sous ou surchauffées, pour obtenir un chauffage uniforme. Les inducteurs linéaires en cables souples sont les plus convenables pour ces essais.

CONCLUSIONS

Le chauffage par induction présente une des plus confortables méthodes pour l'exécution du traitement thermique des constructions soudées. Cependant, l'usage de se mode de chauffage exige une bonne connaissance de plusieurs phénomènes de lesquels dépend le succès final: la qualité de la soudure. Les recommandations qu'on peut conseiller aux utilisateurs de ce procédé de chauffage sont:

- de respecter les rélations fondamentaux données dans les descriptions des modèles différents;
- d'acquerir l'expérience minimale necéssaire en exécutant plusieurs essais sur eprouvettes;
- d'utiliser une source de chauffage par induction bien appropriée au devoir qui devra être executé, surtout au point de vue fréquence et puissance;
- de prévoir une isolation thermique qui reduira les pertes;
- d'adapter la forme de l'inducteur au cordon de la soudure et de le poser non seulement symétriquement par rapport au cordon mais de definir une distance minimale (et constante) entre l'inducteur et la charge;

- d'éviter l'existence des parties vide de l'inducteur parce que dans ce cas la charge de l'inducteur est très inductive. Nous rappelons que la résistance active equivalente est $R_e = R_1 + K^2 R_2$ (k-coefficient de transformation) ce qui veut dire qu'elle croît quand l'inducteur est en "contact" magnétique avec la charge, et que la réactance equivalente est $X_e = X_1 - K^2 X_2$ c'est à dire que le système exige une puissance réactive maximale quand l'inducteur est vide;
- tenir compte de l'existence de l'effet de bord;
- autrement dit, si on ne peut pas obtenir le diagramme de chauffage prescrit, ce qui est le cas quand le courant de l'inducteur ou sa tension atteignent ses valeurs maximales sans que la puissance active soit satisfaisante, il faut chercher la reponse du problème dans les explications données ci-dessus.

BIBLIOGRAPHIE

Dunsky, Ch.V. (1964). Notes du cours d'Electrothermie. Société Coopérative de l'A.E.E.S.,Liège.

Čundev, S.(1977). Electrothermie. Université Cyrille et Méthode, Skopje.

Sluhotzki, A.E. et autres (1981). Installations pour le chauffage par induction. Energoizdat, Leningrad.

SESSION III

MECHANICAL STRESS RELIEVING TREATMENTS/TRAITEMENTS MECANIQUES DE RELAXATION DE CONTRAINTES

III.1	Application of gyrotron for heat treatment of materials (B. E. Paton, V. E. Sklyarevich — USSR/URSS)	141
III.2	Vibratory stabilization of welded constructions — experiments and conclusions (P. Sedek — Poland/Pologne)	145
III.3	Vibratory lowering of residual stresses in weldments (M. Jesensky — Czechoslovakia/Tchécoslovaquie)	153
III.4	Reduction and redistribution of the residual welding stresses through local explosive treatment (S. Hristov, P. Petrov, S. Zvetanov — Bulgaria/Bulgarie)	161
III.5	Investigation on the mechanism of stress relief by explosion treatment — The rule of plastic deformation of metal in the process of explosion (L. Chen, Z. Si, Y. Wang, H. Chen and X. Dong — China/Chine)	169
III.6	Mechanical stress-relief treatment of welded pressure vessels by warm pressure test (K. Kalna — Czechoslovakia/Tchécoslovaquie)	177
III.7	Relaxation overstressing of huge spherical storage vessels repaired by welding (I. Hrivnak, J. Lancos, S. Vejvoda — Czechoslovakia/Tchécoslovaquie)	189

III.1

Application of Gyrotron for Heat Treatment of Materials

B. E. Paton and V. E. Sklyarevich

E. O. Paton Electric Welding Institute, Kiev, USSR

The investigation of the possible application of electromagnetic microwave radiation for heat treatment of materials is of great interest. While interacting with a material, microwave radiation is known to be partially absorbed by it /1, 2/. Here, the material exposed to radiation is heated, this being caused by various loss mechanisms. For semiconducting materials the main ones are exciton absorption and absorption by charge carriers which is due to the transition of electrons and holes to the higher energy band levels, acceleration of charges along the free path length and crystal lattice energy transfer. Energy is also absorbed in ionization or excitation of impurity atoms in the crystal lattice and in interaction of its site charges with the electromagnetic microwave field.

For dielectric materials all the mechanisms of losses are associated with the processes of relaxation (electronic, ionic, dipole) polarization.

In a general case the value of the energy transferred to a material due to electromagnetic radiation is determined by its electrophysical characteristics, i.e. conduction, forbidden gap width, polarization and radiation frequency as well.

Metals are characterized by the significant reflection of microwave radiation from the surface (more than 99%). Penetration inside a material amounts to fractions of a micron due to the skin effect.

Thus, it can be concluded that the electromagnetic microwave radiation is reasonable to use, first of all, for heat treatment of non-metallic materials where other known heat sources (e.g. arc and electron beam) are of a low efficiency. At the same time, it should be noted that electromagnetic radiation, while penetrating these materials to the considerable depth, provides their bulk heating. In many cases this is more efficient and practicable than the surface heating which takes place when using gas and plasma torches.

Lately, due to the availability of the new high-power microwave oscillators the actual prerequisites have been created for the practical application of microwave radiation for the production purposes. In this connection the most promising is gyrotron (cyclotron-resonance maser) developed by a group of Soviet scientists headed by academician Gaponov A.V. /3/. The modern gyrotrons are able to oscilate centimetric, millimetric and submillimetric microwaves with the power of tens of kilowatts (and in the near future - hundreds of kilowatts) and with the efficiency of up to 40% (about an order higher than that of lasers). Radiation is emitted from the device in the form of a beam whose configuration can vary with the help of metal mirrors, i.e. by dissipating the power over a considerable area or by focusing it to the diameter comparable with the radiated wavelength. The beam propagation direction can be easily changed, the beam can be transported for tens of meters without any appreciable losses and can be introduced into the containments with any gas atmosphere composition and pressure.

The E.O.Paton Electric Welding Institute of the Ukrainian SSR Academy of Sciences has developed the technological installation for heat treatment of materials by the microwave radiation. One of the gyrotron types was used as an oscillator, that is the 30 kW gyromonotron which generates the electromagnetic waves of the 83 GHz frequency with the normal distribution of intensity in the beam /3/.

The installation was used for a set of research works of welding, heat treatment and coating and for the determination of the most promising directions in the field of the application of gyrotron radiation as a heat source.

As for welding, the quality joint can be provided in ceramic materials due to the large penetration depth of electromagnetic radiation. In this case the temperature gradient considerably decreases through the weld thickness, thus reducing the probability of cracking. The baffle metal plate can be installed on the weld root side to increase the radiation utilization factor. A portion of radiation which has passed through a workpiece will be reflected from the plate, return to the weld and, thus, increase the uniformity of heating. This method can be used to join members,e.g.of oxide ceramics. Preheating and gradual temperature lowering are carried out only by the defocusing of the electromagnetic radiation.

For ceramic materials which do not admit melting and particularly for those which have the comparatively low absorption factor, i.e. which are uniformly heated through the large volume at a time, the joining was performed in a somewhat different way. Components of a workpiece with well fitted planes are exposed to the radiation and the planes to be joined are preliminary covered with a filler material which is similar to the workpiece material. This method was used to join the specimens of structural silicon nitride-based ceramics. The dimensions of the specimens were 50x30x15 mm, time of heating up to the 1300°C temperature was 3 s, the joining process was performed in a chamber with nitrogen atmosphere under the 60 kPa pressure, the process duration was 12 min. Studies showed that the joint area exhibited the satisfactory homogeneity and the slight decrease (down to 15%) in microhardness.

The gyrotron microwave radiation can be also applied for welding of metals if to use a dielectric (for example, welding flux) as an energy-absorbing element. This principle was taken as a basis for the developed process of joining the metal thick sections under the atmospheric pressure. The fused flux transfers heat to metallic particles of a filler and to the edges of metal being welded, resulting in their melting. The flux grade and percentage are selected so as to provide the maximum absorption of electromagnetic radiation and the physico-chemical reactions which are favourable for the weld quality. Therefore, the process of the electroslag type with weld width of order of the radiated wavelength, i.e. of the dimension of several millimetres, is possible.

Surfacing can be performed by the same principle as well. Filler metal and flux are either fed continuously or preliminary applied in the form of a paste on the surfaces to be deposited. In the latter case it is possible to perform surfacing in hard-to-reach locations and on the inner surfaces of workpieces. Radiation through the interaction with a flux melts the filler metal and the surface of a workpiece being surfaced.

At the same time, with the thorough control of the radiation energy directed onto the ceramic material which is located at the metal surface it is possible to melt the ceramics with the insignificant heating of the metal itself. This enables to apply dielectric coatings with the minimum heating up of workpieces. In the existing processes of enameling, chemical-thermal treatment or, for example, film coating, it is common to heat, first of all, the workpiece being treated. In this case a portion of heat accumulated is spent for the chemical-thermal processes, i.e. enamel melting, film heating, etc. Workpieces are heated in special furnaces up to hundreds of degrees, and heating and cooling take a long time. Here the workpiece metal quality is deteriorated, this often being inadmissible at all, and the geometric dimensions of the structures to be treated are limited.

The application of the gyrotron radiation as a heat source allows in many cases to avoid all these difficulties. The direct heating of enamel films, various powders at the workpiece surfaces up to the temperature required takes practically but seconds and the heat transferred to a workpiece will be negligible.

Thus, for example, the gyrotron radiation was used for deposition of protective films. In this case the film was heated up to the temperature required to form the adhesive contact on the practically cold pipe.

Self-adjusting endothermic technological processes can be performed with the help of the microwave radiation. It becomes feasible if the starting material heated by the radiation possesses the high absorption factor and the final product - the low one. This phenomenon was used to accomplish the boron and silicon nitride powder synthesis reaction.

Similarly, the gyrotron microwave radiation can be used for chemical-thermal treatment of metals. Here the surface of the metal workpiece being treated is covered with powders of various chemical compositions based, for example, on boron or chromium and possessing the comparatively high absorption factor. When powders are melted, boron (chromium) diffuses into metal, their absorption factor drastically decreases, this, in its turn, terminating heating. As a result, borating and chromizing processes can be carried out with the insignificant heating of a workpiece being treated.

When using the microwave radiation as a heat source, the reverse process is possible as well; this is the case when interaction with a material increases with heating. For example, while heating up many materials with the comparatively low absorption factor, the latter, as a rule, drastically increases when the materials are melted. This phenomenon is used to apply refractory non-metallic coatings on ceramic workpieces with the lower melting point. At the starting period of this process a workpiece is heated up with a portion of the microwave radiation which passed through a coating powder poured over the surface. As the powder melts, the transmission of the radiation to the workpiece being coated considerably decreases. Preliminary heating through the substrate provides the coating with good adhesion without any cracks and spalling which are usually observed when using the other heat sources, e.g. indirect arc discharge. The technological processes for ceramic material coating were developed by using this principle.

Therefore, it is apparent that the application of the concentrated high-power electromagnetic radiation within the GHz frequency range allows to improve in principle the existing widespread technological processes and to develop the qualitatively new ones which cannot be performed with any other known heat sources.

REFERENCES

Kovneristy, Yu.K., Lazareva, I.Yu., Ravaev A.A. (1982). Materials absorbing microwave radiation. M., Publ.House "Nauka", 162.
Berezin, V.M., Buryak, V.S., Guttsait,E.M., Marin,V.P. (1985). Microwave electronic devices. M., Publ. House "Vysshaya Shkola", 296.
Pasynkov,V.V. and V.S.Sorokin (1986). Materials for electronics. M., Publ. House "Vysshaya Shkola", 368.

Vibratory Stabilization of Welded Constructions — Experiments and Conclusions

P. Sędek

The Welded Structures Division, Institute of Welding, Gliwice, Poland

ABSTRACT

Currently the stress relief annealing is frequently replaced by vibrational stabilization. In spite of the high efficiency of this process its physical mechanizm is not quite known. In the paper the explanation mechanism of this process basing on research investigation and practical experience has been presented.

KEYWORDS

Dimensional stability; vibratory stabilization; microrelaxation; metal anelasticity; natural ageing; rapid ageing.

INTRODUCTION

During welding, internal stresses resulting from heterogeneous heat cycle occur in a construction. Additional postweld treatments are needed especially in machine building where dimensional stability is often required. Stress relief annealing is such a most known treatment. At the present moment, however, a vibratory stabilization also called vibratory stress relieving is being introduced due to the high energy-consumption of thermal treatments. The vibratory stabilization treatment consists in setting a construction into vibrations of resonance frequency and holding in this condition for some period of time.

CONTEMPORARY STATE OF THE PRIOR ART

Stabilization treatment is not dependent on the installation structure and, in general, it lies in the operation of searching out the resonance frequencies, a proper treatment at a definite time and a control operation in which the comparison is made between after-vibration parameters and those at the beginning of the treatment. High efficiency of the treatment makes it possible to eliminate stress relief annealing.

AN ATTEMPT TO EXPLAIN THE PHENOMENON – EXPERIMENTS

A number of authors are of opinion that high variable stresses are produced during construction vibrations within a range of the resonance frequency, what can induce plastic strains in some construction regions. This results from summing up the internal stresses and stresses created during vibration of the construction. As a result of plasticization of some construction zones, and particularly the regions of welded joints, a reduction in internal stresses takes place /Claxton, 1976; Sagalevich, 1974; Saunders, 1978; Wozney, 1968/. Experiments carried out have shown that stresses generated by vibrations are inconsiderable, and their amplitudes do not exceed the value of 50 MPa. At the same time, it has been found that elements undergo a deformation during vibration /Sędek, 1978, 1980a, 1986/. Measurements of internal stresses have not shown any substantial changes in their distribution and value. Taking into consideration the values of variable stresses due to vibrations and of cyclic yield point, which is usually not less than $0,7\ R_{e0,2}$ for constructional steels, it is difficult to agree on the opinion that the vibration of construction may cause the reduction in internal welding stresses because of summing up the variable stresses and internal stresses.

Experiments carried out on using vibration in relation to welded constructions have shown that vibrated constructions reveal a high dimensional stability after mechanical working. Stress relief annealed constructions exhibit similar shape errors after mechanical working compared with vibration-subjected constructions.

While explaining the dimensional stability phenomenon as a result of using vibrations, it is difficult to be based exclusively on the principle of summing up the external load stresses and internal stresses. It has been found that variable stresses, which are produced in the construction because of vibrations, are insignificant and the stabilization has to proceed as a result of other complex processes. It is suggested that this phenomenon should be associated with the effect of dimensional stability in consequence of natural ageing. Investigations on construction deformations have shown that they are time-dependent after welding. These deformations change with time and their magnitude depends on steel grade.

After 6-month seasoning, the deformation was $0,377 \cdot 10^{-2}\%$ for steel St3 within a zone of heat effect /Sagalevich, 1974/.
Also, measurements taken by the author have shown that constructions undergo a deformation after vibration treatment. Vibration stabilization was tested on the side walls of side-guard manipulators /Sędek, 1980a/. A photograph of the side-guard manipulator element is shown in Fig. 1. A diagram of making strain measurements is illustrated in Fig. 2. The measurements were made by means of a rod gauge on measuring bases in a form of balls welded on the construction /Fig. 3/.
Strains were measured on four elements subjected to vibration treatments and after mechanical working.
The results of these measurements are presented in Table 1.

TABLE 1. Results of the strain measurements on the vibrated parts

Part No	Strains of measuring bases /Nos. 1 to 7/ /$10^{-2}\%$/						
	1	2	3	4	5	6	7
1	0,99	0,80	0,94	1,11	0,78	0,96	0,91
2	0,05	-0,19	0,48	-0,38	0,89	0,35	-2,13
3	2,17	2,24	0,15	-0,07	-0,16	-0,16	-1,97
4	-1,34	-0,13	1,11	-2,94	-0,49	-0,85	-0,96

Table 2 shows the measurement results of strains in vibrated elements which occurred after mechanical working. While mechanical working, measuring bases were damaged in the elements 1 and 2, as well as bases Nos. 2 and 6 in the element 4 and, therefore, not all strain measurements were taken after mechanical working. As can be seen from Table 1, the strain values on vibration are in the range of /-2,94 -2,24/ $10^{-2}\%$, however, those after mechanical working are within /-3,73 - 0,71/ $10^{-2}\%$ /Table 2/.

TABLE 2. Results of the strain measurements on the vibrated parts after machining

Part No	Strains of measuring bases /Nos. 1 to 7/ /$10^{-2}\%$/						
	1	2	3	4	5	6	7
3	-1,18	-3,73	0,05	-0,37	-0,82	-0,61	-2,93
4	0,68	-	0,66	0,71	0,66	-	0,66

Another example of the post-vibration occurrence of strains was a stabilization of mining loader buckets. A shape and main dimensions of the bucket are given in Fig. 4 /Sędek, 1980b/.

Strains were measured by means of rod gauge on bases provided by bearing balls arranged as shown in Fig. 4. The results of measurements of strains that occurred in the elements after vibration and mechanical working are presented in Table 3.

TABLE 3. Results of the strain measurements on the mining loader buckets

Part No	After vibration Strains of measuring bases $/10^{-2}$ %/			After machining Strains of measuring bases $/10^{-2}$ %/		
	1	2	3	1	2	3
1W	0,45	0,45	0,45	-9,49	-4,90	-1,80
2W	2,68	3,12	2,26	7,42	-1,16	-2,80
3W	-8,14	-2,19	-1,32	-1,09	-1,42	5,14
1N	-	-	-	26,31	-1,02	-1,63
1N	-	-	-	16,31	-36,02	3,03

Comparatively, strains were measured on elements worked mechanically without vibration treatment. Designations 1W, 2W, 3W correspond to the elements subjected to vibration and mechanical working, whereas notations 1N, 2N are corresponding to the elements which only were mechanically worked after welding.

On the basis of the results of strain measurements it can be noted that deformations exist after vibration and on mechanical working. Deformations are larger in those elements which are not vibrated. In the case of vibrated elements, strains on mechanical working are larger than those after vibration, however, strains after mechanical working of non-vibrated elements are much greater.

Interesting experiments were carried out by Rappen /1973/. Three welded rings were made from hot-rolled sheet. The rings measured 600 mm in outer diameter and were 150 mm wide and 25 mm thick. One of these rings was vibrated, the second was stress relief annealed, and the third was left in the raw state. All three rings were lathe worked. After mechanical working, measurements were made in order to record circularity errors. Annealed and vibrated rings have been found to be insignificantly deviated from circularity, being contained within a tolerance. The raw ring has shown inadmissible changes in circularity. Deviations in circularity of the annealed and vibrated rings were 0,07 - 0,25 mm and 0,02 - 0,05 mm, respectively.

When all three rings were cut, it has been found that the vibrated and raw rings drew aside by a magnitude of elastic strain, and the stress relief annealed ring preserved its primary shape. Unfortunately, strains were not measured in the ring on vibration.

Also, investigations carried out by Sędek /1983/ had not shown any reduction of internal stresses in welded elements due to vibration.

The final effect in a form of dimensional stability after mechanical working can be best exemplified by vibrating the foundation frames in coal pulverizers for power engineering /Sędek, 1977/. A view of the foundation frame is presented in Fig. 5. The out of flatness of surface, on which power unit was mounted, could not exceed, 0,08 mm per 1 m of length. The results of measurements of flatness in non-vibrated and vibrated frames are given in Table 4.

TABLE 4. Results of the out of flatness measurements on the foundation frames

Frame No	State of frame	Out of flatness /Fig. 5./ /mm/		
		Direction 1	Direction 2	Direction 3
1	unvibrated	-0,267	-0,332	-0,064
2	unvibrated	-0,384	-0,244	-0,135
3	vibrated	0,037	-0,065	0,052
4	vibrated	0,034	-0,024	0,082

It should be mentioned that vibratory stabilization of each frame was performed twice: on welding and roughing. Postweld vibration did not bring good results.

The examples presented illustrate univocally the effectiveness of vibratory treatments resulting in dimensional stability. However, explanation of the stabilization mechanism as a result of using vibratory treatments presents some difficulties. Practically, reduction in welding stresses because of summing up the stresses during construction vibration does not take place due to small stress amplitudes; but strains occur after vibration, and the values of deformations on machining are close to those after stress relief annealing.

It is suggested that explanation of the dimensional stability should be based on microrelaxation effect owing to metal anelasticity. During vibrations, some processes proceed in metal as a result of which the latter undergoes to a small degree a deformation. With inconsiderable plastic strain, the value of modulus of elasticity decreases, what can lead to reducing the level of elastic strain energy stored as a result of the presence of internal stresses /Zener/.

Similar phenomenon takes place in natural ageing, during which constructions spontaneously undergo a deformation with time, and internal stresses practically are not reduced /Bühler, 1962/. One may state that vibratory stabilization is a "rapid ageing", The author makes it clear that this statement has been formulated on the basis of observation of behaviour of constructions subjected to vibration and ageing. No investigations were conducted on phenomena which occured in microregions of metal structure during vibration and ageing.

CONCLUSIONS

Based on investigations and experiments the following conclusions can be drawn:

1. There is no marked reduction in welding stresses during vibratory stabilization of welded constructions.

2. Stabilization can be used only in case of constructions from which dimensional stability is demanded.

3. Vibratory stabilization should not be used in case of constructions where reduction in welding stresses increases operating properties such as: fracture toughness, fatigue strength, stress corrosion resistance.

4. In respect of effectiveness, vibratory stabilization is similar to natural ageing.

REFERENCES

Bühler, H., Pfalzgraf, H.G. /1962/. Untersuchungen über den Abbau von Eigenspannungen in Gusseisen und Stahl durch mechanische Rütteln und Langzeitlagerung im Freien. VDI-Forschungsheft 494, Ausgabe B, Band 28
Claxton, R. A., Saunders, G.G. /1976/. Vibratory stress-relief. The Metallurgist and Materials Technologist, vol 12, 651-656
Rappen, A. /1973/. Möglichkeit der Verzeugsminderung durch Vibration. In A. Rappen /Ed/, Verzugsprobleme durch Eigenspannungen bei der mechanischen Fertigung. Technische Akademie Esslingen, paper 5.
Saunders, G.G., Claxton, R.A. /1978/. VSR- a current state-of-the-art appraisal. In R.W. Nichols /Ed./, Residual stresses in welded construction and their effects, vol 1. The Welding Institute. London, 173-183.
Sagalevich, W.M. /1974/. Methods of replacement the welding distortions and stresses. Mashinostrojenie. Moskva /in Russian/.
Sędek, P. /1977/. Vibration stress-relief of structures coal pulverizers. Instytut Spawalnictwa Report Ha-12, /in Polish/
Sędek, P. /1978/. Vibrational stabilization of sizes of welded structures. Prace Instytutu Spawalnictwa, vol 3, /in Polish/
Sędek, P. /1980a/. Vibrational stabilization of sizes of side wall of side-guard manipulator. Instytut Spawalnictwa Report NK/30/80. /in Polish/.

Sędek, P. /1980b/. Vibrational stabilization of the mining loader bucket. Instytut Spawalnictwa Report Ia-15, /in Polish/.
Sędek, P. /1983/. Können mechanische Schwingungen das Spannungsarmglühen geschweisster Maschinen-elemente ersetzen? Schweissen und Schneiden., vol. 10, 483-486.
Sędek, P. /1986/. Investigation of possibility of vibrational stabilization of selected structures. Instytut Spawalnictwa Report Id-84 /in Polish/
Wozney, G.P., Crawmer, G.R. /1968/. An investigation of vibrational stress-relief in steel. Weld J., vol 9, 411s - 419.
Zener, C. Elasticity and anelasticity of metals. The University of Chicago Press. Chicago Illinois, 144

III.3

Vibratory Lowering of Residual Stresses in Weldments

M. Jesenský

Welding Research Institute Bratislava, CSSR

ABSTRACT

The results of the study of the effect of vibratory stress relieving on mechanical properties of welded joints and the stability of weldment size are outlined.

KEYWORDS

Strength tests; notch and fracture toughness; measurement of the volume of residual stresses in welded joints; measurement of strains, utilization of vibratory stress relieving in practice.

INTRODUCTION

During the welding process residual stresses which usually create three-dimensional stress state are formed in welded joints. In welded joints of unannealed weldments mainly with larger material thicknesses the residual stresses can form un favourable three-dimensional tensile stress state which can decrease strain properties of a joint, lower the dynamic load-carrying capacity of weldments, cause strains in as-machined or as-loaded condition or brittle failure of weldments etc. These drawbacks are solved in manufacture by proposing stress relief annealing of weldments. Majority of weldments is annealed in order to assure the stability of their dimensions and shape during their service life. A considerably lower number of weldments is annealed for other reasons; e.g. to improve mechanical properties of welded joints, to lower hydrogen content in the weld metal and to avoid weldment cracking, to decrease the hardness of transition zone in a welded joint etc.

Expensive stress relief annealing of weldments can be in majority of cases replaced by vibratory stress relieving, later VSR only. Weldments are subjected to vibratory stress relieving on the mechanical accentric vibrator with 1 to 200 Hz revolutions.

From the technical viewpoint vibratory stress relieving can be used as replacement of residual stress relief annealing in majority of selected weldments. The VSR procedure has been introduced and used with much success in Czechoslovakia. At the same time the effect of VSR on mechanical properties of welded joints of weldments and on the hygiene of work was investigated in more detail.

This paper gives a brief description of the advantages of VSR procedure in comparison to stress relief annealing and that of the effect of vibration on some properties of welded joints and weldments having been attained in the course of this study.

MEASURED RESIDUAL STRESSES IN UNVIBRATED AND VIBRATION STRESS RELIEVED WELDED JOINTS

To solve the problem of VSR effect on welded joint properties also the three-dimensional residual stresses were measured. We wanted to find out to what extent the residual stresses are lowered after vibratory stress relieving. The welded specimens 30x650x1100 mm in size were used for the measurement of residual stresses. The longitudinal welded joint type X30 fabricated with N 70 flux and A 107 wire was located in the central part of specimens. One specimen remained unvibrated and the other was subjected to vibration at four resonance levels. The three-dimensional residual stresses were measured in a welded joint in the central part of specimens /Jesenský, 1984/.

The measured three-dimensional stresses in the unvibrated welded joint are shown in Fig. 1a and those in vibratory stress-relieved joint are shown in Fig. 1b.

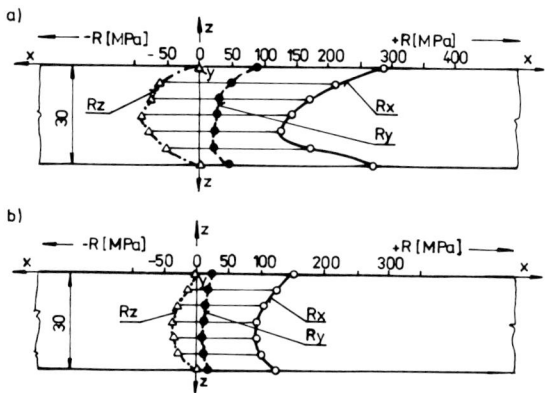

Fig. 1. The course of measured three-dimensional residual stresses in X30 welded joint of Fe 430 D /CSN 41 1449.1/ steel
a - as-welded, b - vibratory stress relieved specimen, R_x - along the weld, R_y - normal to the weld, R_z - along weld thickness, /+/ tension, /-/ pressure

The values of measured residual stresses in these welded joints - being valid also for other welded joints - prove that after vibration the three-dimensional residual stresses on the surface and through material thickness of a welded joint are lower by 40 to 80% in comparison to those in unvibrated welded joints.

For measurement of three-dimensional residual stresses the semi-destructive method which is based on removing of the studied zone in shape of a cylinder ⌀ 30 mm through the joint thickness was used. The foil strain gauges type 1.5/120 LY11 /manufacture by Hottinger Baldwin Messtechnik, West Germany/ cemented on the frame of three-dimensional sensor were used as measuring elements. The sensor is filled up with Belzona mixture into the opening bored through ⌀ 8 mm material thickness. From the measured values prior to and after boring a hole the three-dimensional residual stresses are determined employing graphical-numerical method /Jesenský, 1983/.

Further measurement of surface residual stresses was done e.g. on welded beams made of Fe 510 B /CSN 41 1523.1/ steel, shown in Figs. 2a and 3b. The beams were joined by manual arc process with E 52.33 coated electrode according to CSN standard.

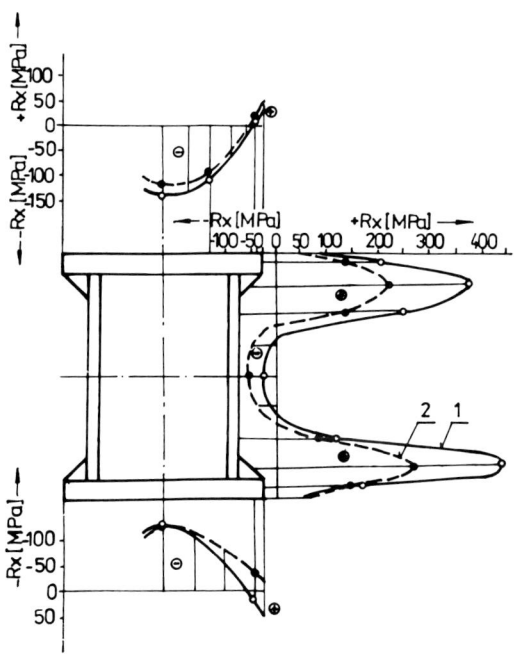

Fig. 2. Measured surface residual stresses in welded beams of Fe 510 B /CSN 41 1523.1/ steel, 1 - stresses in as-welded beam, 2 - stresses after vibratory stress relieving

One beam remained unvibrated and the other was subjected to vibratory stress relieving at four resonance levels. Fig. 2 Shows the course of surface residual stresses in unvibrated and vibratory stress-relieved beam. Also in this case a substantial lowering of residual stresses after vibratory stress relieving can be seen.

The surface residual stresses were measured by tensometric method which consists in cutting prisms 20x20 mm in size along the beam circumference. The HBM strain gauges type 1.5/120 LY 11 which were cemented on the material surface in the shpe of rosette of 0 °, 90 ° were used as measuring elements

MEASUREMENT OF STRAINS IN WELDED BEAMS AFTER PLANING

During the study of this problem except for other measurements also strains in the above-mentioned welded beams shown in Fig. 3 were measured.

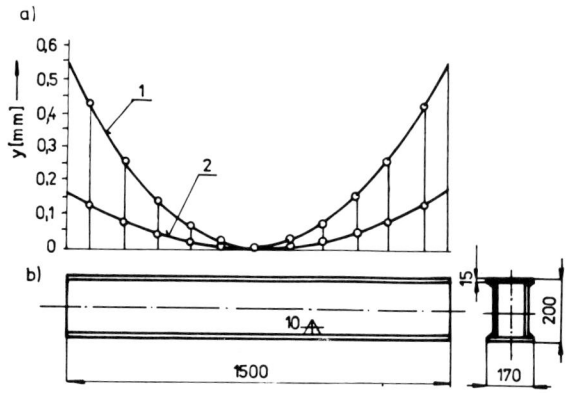

Fig. 3. Measured strains "y" of welded beams after machining one flange 5 mm in depth; a - longitudinal strains in: 1 - as-welded beam, 2 - vibratory stress relieved beam, b - MMA welded beam of Fe 510 B steel using E 52.33 electrode according to CSN Standard

One beam was untreated. The other beam was vibrated to lower residual stresses and to stabilize dimensions at four max. 120 Hz resonance levels. The strains "y" /Fig. 3a/ in the untreated flange along the beam were measured in the unvibrated and vibratory stress-relieved beam by aid of deflectometer. After gradual machining - 2+2+1 = 5 mm planing from one flange - the total strains - deflections "y" /Fig. 3a/ were measured. The course of strains designed 1 - for as-welded beam, that designated 2x corresponds to vibratory stress-relieved beam.

The measured values in Fig. 3a prove that after vibratory stress relieving of the beam the residual stresses were lowered and the shape was stabilized /Jesenský, 1984/. Substantially higher strains were measured in the unvibrated beam. At the same time it has been pointed out that the measuring system was calibrated and its maximum deviation between single measurements represented ± 0.15 mm.

STRENGTH AND YIELD STRENGTH OF MATERIAL OF UNVIBRATED AND VIBRATORY STRESS-RELIEVED WELDED JOINTS

For the solution of this task the strength tests of welded joints in different steels with strength values varying between 370 and 700 MPa were performed. The measured values of welded joints in bars have proved that vibration exerts a negligible effect on yield strength and tensile strength of welded joint material. A slight increase in yield strength, practically within the scatter of measured values /Haramia, 1982/ was proved in vibratory stress-relieved joints.

NOTCH AND FRACTURE TOUGHNESS OF UNVIBRATED AND VIBRATORY STRESS-RELIEVED WELDED JOINTS

The samples for measurement of notch and fracture toughness were fabricated from Fe 430 D /CSN 41 L449.1/ steel welded joints from above mentioned specimens 30 mm in thickness.

The course of measured values of notch toughness in dependence on temperature /measured by Haramia, 1982/ is shown in Fig. 4.

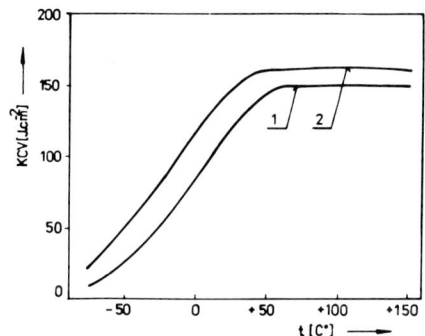

Fig. 4. The course of notch toughness KCV in dependence on temperature measured in specimens with X30 joint in Fe 430 D steel; 1 - as-welded, 2 - VSR specimen

The course of fracture toughness in dependence on temperature /Haramia, 1982/ is shown in Fig. 5.

The measured values in Figs. 4 and 5 and those of other welded joints have proved that the notch and fracture toughness values of vibratory stress-relieved joints are higher than those of unvibrated joints.

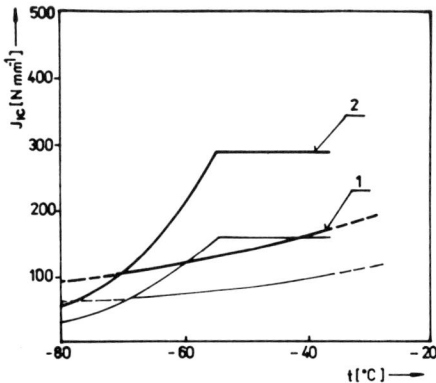

Fig. 5. The course of notch toughness J_{IC} on temperature measured in specimens with X30 joint in Fe 430 D steel; 1 - as-welded, 2 - VSR specimen

THE EFFECT OF VIBRATORY STRESS RELIEVING ON FATIGUE STRENGTH

To solve the problem of VSR application the fatigue strength of welded joints and weldments in dissimilar steels was studied.

Fig. 6 shows the results of fatigue tests of welded specimens 30 mm in thickness with X30 welded joint in Fe 360 B steel /Fig. 6a/ fabricated by manual arc process employing E 44.83 coated electrode and that in Fe 430 D steel fabricated by submerged arc process employing N 70 flux and A 107 wire /Fig. 6b/.

The measured values in Fig. 6 prove that fatigue strength in vibratory stress relieved joints was not decreased /dependence designated with No. 2/. Similar results were attained also in other welded joints of different steels and in welded beams. The vibratory stress relieving of welded joints and weldments does not cause decrease in their fatigue strength.

HYGIENE OF WORK AT VIBRATORY STRESS RELIEVING OF WELDMENTS

By the introduction of VSR into manufacture also the tasks of hygiene of work with respect to higher noisiness by vibratory stress relieving of weldments were solved. The guide-lines for vibratory stress relieving of weldments and hygiene of work /Jesenský, Horný, Vránka, 1986/ have been elaborated. Except for specifications of vibratory stress relieving of parts this guide-line also contains the measures of safety and hygiene of work in accordance with the CSN Standard.

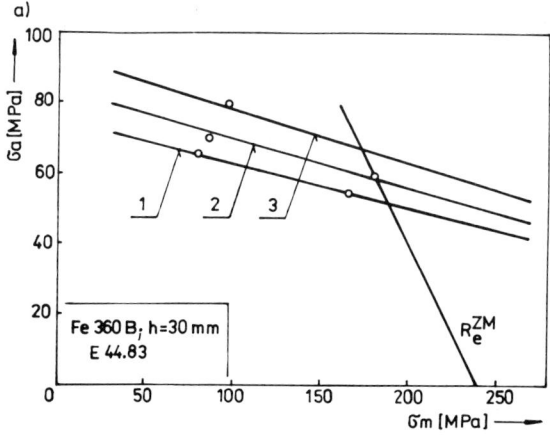
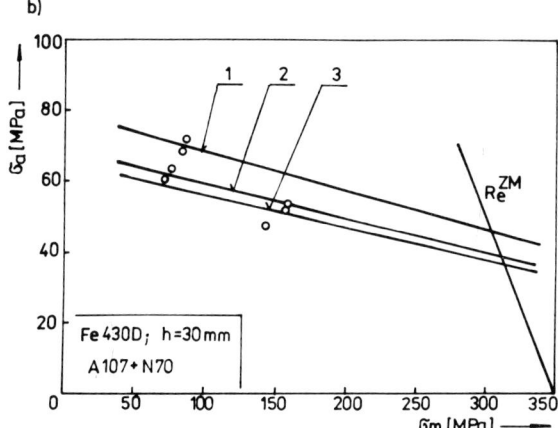

Fig. 6. Measured values of fatigue strength in welded joints; a - dependence of stress amplitude on mean stress in Fe 360 B steel specimens; b - the same dependence in Fe 430 D specimens; 1 - as-welded condition, 2 - after vibratory stress relieving; 3 - after stress relief annealing; R_e - yield strength of base metal

CONCLUSIONS

Based on hitherto attained research results and experience in practical use of VSR weldments the introduction of VSR into practice can be recommended. The vibratory stress relieving can be employed for stabilization of the size of suitable weldments prior to their machining and servicing as a replacement of stress relief annealing. Much experience and good results were attained with VSR in our manufacturing plants.

The VSR process is used for lowering of residual stresses and stabilization of the size of different weldments such as frames of forming machines, machine frames, grey cast iron castings, etc. which were up to now subjected to stress relief annealing.

According to hitherto obtained findings the shape and size of suitable VSR weldments were sufficiently stabilized after vibratory stress relieving and in exploitation. As a fact, VSR does not negatively affect the static and dynamic strength of welded joints and weldments, fracture and notch toughness and homogeneity of welded joints.

Based on the attained data the implementation of VSR procedure as replacement of stress relief annealing for the stabilization of size of weldments, castings and forgings leads to high savings of production costs to our national economy. The saving of thermal energy has to be emphasized first of all because in VSR procedure it does not exceed 3% of energy required for annealing of weldments and in average 5% of production costs.

REFERENCES

Haramia, F.:1982, Progress Report - VUZ Bratislava, No. 1508/206.
Jesenský, M.: 1984, Compilation - VUZ Bratislava, 57-61.
Jesenský, M.: 1983, Residual stress measuring /in Slovak/ - VUZ Bratislava.

III.4

Reduction and Redistribution of the Residual Welding Stresses through Local Explosive Treatment

S. Hristov*, P. Petrov* and S. Zvetanov**

*High Technical University "A. Kantchev", Russe, Bulgaria
**Engineering Works "Dunarit", Russe, Bulgaria

ABSTRACT

Local explosive treatment was used for the reduction and redistribution of the residual welding stresses. Its application considerably improved the reliability and the efficiency of welded joints and constructions. The effect of the most important technological factors (type of detonating cord, transfer medium, mode of arrangement of detonating cords and quantity of explosive) upon the efficiency of the local explosive treatment was investigated in the experiments described.

KEYWORDS

Local explosive treatment; explosive system; transfer medium; residual welding stresses; limit of effective increase of the quantity of explosive.

INTRODUCTION

On arc welding in the connections always arise residual welding stresses (RWS). They usually have considerable values and in weld zone they can even reach the yield limit. In certain conditions RWS have negative effect upon reliability and efficiency of welded joints and constructions (Vinokurov, 1984). That call for serious attention to the problem of reduction of RWS. Not long ago in Bulgaria the only method applied for this purpose was thermal treatment (high temperature annealing at 600-750°C). But this has a lot of disadvantages, which make it irrational and difficult to apply, in particular when it concerns constructions with a large mass or large overal dimensions. Due to these reasons other possibilities for reduction and redistribution of RWS are searched for. An alternative method is the local explosive treatment (LET) of the welded joints (Kudinov, 1981).

EXPERIMENTAL PROCEDURE FOR LET

The specimens which are submitted to LET represent connections with double butt welded seam. They are obtained through automatic submerged arc welding. The explosive system involves a scheme on which are disposed detonating cords upon the welded connection, the transfer medium, detonators and mean of fixing the detonating cords. The basic schemes for the arrangement of the detonating cords (Fig. 1) are two: unilateral (scheme I) and bilateral (scheme II). Moreover some varieties of the unilateral scheme have been applied throughout the researches which are shown in Fig. 1 too (schemes III and IV).

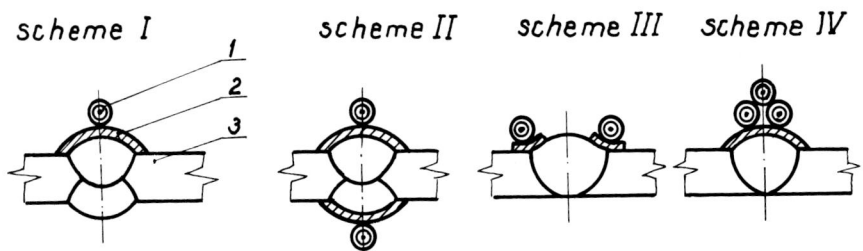

Fig. 1. Schemes for the arrangement of the detonating cords (1-detonating cord, 2-transfer medium, 3-welded joint).

For the evaluation of the efficiency of the process three specific criteria (Kw, Km, Kf) are introduced. Criterion Kw gives the relative reduction of the longitudinal RWS acting in the seam metal. Criterion Km shows the relative reduction of the maximum of the tensile RWS in the welded joint and criterion Kf gives the relative reduction of the tensile area of RWS diagram (Hristov, 1983). To use these criteria it is necessary to know the size and distribution of RWS before and after LET. A great number of methods are known, with the help of which the RWS can be found (Kasatkin, 1976). The comparative analysis made has shown that the RWS should be defined by strain measurements.

TYPE OF DETONATING CORD

Detonating cords are the basic explosive for the treatment of welded constructions. But as intended for other purposes it is necessary to estimate the possibilities of different types of detonating cords and the efficiency of their use as a means of reduction and redistribution of the RWS. The subject of research were a water resisting detonating cord-C, geologocal detonating cord with cover of natural material-F, a geological detonating cord with cover of synthetics-H, a reinforced detonating cord-E, a detonating cord with lead mantle-A and an antifire -damp detonating cord-I. The results of the experiments are summarized in Fig. 2. All welded connections are treated with one detonating cord according to scheme I. From the values of criterion Kw it follows that in all specimens except those of series C the tensile stresses in the seam are almost completely eliminated. Even in the specimens of series H they are converted into compressive ones. The peak values of the tensile stresses which before the LET were in the middle of the seam after it they move away from the seam. Because of the small amount of explosive the peak values are not considerably reduced. The lowest values of criterion Km are obtained with the specimens of series C and comparatively high with the series H and I. Analogical results are obtained through evaluating with criterion Kf. According to this criterion the process efficiency is better and with the specimens of series H and I it goes up to 60%.

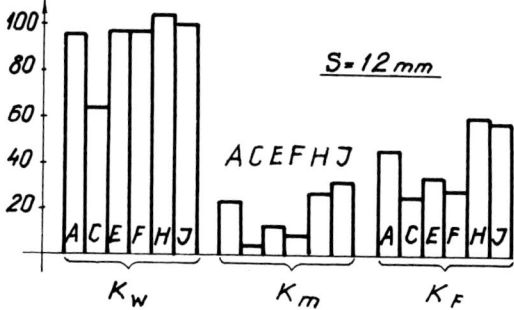

Fig. 2. Effect of the detonating cord type on the LET efficiency.

In conclusion in connection with the problem of the reduction and redistribution of RWS in welded constructions we can recommend the usage of the detonating cords with which series H and I were treated.

TYPE OF TRANSFER MEDIUM

The transfer medium exerts a considerable effect upon the efficiency of the process. It effects in two ways. First through it the contact area between the detonating cord and the threaded connection is improved. Second the homogeneity of the medium, transfering the energy of the explosive is improved also. If there is no transfer medium then except for a narrow line of contact the explosive energy is transferred through air. And this is a material, whose properties are radically different both from the cover of the detonating cord and from the metal of the welded connection the losses are great and the explosive energy coefficient utilization drops abruptly. Also if there is no transfer medium a constant good contact along the seam cannot be guaranteed. Because of the rigidity of the detonating cord and the inevitable curves in some points it separates from the metal, the explosive effect is sharply weakened and the required reduction of RWS is not obtained. That is why the transfer medium must guarantee good adhesion of the detonating cord to the surface of the treated joint with stable contact along the seam. For these reasons it was necessary to study the behaviour of various materials as transfer mediums under the conditions of explosive treatment. The glues "Hemopren" and "Proma" (a), the glue "Proma" with talcum powder (b), or aluminium oxide (c), plasticine (d)

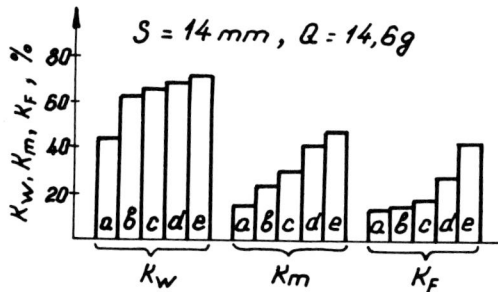

Fig. 3. Effect of the transfer medium type on the LET efficiency.

and putty (e) were tested. They all possess good adhesive properties and ensure reliable fixing of the detonating cord to the treated surface. Their density is different being lowest in the glues and highest in the plasticine and putty. The results of the experiments are shown in Fig. 3. From this it is clear that the best transfer media are plasticine and putty, i.e. the media with the highest density.

MODE OF ARRANGEMENT OF THE DETONATING CORDS

If we assume that the necessary quantity of explosive is determined in advance then the question arises of how the detonating cords should be arranged on the welded connections. It is possible to use a smaller number of detonating cords with a more powerful charge or a greater number of detonating cords with a smaller quantity of explosive in them. In Fig. 4. the results of the LET of welded connections with different thicknesses are shown. Detonating cords are arranged according to scheme I. The figures in the columns of the hystogram show the number of the detonating cords with which the experiment was carried out. Specimens of welded connections 12 mm thick were treated with an equal quantity of explosive but in the first series it was equally distributed in two detonating cords. A considerable increase of the efficiency of the process was obtained in the second case especially with criteria Kf and Km.

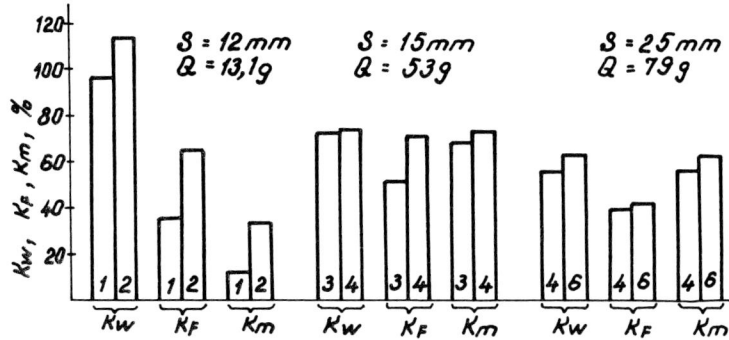

Fig. 4. Effect of the number of the detonating cords on the LET efficiency.

In the similar way it was proceeded with specimens of welded connections with the thickness of 15 mm and 25 mm. Because of the greater thickness more detonating cords were used.
In three thicknesses more effective reduction and redistribution of RWS is obtained when a greater number of detonating cords are used. In other words when the same quantity of explosive is more uniformly distributed over the surface of the treated connection the results are better.
Another experiment was conducted in the same direction. Specimens of welded connections 18 mm thick were treated with 3 and with 5 detonating cords, simultaneously primed. In the first series they were mounted one next to the other according to scheme I, and in the second series in two layers, according to scheme IV. The experimental data obtained (Fig. 5) confirm the conclusion, that it is more rational to use a greater number of detonating cords, containing as little explosive as possible and distributed in an uniform layer on

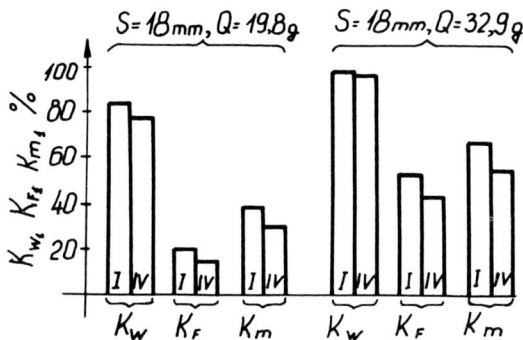

Fig. 5. Values of criteria Kw, Kf and Km in one layer (I) and two layers (IV) arrangement of the detonating cords.

ABSTRACT

Local explosive treatment was used for the reduction and redist-

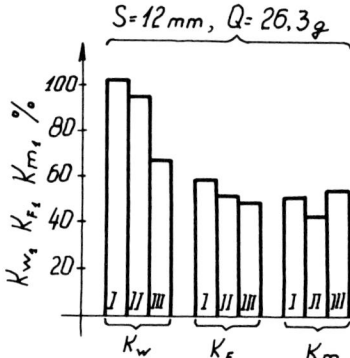

Fig. 6. Comparison of the efficiency of the schemes I. II and III.

the surface of the treated connection.
In Fig. 6 experimental data are compared, obtained by LET of specimens of welded connections with thickness 12 mm according to three different schemes: I, II and III (Fig. 1). All experiments were carried out with the same quantity of explosive and the same number of detonating cords. In all the specimens the rate of decrease and redistribution of RWS according to criteria Kf and Km is almost the same. The difference in values of the criteria does not exceed 11%. The treatment according to scheme III leads to comparativly slight reduction of RWS in seam metal. In this scheme criterion Kw is with 30% and 35% lower than in scheme II and scheme I. Therefore, when one is aiming at reduction of the RWS mainly in the seam one should work according to scheme I or scheme II. Scheme III should be applied when redistribution of RWS with minimum peak values in the heat affected zone is required. Scheme II gives a possibility of most uniform reduction of the RWS in both surfaces of the welded connections.

QUANTITY OF EXPLOSIVE

The specimens of the welded connections are submitted to LET according to scheme I. This scheme was chosen, because it was most effective. The effect of the quantity of explosive Q on the rate of reduction and redistribution of RWS is shown in Fig. 7.

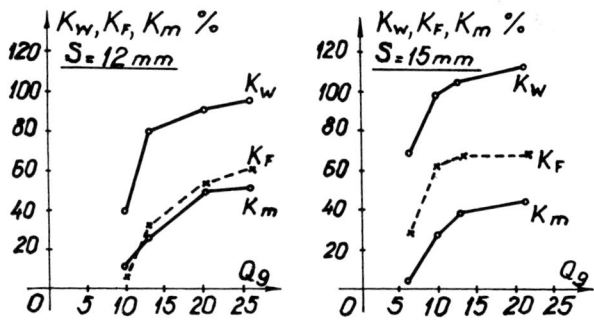

Fig. 7. Effect of the quantity of explosive Q in LET.

Naturally with the increase of Q the values of criteria Kw, Kf and Km also increase. But the dependence is not linear. In the range of the small quantities of explosive their increase leads to proportionally equivalent improvement of the effect of LET. With the further increase of Q the effect of the LET gradually fails. Above certain value of Q, depending on the concrete conditions, its further increase is practically useless. We call this value of Q a limit of effective increase and mark it Ql.
A series of factors effect the magnitude of Ql: transfer medium, thickness of treated connection ets. Thus for example for the specimens 12 mm thick the glue "Proma" is used as a transfer medium. The magnetude of Ql according to criterion Kw is 13 g. and according to criteria Kf and Km - 20 g. The transfer medium in LET of specimens 15 mm thick is putty. According to the three criteria Ql is equal to 13 g. A result is obtained which is opposite to the expected one. Because of the greater thickness of these specimens the value of Ql should be higher, than in the specimens 12 mm thick. But obviously the choice of a more adequate transfer medium has lead to a more effective usage of explosive energy and a lower value of Ql is obtained.
The pattern of the dependences between Q and the criteria Kw, Kf and Km as well as the knowledge of the values of Ql have an important practical significance, because they make it possible to select correctly the quantity of explosive necessary for LET of a given construction.

MECHANICAL PROPERTIES OF THE EXPLOSIVELY TREATED JOINTS

To evaluate the fffect of the LET on the mechanical properties of the welded connections and on their reliability and efficiency a variety of comparative tests were carried out. Specimens of the same kind of steel are welded under the same conditions. Half of them are submitted to LET as given conditions and the other half is not treated additionally. Standard samples for tensile test, bending test, notch impact test and fatigue failure test were made from them. Testing in cyclic loads is carried out according to the scheme "bending and rotation" and symmetric cycle of alternative stresses. The results from the test are given in Fig. 8. The fatigue curves are obtained equidistant and are

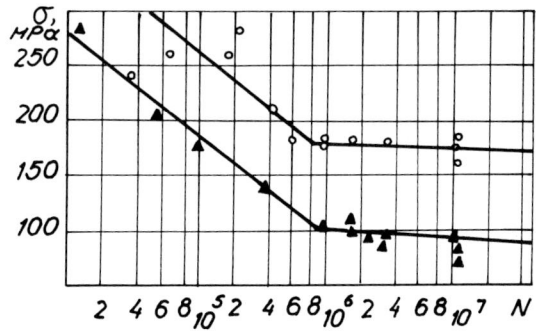

Fig. 8. Effect of the LET on the fatigue strength.

approximated with two straight lines. The fatigue strength is about 95% higher in explosively treated welded connections. The results from the other tests show that the mechanical properties of the explosively treated and not treated welded connections are practically equal with the exception of notch ductility. It is considerably higher in welded connections submitted to LET (Hristov,1981).

APPLICATION OF THE METHOD

The method of reduction and redistribution of RWS through LET has found application in heavy machine-building, in transport machine-building and in energetics.
In heavy machine-building the method is applied in the treatment of units of metal constructions of various modifications of excavators, stone-crushers and other equipment. It increases their efficiency and reliability.
A technology was developed for the reduction and redistribution of RWS in tank waggons used for the transportation of chemical products (Hristov, 1986). The reduction of the RWS obtained guarantees practically equal corrosion resistance of the welded connections and the basic metal which is an important condition for the reliability of the tank waggons.
Positive results are obtained also in LET of welded steel constructions designed for atomic power engineering.

CONCLUSION

The local explosive treatment is a modern highly efficient method for the reduction and redistribution of residual welding stresses particularly suitable for the treatment of metal constructions with large over-all dimensions. Its application considerably increases the reliability and the efficiency of welded connections and constructions.
By modifying the scheme of the local explosive treatment, the type of transfer medium, the number and the type of the detonating cords, the process of LET can be controlled in wide ranges. This gives a possibility for every required reduction and redistribution of residual welding stresses to be achieved.

REFERENCES

Hristov, S. H., D. G. Nikolov, and P. S. Petrov (1981). Reducing and redistribution of the residual welding stresses through explosion processing. Zavaryavane, number 2, 20-22.
Hristov, S. H., and co-workers (1983). On the effect of the explosive for reducing the residual welding stresses in butt welded joints. Zavaryavane, number 1-2, 6-8, 19.
Hristov, S. H., and co-workers (1986). Reduction of residual welding stresses of big-sized tanks by means of explosion. Proceedings of the ninth international conference on high energy rate fabrication. Academy of Sciences of the USSR, Sibirian division, Novosibirsk, 18-22.
Kasatkin, B. S, and co-workers (1976). Experimental methods for investigation of the deformations and stresses. Naukova dumka,Kiev.
Kudinov, V. M., and co-workers (1981). Mechanism of residual stress relieving by explosion treatment. 7-th internaional conference on high energy rate fabrication, Leeds University, Leeds, 23-27.
Vinokurov, V. A., and A. G. Grigoryants (1984). Theory of the welding strains and stresses, Mashinostroenie, Moscow.

Investigation on the Mechanism of Stress Relief by Explosion Treatment — the Rule of Plastic Deformation of Metal in the Process of Explosion

L. Chen, Z. Si, Y. Wang, H. Chen and X. Dong

Institute of Metal Research, Chinese Academy of Sciences, Shenyang, PRC

ABSTRACT

This paper describes the rule of metal deformation in the process of stress relief by measuring residual stress (σ_R) and plastic strain of metal (ε_p) after explosion treatment on a three-bar frame designed by authors. The experimental results show that there exists a relationship between plastic strain and original elastic strain (ε_e):

$$\varepsilon_p = \varepsilon_e + C$$

where C is a constant. Its value and sign depend on the explosion condition. The authors propose a mechanism of stress relief through plastic deformation which is induced by the residual stress during the explosion treatment.

KEYWORDS

Stress relief by explosion treatment; hard explosion; moderate explosion; soft explosion; residual stress; plastic deformation.

INTRODUCTION

All processes of residual stresses relief produce plastic deformation which counteracts the preexisting elastic deformation. The subject of this investigation is to study the rule of transformation from elastic strain to plastic strain.

It is well known that residual stress relief by annealing is usually realized by metal creeping at a high temperature. The amount of creep increases with the increase of the temperature and duration of heat treatment. When the quantity of the plastic strain produced by metal creeping is equal to that of the original elastic strain, the residual stress will be relieved and the creeping process will stop. So long as the temperature and the duration of the treatment are sufficient, the plastic strain (ε_p) produced by the creep is always equal to the original residual elastic strain existing in the zone during the annealing process, no matter what the magnitude and sign of the residual stresses are. That is:

$$\varepsilon_p = \varepsilon_e = \sigma_R/E \tag{1}$$

Theoretically speaking, residual stress can be eliminated completely by annealing, but practically speaking, it can never be entirely eliminated due to nonuniformity of heating and cooling during annealing, especially in a large welded structure. The resudial stress in a welded structure made of dissimilar metals with different coefficients of thermal expansion also cannot be eliminated by annealing.

Explosion treatment of residual stress is a new technique developed in recent years. This new technique opens up a possible way of relieving residual stress in the large welded structures. Quite a lot of works dealing with this explosion technology and its effect on the relief of residual stress of butt welded test plates have been done at home and abroad (Han, 1984; Kudinov, 1981, 1982;(Кудинов,1976;Труфяков,1974). However, so far only a few studies on the mechanism of relieving residual stress by explosion treatment have been reported(Петушков1974,1982), and the report dealing with either the quantative measurement or the rule of plastic deformation during explosion has not been seen yet. The problem of how to use explosion treatment to eliminate residual stress effectively in a large complicated welded structure, in which the distribution of original residual stress is difficult to measure accurately, is still not solved both theoretically and practically. There are some doubts whether the explosion treatment would decrease the residual stress in a structure. The aim of this paper is to study the rule of plastic deformation by explosion in a plane on the basis of the quantitative measurement of plastic deformation and to afford theoretical grounds for the application of explosion treatment to relieving residual stress in the large complicated welded structures.

MEASUREMENT OF PLASTIC STRAIN (ε_p) PRODUCED BY EXPLOSION AT DIFFERENT LEVELS OF ORIGINAL RESIDUAL STRESSES (σ_R)

Method of Measurement

Principle of measuring method. It is difficult to establish a relationship between ε_p and σ_R quantitatively in an actual welded plate due to distribution of original residual stresses in each part of the welded plate. When the applied force P at both ends of the test plate is changed as shown in Fig.1, different levels of stresses in the plate can be simulated. Test plates with different levels of stresses are treated by explosion with different parameters, and then the amount of plastic deformation ΔL is measured in a fixed gauge length L. Average plastic strain in a gauge length L is:

$$\varepsilon_p = \Delta L/L$$

Test material and loaded frame. The material for the test plate, which is used for the measurement of plastic deformation produced by explosion, is 16Mn steel. Its chemical composition is: 0.16%C, 0.40-0.60%Si, 1.30-1.60%Mn. Its yield strength is 350 MPa. By changing the welding sequence and welding parameters of load welds, the required load will be applied to the specimen through contraction of the weld by the frame as shown in Fig.2.

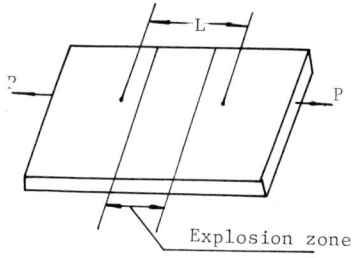

Fig.1 Principle of quantitative measurement of ε_p and σ_R.

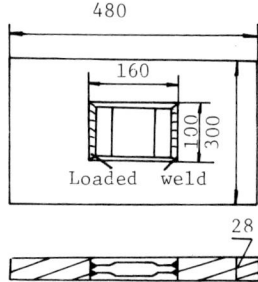

Fig. 2(a). Loaded frame for simulating tensile stress.

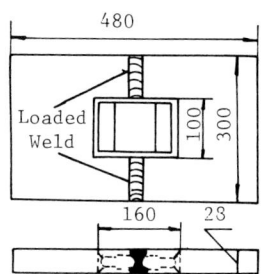

Fig. 2(b). Loaded frame for simulating compressive stress.

Measurement of stress and strain. A mechanical extensometer was used for measurement of the preloading stress and plastic deformation produced by explosion. Let L_o be the original gauge length of the measured specimen, L_b be the gauge length when the specimen is loaded, and L_a be the gauge length after the specimen has been treated by explosion and unloaded, then the original residual stress (σ_R) of the simulated specimen is:

$$\sigma_R = \left(\frac{L_b - L_a}{L_o} \right) E \quad (2)$$

The average plastic strain (ε_p) produced by explosion in a gauge length of 70 mm is:

$$\varepsilon_p = (L_a - L_o) / L_o \quad (3)$$

The original residual stress and plastic strain produced by explosion of each specimen are the average values of six pairs of measured marks, which are obtained from equations (2) and (3).

Explosion condition. Rubber explosive is used for all explosion in this work. The explosion rates are in a range of 3800-5700 m/s. Cushions of different materials with thicknesses changing within 10 mm are inserted between the explosive and the test plate.

Results of Measurement

The measured results of plastic strains produced by different explosion conditions are shown in Fig.3. It is obvious that an extended plastic deformation is produced in the whole zone of residual tensile stress under all explosion conditions. This will undoubtedly lead to lowering the residual tensile stress. It is of interest to note that the explosion will produce a contract plastic deformation in the metal in the case of a residual compressive stress. Therefore the explosion treatment will also lead to lowering the residual stress even in the zone of residual compressive stress.

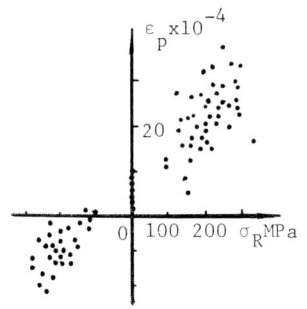 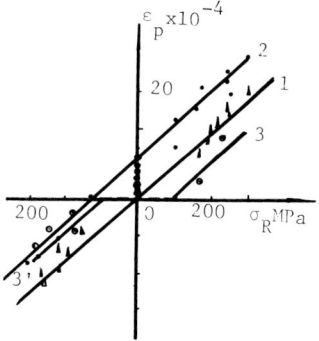

Fig.3. ε_p and σ_R under different explosion conditions.

Fig. 4. Three typical explosion conditions.

All explosion conditions in the tests can be divided into three categories (Fig.4).
The 1st category of explosion shows that the relationship between plastic strain (ε_p) produced by explosion and original residual stress (σ_R) in the metal is represented by a straight line passing through the origin of the coordinate (Fig.4, line 1). The explosion of first category, which is called moderate explosion, is a critical explosion condition. The action of relieving stress in the residual tensile stress zone and that in the compressive stress zone are all the same under a moderate explosion.
The 2nd category of explosion, which is called hard explosion, is in favour of developing the extended plastic deformation. The relationship between ε_p and σ_R is represented by a straight line 2 as shown in Fig.4. It is obvious that this is due to the result of explosion with higher explosion pressure. Hard explosion condition is a useful means to produce a certain amount of residual compressive stress in the welding zone to improve property of the welded joint.
The 3rd category of explosion, the explosion pressure of which is lower than that of the 1st category, is called soft explosion. The relationship between ε_p and σ_R is represented by a deflection line 3 as shown in Fig.4. The soft explosion is not an ideal explosion condition in relieving residual stress.

ANALYSES OF TEST RESULTS AND DISCUSSION

The moderate explosion condition shows that the plastic strain produced by explosion is entirely induced by the residual stress since the relationship of ε_p and σ_R is a straight line passing through the origin of the rectangular coordinate as shown in Fig.4. The res..stance of deformation to the residual stress is zero and the deformational behaviour of material is similar to fluid. Under this explosion condition, an extended plastic deformation under the action of residual tensile stress or a contract plastic deformation under the action of residual compressive stress will be produced as a result of metal flow. The residual stress decreases correspondingly with the development of plastic deformation till complete elimination of residual stress, and thereafter the plastic flow will stop. It is obvious that the rule of plastic deformation of metal under moderate explosion condition is the same as that of stress relieving by annealing. Therefore, the finally formed plastic strain ε_{pm} should be equal to the originally existing elastic strain ε_e under the moderate explosion condition, that is:

$$\varepsilon_{pm} = \varepsilon_e = \sigma_R/E \qquad (4)$$

in which both compressive plastic strain ε_p and compressive elastic strain are negative values while the extended elastic strain and plastic strain are positive values. The condition of moderate explosion is represented by a straight line passing through the origin of ε_p-ε_e rectangular coordinates and its slope is 45 degrees as shown in Fig.5.

Hard explosion is more violent than moderate explosion. Explosion exerts pressure on metal, which not only causes metallic material to have a deforming character like fluid but also produces a certain amount of two dimensional flow on a plane in the metal due to excessive pressure of shock wave. Therefore, a definite amount of extended plastic (C) will be produced. C is a function of the explosion parameters and property of explosion treated material and is independent of the original residual stress. It is a constant under a given explosion condition and explosion treated material, and is always a positive value.

a. hard
b. moderate
c. soft
d. quasi-hard

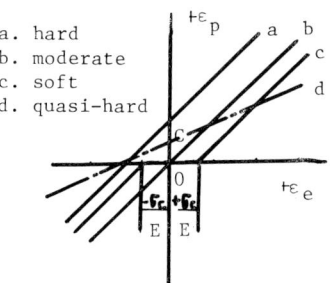

Fig.5. Mechanism of plastic flow induced by residual stress.

The total plastic strain (ε_{ph}) produced in the explosion treated metal consists of two parts under the hard explosion condition. The 1st part is the plastic strain produced by plastic flow, which is induced by residual stress. Its value is equal to ε_e. The 2nd part is the extended plastic strain C produced by metal flow due to excessive pressure of shock wave. Therefore, the total plastic strain can be expressed as follows:

$$\varepsilon_{ph} = \varepsilon_e + C \qquad (5)$$

Soft explosion is less violent than the moderate explosion. The metal exploded by soft explosion has not achieved the fluid character entirely because of insufficient pressure of shock wave, hence there still exists a definite resistance to deformation under the action of the residual stress. Metal flow and the corresponding plastic strain occur only when the explosion treated metal has a high value of residual tensile or compressive stress. The plastic flow will stop when the value of residual stress is lower than a certain critical value σ_c. The plastic (ε_{ps}) produced by soft explosion can be expressed as follows:

$$\varepsilon_{ps} = \varepsilon_e \pm C' \qquad (6)$$

When the residual stress is tensile, formula (6) has a minus sign, and it will have a plus sign when the residual stress is compressive. Soft explosion is represented by two deflection lines parallel to the line representing the moderate explosion conditions, the intercepts of two parallel lines at abscissa are $\pm C'$ respectively.

To sum up the above three explosion conditions, the mathematic expression of the plastic strain produced by explosion can be written as follows:

$$\varepsilon_p = \varepsilon_e + C \qquad (7)$$

Whether formula (7) represents moderate or hard explosion condition depends on that C is equal or greater than zero. When the signs of C and ε_e are opposite, formula (7) will represent the soft explosion condition.

To establish a mathematical relationship among C, explosion parameter and property of explosion treated work piece is an important subject of theoretical investigation on residual stress relief by explosion. Owing to the complexity of the explosion state equation under the condition of three-dimensional diffusion, it is difficult to get result for practical application at present. On the contrary, it is quite simple and practical to measure C for the need of engineering application. C is a plastic strain produced by explosion in test material with no residual stress, that is, $C=\varepsilon_p$ when $\varepsilon_e=0$.

Comparing Fig.4 with Fig.5, it may be seen that there is a certain difference between the result of actual measurement and the plastic strain predicted by the mechanism of plastic flow induced by residual stress. In the case of moderate explosion condition, this difference is shown in Fig.6.

a. experimental results
b. theoretical analyses

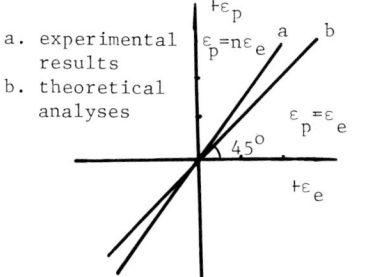
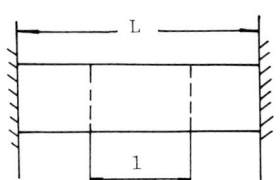

Fig.6. Deviation of experimental results from theoretical analyses.

Fig.7. Analytical model.

The average measured plastic strain (ε_p) in a gauge length of 70 mm is:

$$\varepsilon_p = n\, \varepsilon_e \qquad (8)$$

where n=1.4 for the test conditions.

The above mentioned deviation results from the difference between the test condition and the conditions of equations 4, 5, 6 and 7, which are established by the mechanism of plastic flow induced by the residual stress. The conditions of equations 4, 5, 6 and 7 are based on that the whole residual stress zone is treated by explosion simultaneously just like annealing a whole workpiece in relieving residual stress. But the test condition used in this work is that the explosion is done in the partial zone of residual stress like partial annealing treatment in relieving stress. This can be explained quantitatively by the following example. Let L be the length of residual stress zone (Fig. 7), σ_R the residual stress, then elastic strain $\varepsilon_R = \sigma_R/E$. Plastic strain ε_p produced in range 1 is equal to original elastic strain ε_e, when the whole range L is treated either by explosion (provided that both ends are fixed), or by annealing (provided that both ends of test piece can extend or contract freely with the change of temperature) in the relief of residual stress. The produced plastic strain ε_p in the gauge length 1 is greater than the original elastic strain ε_e, if only range 1 is treated by annealing or by explosion. Its value can be expressed as follows:

$$\varepsilon_p = 1 + (L-1)a/1 \, \varepsilon_e \qquad (9)$$

where a is an equilibrium coefficient of the process. When the temperature is high enough and the time at the temperature is kept long enough in the partial annealing treatment in relieving stress, a is equal to one. In this case, the plastic strain in the garge length 1 may reach the ultimate value $(L/1)\varepsilon_e$. However, the average plastic strain in the whole range L is still equal to ε_e. Owing to the short time of explosion process in the case of partial explosion treatment, the equilibrium coefficient of process a is less than one. Hence ε_p in the gauge length 1 does not reach the ultimate value $(L/1)\varepsilon_e$, but it is greater than ε_e. The value of the average plastic strain measured in the whole range L should be less than ε_e in such a case. In practical application, the explosion is usually done in a partial zone of residual stress. However, it is not the quantity of the local elastic strain at the explosion position, but the state of average plastic strain in the whole residual stress zone to determine the residual stress relieving effect of the whole work piece. The above mentioned explosion is called the fourth explosion condition, i.e. quasi-hard explosion that can be applied practically as shown in Fig. 5. The slope of the straight line representing the relationship between ε_p and ε_e under the quasi-hard explosion condition can be adjusted by changing the proportion of the exploded zone to the unexploded zone.

CONCLUSIONS

The rule of plastic flow of a metal is similar to that of fluid when the metal is treated under sufficient intensity of explosion. The residual stress induces the metal to produce a plastic strain that is equal to the original residual elastic strain ε_e in the explosion process. An excessive explosion pressure may cause the metal to produce an extended plastic strain C. The total amount of plastic strain ε_p formed in the process of explosion is:

$$\varepsilon_p = \varepsilon_e + C$$

where there are three typical sonditions.
When C=0, it is the case of moderate explosion. The plastic strain produced is entirely dependent on the original residual stress. The result of explosion is that the original elastic strain is completely converted into the plastic strain and that the residual stress is completely eliminated.
When C>0, it is the case of hard explosion. The result of explosion is that it is possible to produce a definite residual compressive stress in the explosion treated zone.
When the signs of C and original elastic strain are opposite, it is the case

of soft explosion. The result of this explosion is that there exists a definite amount of original residual stress in the explosion zone.

REFERENCES

Han, G., Chen, L., Dong, X. and Si, Z. (1984). Investigation on the possibility of relieving residual stress by explosion treatment in the butt weldments of medium and thick section. Proceeding of International Welding Conference, The Welding Institution of the Chinese Mechanical Engineering Society, X-4-1-X-4-6.
Kudinov, V.M., Petushkov, V.G. and Soskov, A.A. (1981). The residual stresses in explosion treated circumferential welds of pipes. Welding Research Abroad, 27 No. 3, 38-39.
Kudinov, V.M. and co-workers (1982). Mechanism of residual stress relieving by explosion treatment in 7th international conference high energy rate fabrication, University of Leeds, Leeds.
Кудинов,В.М., Труфяков,В.И., Петушков,В.Г., Березина,Н.В. и Михеев,П.П.,(1976). Параметры зарядов взрывчатого вешества для снятия остаточных напряжений в сварных стыковых соединениях, Автоматическая Сварка, №1, 46-49.
Петушков,В.Т., Кудинов,В.М., Березина,Н.В., (1974). Механизм перераспределений остаточных напряжений при взрывном нагружении, Автоматическая Сварка, №3, 37-39.
Петушков,В.Г.,(1982). О механизме снижения остаточных напряжений обработкой взрывном, Автоматическая Сварка, №4, 1-4.
Труфяков,В.И., Михеев,П.П., Кудинов,В.М., Петушков,В.Г., Березина Н.В., Гуша,О.И. и Лебедев,В.К.,(1974). Повышение сопротивления усталости сварных соединений взрывным нагружением, Автоматическая Сварка, №9, 29-32.

III.6

Mechanical Stress-relief Treatment of Welded Pressure Vessels by Warm Pressure Test

K. Kálna,

Welding Research Institute, Bratislava, CSSR

ABSTRACT

Large pressure vessels and storage tanks can be hardly stress-relief heat treated. However, residual stresses in welded joints can be reduced significantly by means of warm pressure test. In that case steels and welding consumables should be chosen with respect to achieve sufficient fracture toughness in both base metal and welded joints. The fracture toughness - J_{IC} - criterion should be assessed considering actual level of residual stresses - S_r - in welded joints. The results of residual stress measurements performed on welded plates /with the thickness ranging from 100 to 200 mm/ as well as on four model pressure vessels /with the wall thickness 45 and 120 mm/ in both as-welded and warm prestressed condition, have proved the benefit of the warm prestressing method. Increased resistance of pressure vessels treated by this method to brittle failure has been proved in practice.

KEYWORDS

Pressure vessel; brittle failure; residual stress; warm prestressing; warm pressure test.

INTRODUCTION

In fabrication of welded pressure vessels the stress-relief heat treatment represents a costly and time consuming operation for which high capacity furnaces and a lot of thermal energy are needed. The surfaces of heat treated structures must be usually cleaned. Large storage tanks with the diameter over 10 m can be heat treated locally only, however, this will substantially increase the fabrication costs. Moreover, with local heat treatment residual stresses, eventhough of lower level, still remain, particularly in the areas distant from the locally treated parts.

It is known that stress-relief heat treatment may exhibit some
negative effects on the properties of welded joints and, on
the other hand, that non stress-relief heat treated welded
structures may operate safely even in severe service conditions.

The complex effect of stress-relief heat treatment can be considered from the following aspects:
a/ reduction of residual stresses, resulting in:
- increased resistance to brittle failure
- increased resistance to stress corrosion cracking
- usually decreased resistance to fatigue failure
- decreased distortions after mechanical machining
b/ tempering of the hardened phases and stabilisation of microstructure in the heat affected zone or in the weld metal
c/ decreased both yield and tensile strength of the weld metal
d/ possible distortion of weldment
e/ possible occurence of reheat cracking

According to the different existing pressure vessel and piping
codes the stress relieving heat treatment of welded joints is
obligatory to be applied beginning from a certain minimum wall
thickness. In case of CMn steels stress-relief treatment if
required from 25 to 50 mm range /Bouhalier, 1984/. On the
other hand, the codes for bridges and similar civil engineering steel structures do not require stress-relief heat treatment of welded joints. As it follows from this statement, the
necessity of heat treatment of welded joints in steel structures including pressure equipment should be analysed more
exactly for each individual case /Kálna, 1934, 1986/.

REVIEW OF PRESENT KNOWLEDGE ON MECHANICAL STRESS RELIEF
ITS EFFECT ON FRACTURE TOUGHNESS OF WELDED JOINTS

Early knowledge about the effect of residual stresses on
brittle failure of welded joints summarized Kihara /1962/ in
the IIW-X-29-62 document and is presented schematically in
Fig. 1 as temperature dependances of the yield point - R_e - ,
the threshold value of brittle fracture stress - S_{lim} - and
the course of the crack-arrest temperature /CAT/.

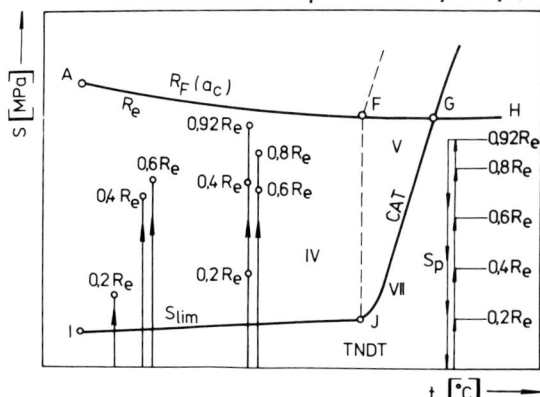

Fig. 1. Effect of warm prestressing on brittle
fracture of stress-relief heat treated
welded joints

At temperatures below the TNDT the specimens in the stress relieved condition containing a crack fail at the fracture stress equal to the yield point (R_F (a_c) = R_e). The same specimens but not stress relieved fail at the fracture stress $R_F \geq S_{lim}$ /the area IV/. The level of S_{lim} is very low /30 to 50 MPa/, thus it cannot be adapted for calculations as the allowable stress. However, as far as the crack containing non stress relieved specimen was prestressed at the level S_p = (0,20 ÷ 0.90)R_e at temperature over the CAT /area VII/ and then reloaded up to rupture at lower temperatures /area IV/, the fracture stress reached value equal or higher than S_p . ($R_F \geq S_p$). In Fig. 1 the level of prestressing is indicated at the symbols /empty circles/ of experimentally obtained values of fracture stress /e.g. 0.4 R_e/.

In spite of some simplifications introduced into the scheme in Fig. 1 it is valied up to now. In fact, the method of mechanical reduction of residual stresses has been practically used for very long time, e.g. by preloading tests of bridges, pressure tests of vessels etc. However, as it follows from Fig. 1 such preloading tests should be performed at elevated temperatures, i.e. above the CAT.

The knowledge about the beneficial influence of warm prestressing was utilized by the author of this paper for the preparation of heavy section specimens for the Robertson test at Skoda Works since 1963. With this purpose to the test plates of 1200 mm in width and 100, 150 and 200 mm in thickness fixing ends of the same thickness were electroslag welded and then prestressed at the temperature +80 to +100 °C to the stress level by 10 to 25% higher than the prescribed loading stress. In this way the load-bearing capacity at room temperature of these welds could be increased by about 25% as compared with the prestressing level. The validity of the beneficial effect of warm prestressing on the fracture stress has been verified by more than 100 heavy section wide plate test specimens prepared in this way for testing on the 80 MN test machine at Skoda Works /Kálna, 1966/.

With the aim to study the mechanism of warm prestressing the level of residual stresses was measured on prestressed welded wide plate test specimens. The results obtained are given in Table 1. The base metal of test plates was a CMn pressure vessel steel with R_e = 240 MPa and R_m = 440 MPa. The welding wire A215 /MnMoNiV type/ 3.15 mm in dia. and the basic flux VUZ-4F were used for electroslag welding. It follows from the data in Table 1 that the level of residual stresses in ES welded joints decreased with increasing prestressing level. It is also obvious that the maximum reduction of S_r was obtained in the y-axis direction which is the prestressing load direction. Considerable scatter band of experimental data indicated the necessity to precise the methods of residual stress measurements as well as to determine the residual stresses also in the thickness direction of welded joints.

More recent experience on the effect of prestressing on the fracture toughness reported Nichols /1966/ in the IIW X-409-67E document with the conclusion that during prestressing local stresses may reach a value higher than R_e of the material in the vicinity of the crack tip. By such redistribution of residual stresses as well as by decreasing of the crack tip acuity increased resistance to brittle failure of the structure is secured. Consequently, repeated pressure tests of pressure vessels during their service life have been proposed as a means of improving the fracture mechanics characteristics of the vessel, considering acting stresses, fatigue crack growth rates and the critical crack size a_c values /assessed by the CTOD method/ for a given structure.

However, replacement of relaxation heat treatment by the mechanical prestressing procedure applied to welded pressure vessels Nichols limited by the following conditions:
- 25 mm maximum wall thickness
- applicable only for pressure vessel CMn killed steels with the ratio $R_e/R_m \leq 0.8$
- welded joints must not be situated into the regions with high stress concentration
- double V and single V butt joints fabricated with backed root are permitted only
- defect free welded joints must be ensured
- mechanical prestressing must not be applied to storage tanks for liquid gases with the boiling point below -29 °C.

Some recent experience on warm prestressing method, reported by Pickles and Cowan /1983/ is presented in Fig. 2.

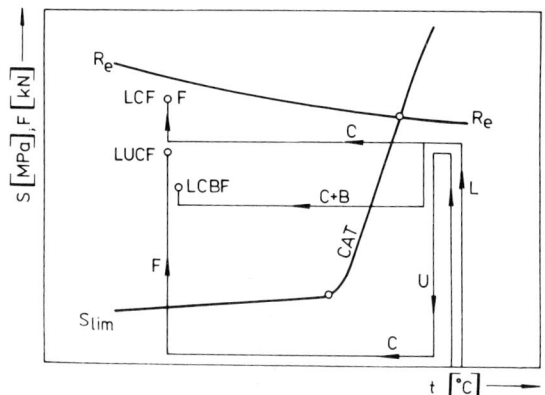

Fig. 2. Effect of warm prestressing on fracture **stress** of steel

As it follows from Fig. 2, to procedures of prestressing can be applied:
LUCF: loading - unloading - cooling - fracturing
LCF : loading - cooling - fracturing
where loading is performed at temperatures above the CAT for a given steel.

The LCF cycle is considered as a more effective compared to the LUCF one because with it the fracture occurs at higher stress or at higher fracture toughness K_{CJ}. If embrittlement of the stretched zone at the crack tip takes place during cooling /dure to e.g. ageing or radiation damage/ - see the LCBF cycle - the fracture occurs at lower stress or lower K_{CJ} value. Tests were performed applying three-point-bend and compact tension test specimens 25 to 75 mm thick. Laboratory tests were supplemented by warm pressure tests of model pressure vessels /1524 mm in dia./ made of a CMn steel /0.36% C/ with rather high CAT values; the prestressing temperature was +62 °C.

Nakamura and others /1981/ studied the influence of static and cyclic prestressing on the fracture toughness of A 533 B-1 reactor pressure vessel steel. The results of the test series performed with static prestressing /K_p=80 MPa$\sqrt{m}$/ are presented graphically in Fig. 3 where the solid line represents the temperature dependence of the fracture toughness of the steel not influenced by prestressing of the test specimens.

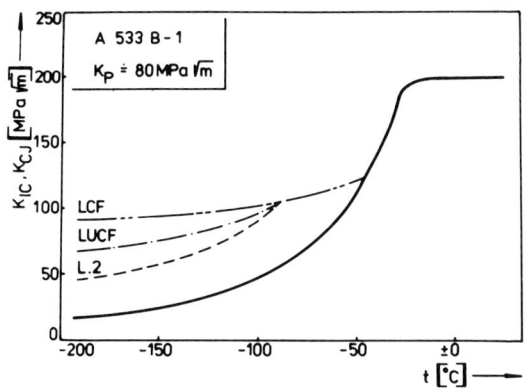

Fig. 3. Temperature dependence of fracture toughness; prestressing with K_p = 80 MPa $\sqrt{m}$

The L.2 line represents the test results obtained on prestressed /symbol L/ and stress relieved /symbol .2/ specimens. Symbols of the other curves mean:
L - loading; C - cooling; U - unloading; F - fracturing.
In the range of low testing temperatures maximum fracture toughness values were obtained with the LCF cycle, i.e. without unloading before cooling of the test specimens. However, at the test temperatures higher than -50 °C the prestressing did not show any effect on K_{CJ} value. This phenomenon can be ascribed to the occurence of a stretched zone at the crack tip during prestressing /r_L/ and fracturing /r_F/. At low test temperatures $r_L > r_F$, the prestressing increases the K_{IC} value, whereas at higher test temperatures $r_L < r_F$ and prestressing does not influence the K_{IC} values.

The warm prestressing is most important as a means for decreasing of residual stresses in welded joints and thus increasing their resistance to brittle failure. Moreover, further benefits are observed such as decreasing of complementary bending stresses arising due to geometrical imperfections of welded joints /ovality, linear and angular misalignments etc./. During loading the strain concentration takes place in the root of a defect, being transformed into compressive stress after unloading, thus arresting further growth of that defect. From this point of view the warm pressure test is advantageous also for heat treated pressure vessels in which - in spite of stress-relief heat treatment - non negligible residual stresses as well as shape imperfections may occur during cooling down. By warm pressure test more uniform loading state and increased safety against brittle failure are achieved.

TESTS OF MODEL PRESSURE VESSELS

With the aim to verify the reduction of residual stress level due to warm prestressing as well as the serviceability of heavy section pressure vessels without stress relief heat treatment a series of tests on model pressure vessels was performed at the WRI, Bratislava. Some of experimental data obtained by the tests on pressure vessels fabricated of CMn steel with R_e = 350 MPa are summarized in Table 2. Two model vessels, made of Fe 510 E steel grade were 1200, 1500 mm resp. in diameter and 45 mm in wall thickness. Longitudinal as well as circumferential welds were fabricated by MMA and SAW processes. With respect to low design temperatures /-40 and -60 °C/ the ele-ctrodes and welding wires containing up to 2,5% Ni were used. The sketch of this model pressure vessel is shown in Fig. 4. Other two model vessels represented the pressure head of a high pressure heat exchanger designed for the lowest service temperature -20 °C /without internal pressure/ and the highest service temperature of +225 °C at the 15.8 MPa.

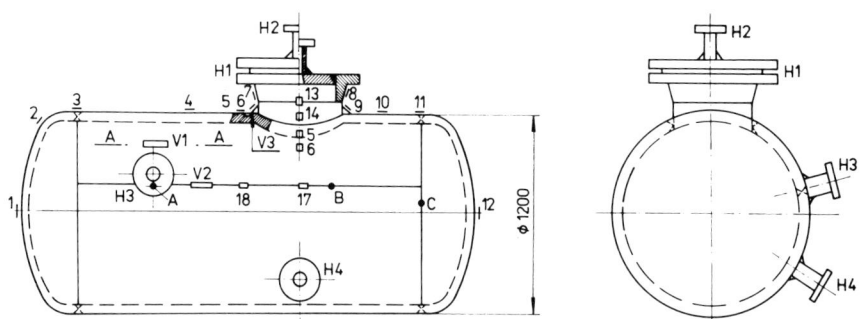

Fig. 4. Dimensional sketch of the 1200 mm dia. model pressure vessel

The dimensional sketch of these models is presented in Fig. 5 together with the diagram of distribution of tangentianal and meridian /S_m/ stresses recorded during prestressing at p=30 MPa

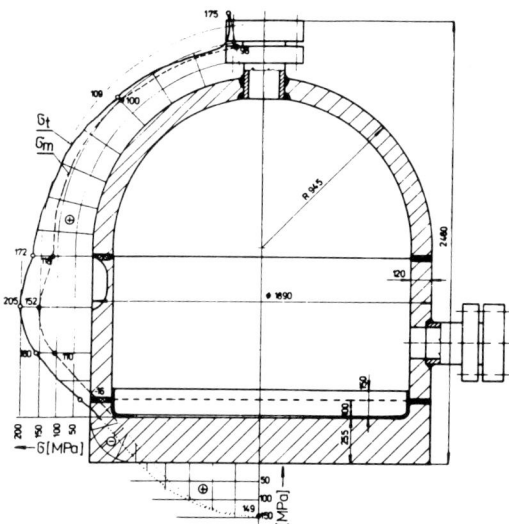

Fig. 5. Dimensional sketch of high-pressure heat exchanger and diagram of stress distribution recorded during the pre-stressing, p = 30 MPa

Two pressure vessel steels, one CMn type with R_e = 300 MPa and one CMnMo type with R_e = 420 MPa were used as the base metals. As it follows from Fig. 5, two circumferential welds joining the cylindrical shell with hemispherical head 120 mm in thickness on one side and with a model tube sheet 255 mm in thickness from the other side were fabricated by SAW process, using welding wires and fluxes adequate to the base metals and with respect to the desired fracture toughness level in the weld metal.

Similar testing programs were applied to all four model vessels. Therefore testing of one model vessel only /according to Fig. 4/ is described hereafter. The test parameters and the results obtained with the rest of model vessels are contained in Table 2.

After welding the volume residual stresses were measured /Jesensky, 1983/ in the longitudinal welded joint. The diagram of through-thickness residual stresses is presented in Fig. 7a. The hole remaining due to the stress measurement method applied was sealed by the flanged nozzle H3, whereas the H2 and H4 nozzles represent the inlet and outlet holes for the cooling liquid. Three artificial notches /V1, V2, V3/ were located into the base metal /V1/, into the longitudinal weld /V2/ and in the weld joining the manhole with the shell /V3/, as shown in Fig. 4 were machined.

The loading diagram is presented in Fig. 6 and consists of the following phases:
- 8 load-unload cycles of warm prestressing at +34 $^{\circ}$C temperature /i.e. above the CAT value/ and internal pressure

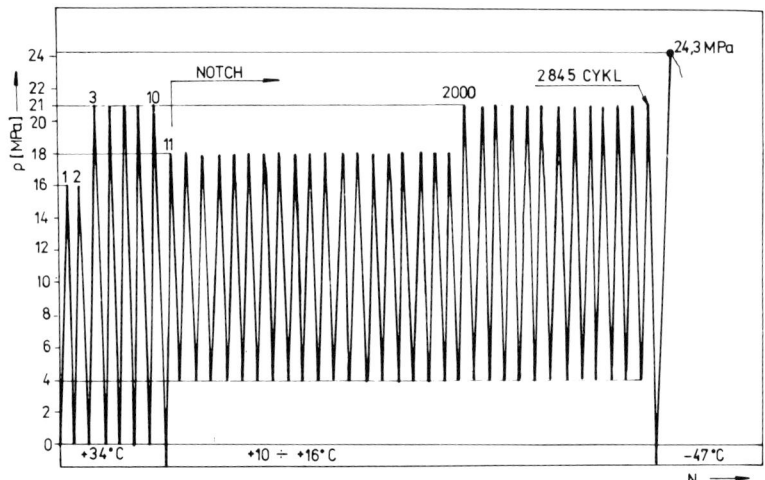

Fig. 6. L₀ading course of 1200 mm dia. model pressure vessel

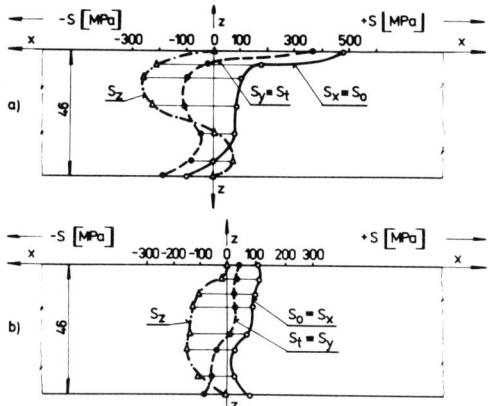

Fig. 7. Distribution of through-thickness residual stresses in longitudinal SA welded joint; vessel 1200 mm in dia; a - as-welded state; point A; b - after warm pressure test; point B

p=21 MPa /higher pressure could not be obtained due to a damaged sealing of the manhole flange
- 2000 cycles perfored at ambient temperature in the pressure range 4 to 18 MPa
- 845 cycles performed at ambient temperature in the pressure range 4 to 21 MPa
- fracturing at -43 to -47 °C temperature

The destruction occured at 24.3 MPa of internal pressure, corresponding with the membrane stress of 293 MPa.

A part of the longitudinal weld /from the location "B" shown in Fig. 4/ was used for residual stress measurements, the diagram of which is presented in Fig. 7b and corresponds to the location B shown in Fig. 4. As it follows from these measurements, the tangential residual stresses S_{ry} decreased from the level 370 MPa after welding to the level 40 MPa after warm pressure testing and destruction, i.e. by about 90% of the original level. More details on this test were published /Kálna, 1986/. Similarly, Figs 8a and 8b show the residual stress levels in as-welded condition of the reference weld and in the circumferential SA weld of the model pressure vessel made of CMn steel after warm pressure test.

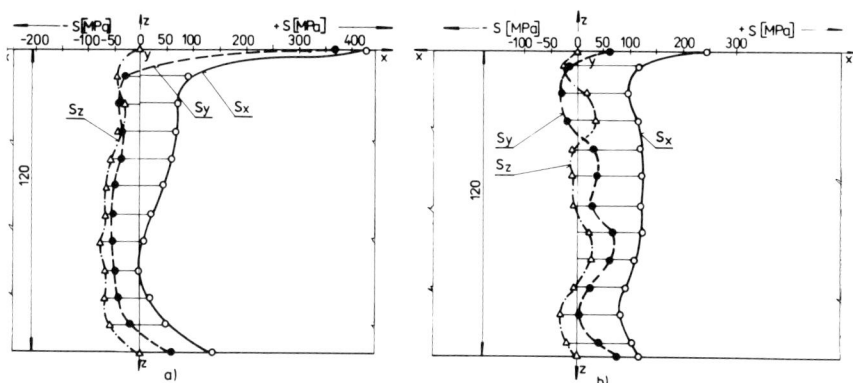

Fig. 8. Distribution of through-thickness residual stresses in circumferential SA welded joint; vessel 1890 mm; a - reference specimen, as-welded state; b - vessel after warm pressure test

Again, the residual stresses decreased by about 44% in the tangential and by about 82% in axial directions. The above examples as well as the data given in Table 2 show unambiguously the benefits of warm prestressing in terms of decreasing the peaks of residual stresses and their more advantageous distribution in welded joints.

CONCLUSIONS

The warm pressure test increases the resistance to brittle failure of non stress-relief heat treated pressure vessels, mainly due to:
- decreasing of residual stress level /S_r/ creasing being dependent on the level of prestressing as well as on the number of prestress loading cycles;
- improving the shape imperfections of welded joints and of the vessel, thus decreasing the level of complementary bending stresses in the wall of the pressure vessel;
- creating a stretched zone at the tips of cracks and/or sharp pointed defects, thus stopping their growth.

With the aim to decrease the residual stresses, the warm pressure test may be used under the following conditions only:
- the test temperature should be higher than that of CAT;
- the base metal as well as the metal of welded joints should ensure sufficiently high plastic properties /i.e. A_5, Z, KV/ and should not contain inadmissible defects;
- the notch toughness of base and weld metals at the lowest service temperatures of the pressure vessel should correspond with a criterial value derived from the J_{IC} fracture toughness by means of fracture mechanics approaches;
- after warm pressure test a full NDT of welded joints should be performed;
- for the construction of storage tanks destined for liquified gases and where stress-corrosion-cracking must be avoided, microalloyed CMn steels and adequate welding consumables should be preferably employed.

REFERENCES

Bouhelier, C. /1984/. Heat treatment in pressure vessel technology. Fifth Internat. Conf.: Pressure Vessel Technology. Vol. II. San Francisco.
Jesenský, M. /1983/. Residual Stress Measurement. VÚZ Bratislava.
Kálna, K. /1966/. Analysis of Size Factor at Brittle Fracture of Al Pressure Vessel Steel. UZÚ ŠKODA Plzeň. Thesis.
Kálna, K., M. Jesenský and P. Polák /1981/. Welding News. 31, 87-97.
Kálna, K. /1984/. Determination of heat treatment requirements of welded joints by fracture mechanics analysis. 9th Conference on Welding, OMIKK - TECHNOINFORM Budapest.
Kálna, K., M. Jesenský, P. Polák and K. Ulrich /1986/. Welding News. 36, 49-59.
Kihara, H. /1962/. Recent studies in Japan on brittle fracture of welded steel structure under low applied stress level. Doc. IIW No. X-291-62.
Nakamura, H., H. Kobayashi, T. Kodaisa /1981/. On the effect of pre-loading on the fracture toughness of A 533 B-1 steel. Advances in Fracture Research. Vol. II. Cannes.
Nichols, R.W. /1967/. The use of overstressing techniques to reduce the risk of subsequent brittle fracture. Doc. IIW No. X-409-67E.
Pickles, B.W., A. Cowan /1983/. Int. J. Pres. Ves. and Piping. 14, 95-131.

TABLE 1. Surface residual stresses in electroslag welded joint of thick plates as weld state and after warm preloading

Plate cross section	mm	100 × 1200		150 × 1200		200 × 1200	
Warm preloading stress		0.70 R_e		0.85 R_e		1.15 R_e	
S_r ori-entation	to weld axes	S_{rx}	S_{ry}	S_{rx}	S_{ry}	S_{rx}	S_{ry}
	to preloading direction	R_e	R_e				
S_r as weld state	MPa	100÷140	60÷100	50÷100	200÷250	170÷200	280÷360
S_r after preloading	MPa	60	70	30÷40	+150÷/-20/	150÷180	+15÷/-50/
Decrease ratio	%	60	70	60	40	10	100

TABLE 2. Surface residual stress in pressure vessel welded joints as weld state and after warm pressure test

Steel		Fe510E	Fe510E	20Mn5	CMnMo
SA welded joint		A234+N70	A234+F205	A202+F102	A234+F205
Yield stress BM/WM	MPa	420/570	400/510	300/600	425/540
Vessel: dia/thickness	mm	1200/46	1500/45	1890/120	1890/120
Service/test pressure	MPa	18x1.17=21	11x1.5=16.5	15.7x1.9=30	15.7x1.5=24
Pressure test — Load cycles	-	8	10	2	2
Pressure test — Temperature	°C	+34	+33	+55	+58
Pressure test — Stress S_x/S_y	MPa	250/125	260/130	206/103	165/83
Repeat Load — Pressure	MPa	18 21	15 18	24	16 20 23
Repeat Load — Number of cycles	-	2000 845	2000 3760	2770	3180 520 3280
Failure — Pressure	MPa	24.3	22.2	33.5	26.1
Failure — Temperature	°C	-47	-34	-22	-25
Weld location		longitudinal	longitudinal	circumferen.	circumferen.
S_r direction to weld axis	-	S_{rx} S_{ry}	S_{rx} S_{ry}	S_{rx} S_{ry}	S_{rx} S_{ry}
S_r after welding	MPa	480 370	390 180	430 370	375 80
S_r after test	MPa	110 40	80 20	240 65	230 125
Decrease ratio	%	77 90	80 90	44 82	39 -

III.7

Relaxation Overstressing of Huge Spherical Storage Vessels Repaired by Welding

I. Hrivňák*, J. Láncoš** and S. Vejvoda**

*Welding Research Institute, Bratislava, Czechoslovakia
**The Institute of Applied Mechanics, Vítkovice, Brno, Czechoslovakia

ABSTRACT

A series of spherical storage tanks made from low-alloy Ni-Cu-
-V TTStE47 steel was damaged and had to be repaired by welding.
The wall thickness of tanks was 27-34.5 mm and their volume
ranged between 1500 and 3300 cubic meters. The vessels were
used for storage of liquified petrochemical gases including
ethylene. The overhead tanks were max. 18.50 m in diameter.
The operating pressure was 1.85 and/or 2.0 MPa and the operating temperature -30 up to +40°C. There were two kinds of
welds in the tanks: manual metal arc welds made with Tenacido
60 electrodes and submerged arc welds produced by S2Ni2 wires
and basic flux LW 320. MMA welds were the repaired ones. The
level of residual stresses was measured by three methods: destructive method on repaired samples, drilling method and "in
situ" measurement during repair welding. To reduce the internal stresses induced to vessels by repair welding three successive pressure cycles up to the yield stress of the base metal were used. The results were good enough to put the vessels
back to the operation.

KEYWORDS

Storage tank; liquified gases; repair welding; internal stresses induced by welding; pressure test; successive pressure
cycles; low-alloy steel.

INTRODUCTION

In previous works (Hrivňák, 1985a,1986) a review of conditions
of relaxation annealing of welded joints in carbon and microalloyed steels was outlined. Further on, it was pointed out
that there also exist other procedures of relaxation treatment
of welded joints, pressure vessels and structures. They are

based preliminarily on overstressing or application of stress loading cycles on the structure or pressure vessel to the yield strength or in austenitic steels with unexpressive yield strength loading to 2-4% strain of base metal or on mechanical vibration treatment or that within ultrasonic frequencies. This work presents results of measurements with application of overpressure cycles in relaxation treatment of huge storage vessels for liquified gases.

A series of huge spherical vessels for liquified gases was repair welded. The 1500-3300 m^3 vessels were made of TTStE47 steel. The operating pressure in vessels was 1.8-2.0 MPa and the operating temperature varied between ambient temperature and -30°C. The wall thickness of spherical storage tanks was 27.0-34.5 mm. The overhead storage tanks were max. 18.50 m in diameter. Storage tanks were fabricated by manual arc and submerged arc welding procedures. With the exception of a storage tank having 34.5 mm wall thickness (3300 m^3 volume) the others were not annealed after assembly. The mentioned storage tank was annealed after assembly at 540-580°C.

The nominal chemical composition of TTStE47 steel according to TUV Werkstoffblatt 35 702.81 is as follows: max. 0.17% C, 1.00 -1.60% Mn, 0.05-0.55% Si, max. 0.35% S, max. 0.025% P, 0.47- -0.75% Ni, 0.06-0.20% V, max. 0.05% Nb, 0.45-0.75% Cu. The real chemical composition of several analysed melts agreed well with the Standard composition. Carbon content varied within the upper limit, Mn content within 1.2-1.4%, Ni content within 0.55-0.66%, V content was approx. 0.15% and Cu content within 0.40-0.50%. The steel purity was good - P content $\sim$ 0.015% and S $\sim$ 0.01%. The required mechanical properties were: R_e min. 450 MPa, R_m 560-730 MPa and A_5 min. 17%. The real values of longitudinal and transverse yield strength R_e were 480 -490 MPa, the strength R_m 628-636 MPa, transverse ductility 29 -30% and transverse reduction in area was min. 60%. Notch toughness of base metal was assured at -40°C (however, DVM notch was present). KV pieces gave sufficient values at -40 to -50°C also in transverse direction. According to 50% ductile fracture criterion the transition temperatures varied between -5 and -40°C.

The analysis of TTStE47 steel weldability proved that the value of carbon equivalent according to IIW was CE = 0.46-0.475% and the value of cracking parameter P_{CM} according to Ito and Bessyo was P_{CM} = 0.26-0.28%. For welding 27 mm thicknesses with basic electrodes 190-200°C preheat was calculated according to Ito-Bessyo concept. According to Subcommission IX-G concept (Hrivňák, 1985b) the preheat temperature represents 112°C and according to Séferien (1962) it ought to be 121°C.

The calculated maximum real hardness of underbead zone was

380 HV (Hrivňák, 1985b) and the maximum calculated hardness of pure martensitic structure was 414 HV. Further calculation proved $\sim$ 300 HV hardness at cooling time $\Delta t_{8/5} \approx 30$ s. In storage tanks manual welds were repaired. The total length of welds was 340-450 m, manual welds represented at about 60%. In one storage tank cracks in transition zone 11.0-193 m in length (in average 120 to 460 cracks) were observed in manual welds. The repair welding procedure was elaborated in the Welding Research Institute (1985). As the stress distribution in the storage tanks was changed due to a great number of repair welds we have decided to apply the mechanical overstressing relaxation treatment. Three pressure cycles following the required pressure tests were used. At these pressure cycles the maximum pressure was chosen developing the membrane stresses in the wall just below the yield point of the parent material or developing resulting membrane stresses on the level of yield stress (weld area). By this treatment a plastic flow and relaxation of high stresses occur in the areas where the condition of plasticity is fulfilled. Above all, it has to be stressed here that 100% ultrasonic inspection with normal and tangential probes and the MPA (magnetic particle analysis) were performed after repair welding of cracks. The pressurized storage tanks were free of defects.

MEASUREMENT OF RESIDUAL STRESSES IN ORIGINAL AND REPAIRED WELDED JOINTS

The surface residual stresses were measured by three independent methods:
1. Method of sampling a material column by foil strain gauges (Vejvoda and Láncoš, 1985c). Having measured the initial values the measured areas were milled to 33 x 33 m prism-shaped pieces. Both original and repaired welds were measured in 36 mm distance on both sides of welded joints.
2. Drilling method based on the theory of stress concentration and strain around circular hole and their strain gauge sensing. The holes were drilled in weld axis and in 30 mm distance from weld axis by a 6 mm long drill made of hard metal directly on repaired surface of containers (Vejvoda and Láncoš, 1985a).
3. "In-situ" strain gauge measurement during repair welding of the defect (Jaroš, 1986).

The values measured by the first method were practically the same in both original and repaired welds. Maximum residual stresses in weld axis direction were $\sigma_x \approx 270$ MPa. Maximum residual stresses in normal direction to the weld were lower in both original and repaired welds with the upper limit $\sigma_y \approx 200$ MPa.

The method of drilling a hole (modified Mathar method) was used for measurement of surface residual stresses on annealed storage tank 34.5 mm in wall thickness. After the first roading holes $\emptyset$ 3 mm were drilled 3 mm in depth at first and then strains were measured. Having drilled holes 6 mm in diameter into 6 mm depth strains were measured again. The modulus of steel elasticity E = 205000 MPa, Poisson's ratio μ = 0.3 and yield strength R_e = 450 MPa were used for calculation of residual stresses.

In the vicinity of original weld after annealing and original pressure test the residual stresses (σ'_I = 180 MPa, σ'_x = 80 MPa, σ'_y = 176 MPa) were measured. A very low residual stress σ = = -11 MPa was measured in a sufficient distance (570 mm) from welds on the inner surface of a vessel wall.

In the zone of repair welded defect (without subsequent annealing) the values $\bar{\sigma}_I$ = 341 MPa, $\bar{\sigma}_x$ = 260 MPa, $\bar{\sigma}_y$ = 135 MPa were measured.

The third measurement method was employed during repair of a stress-relieved pressure tank. The length of repaired weld was $\sim$370 mm, its depth was $\sim$23 mm. Strain gauges together with thermocouples were located in 50 mm distance from weld edges. Figure 1 shows the flow chart of strain gauge and thermocouple location (Jaroš, 1986). The values of measured residual stresses are shown in Fig. 2 (Jaroš, 1986). Whereas sufficiently low stresses (104-148 MPa) were induced normal to the weld in 50 mm distance, they are substantially higher in the zone of weld ends. From performed tests and measurements it can be concluded here that each repair welding of welded joints introduces additional residual stresses into the wall of a vessel and their amount together with orientation depend on size and orientation of the repaired weld section under given thermal conditions of welding. On the contrary, the original welds have already had a lower level of residual stresses due to relaxation in the initial pressure test at the beginning of their service.

CALCULATION OF PRESSURE TEST CONDITIONS

The temperature of a vessel wall at pressure test ought to be higher than the transition temperature of ductile-brittle fracture. To evaluate brittle fracture resistance the DT impact bend tests on 25 x 40 x 80 mm specimens with a pressed notch in transverse direction were performed on both original and repaired welded joints. The TDTE and TDT 50% temperatures were determined (Kálna, 1984). Further on, the static through-thickness fracture toughness test K_{CJ} in the HAZ and weld metal of both the original and repaired welded joint, dynamic fracture toughness test (Holzmann, 1985) and fatigue crack growth rate tests of base metal and the HAZ were used as well. At -40°C KV notch toughness above 27 J was measured in all cases. At this temperature 63-76 J notch toughness in rolling direction and 27-43 J value in transverse direction were measured.

TDTE -21 to -7°C and TDT -20 to -8°C temperatures were deter-

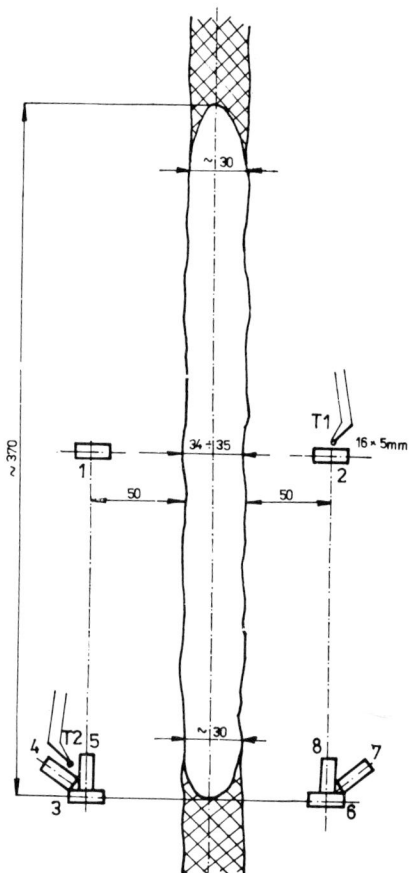

Fig. 1. Flow chart of strain gauge and thermocouple location

mined in static fracture toughness tests.

At -40°C the minimum K_{CJ} value of 118 MPa$\sqrt{m}$ for base metal and 55 MPa$\sqrt{m}$ for the original heat affected zone were measured. The original weld metal had minimum value K_{CJ} = 70 MPa$\sqrt{m}$ at this temperature.

From performed measurements the minimum temperature of vessel wall at pressure test was determined so as to prevent cleavage fracture initiation. The crack arrest temperature for base metal was +22°C (Kálna, 1984). With respect to accummulated energy by the storage tank the testing temperature of 30-40°C of pressure water was selected.

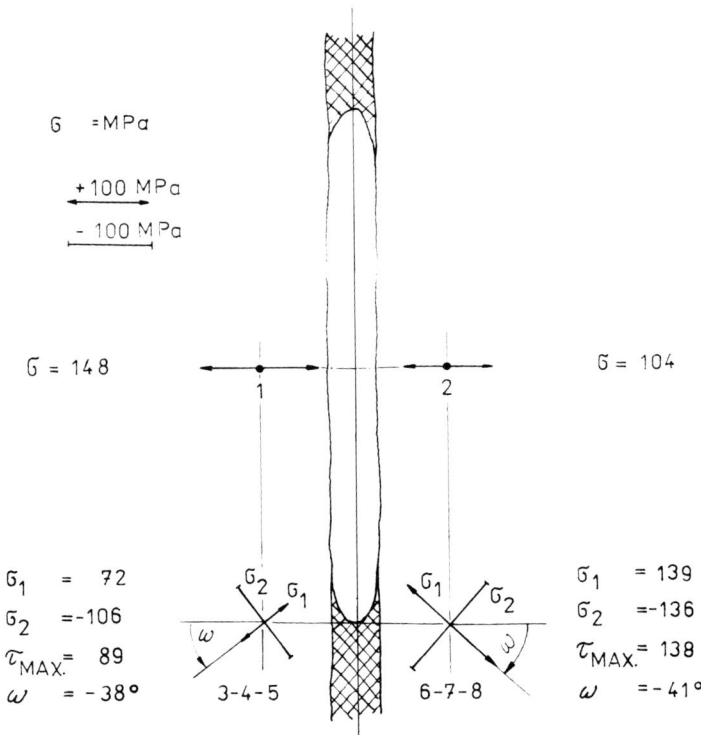

Fig. 2. The values of additional stresses induced in weld zone by repair welding

Except for prescribed pressure test at p = 2.34 MPa eventually p = 2.60 MPa specified pressure for a storage tank 34.5 mm in wall thickness, three pressure cycles with maximum 2.60 MPa eventually 2.85 MPa pressure were induced. The safety factor for membrane stress as a result of applied overpressure represented 1.17 at the guaranteed yield strength R_e = 450 MPa. The pressure rise time to 1 pressure cycle was 30-60 minutes, dwell time of a storage tank at maximum pressure was 10 minutes and pressure fall down to 0.1 MPa was at about 30 minutes. If pressure rose from 0.1 MPa to 2.60 eventually 2.85 MPa the volume of storage tanks was increased by 4-8 m^3.

During demineralized water overstressing the temperature of vessel wall was measured at least on six points.

MEASUREMENT OF RESIDUAL STRESSES AT PRESSURE TESTS

To determine the redistribution of stresses during application of three successive pressure cycles the stresses on the outer surface of three vessels were sensed by resistance strain gauges. In one vessel the strain gauges were located in the area of a repaired and unrepaired weld (Láncoš and Vejvoda, 1985b). The above mentioned values for original and repaired welds were considered for initial residual stresses. On the whole, 18 strain gauge clips were located in the following way: two clips were in the axis of the old weld and two of them in the axis of the repaired one. Two clips were located on either side of weld in the transition zone and then in base metal 30 mm from weld.

Table 1 gives a survey of the evaluated residual stresses in discharged pressure-free vessel. For comparison the lowest line of the table gives residual stress values in base metal in 50 mm distance from the measured weld area.

TABLE 1 Residual Stresses in Discharged Pressure-Free Vessel

Specification of measurement location		Residual stresses (MPa)		
		x	y	xy
In weld axis	Original weld	24.8 34.1	81.6 103.5	-27.3 -30.9
	Repaired weld	16.5 50.9	86.4 99.2	-32.8 -60.5
Transition zone	Original weld	48.8 69.9	112.2 143.7	-17.7 -23.4
	Repaired weld	14.9 110.8	71.0 118.0	10.9 -113.7
Base metal 30 mm from transition zone	Original weld	24.9 29.1	-16.4 -42.3	-25.0 15.1
	Repaired weld	100.3 117.6	-6.0 -99.9	-4.9 -80.3
Base metal 500 mm from weld		-2.8	-3.7	3.5

From performed measurements two significant conclusions can be drawn:
1. After three pressure cycles the level of residual stresses is decreased in both repaired and original welds. As a fact, the highest decrease was recorded after first cycle. At the second and third cycles the values of further decrease drop substantially.
2. Residual stress after three pressure cycles attains max. 143.7 MPa normal to weld axis and 110.8 MPa in weld axis. No expressive difference between values of residual stresses in original and repaired welds was observed. Both values

(in weld direction and normal to weld axis) are, however, expressively lower than those prior to prestressing.

After three pressure cycles of another storage tank subjected to overstressing residual stresses in transition zone along weld axis were $\tilde{\sigma}_x$ = 124.6-187.5 MPa and those normal to weld were 102-162.9 MPa. These values are in order the same as in previous measurement. Discussing the total course of measurements it has to be mentioned here (Vejvoda and Láncoš, 1985c) that after unloading from the first pressure test and subsequent two pressure cycles the dependences $\sigma = f(\varepsilon_t)$ and $p = f(\varepsilon_t)$ in weld zone were practically linear. Only in one case the dependence $p = f(\varepsilon_t)$ was linear already after unloading from the third pressure cycle. Except for membrane stresses σ_m also bend stresses σ_b = 0.2-0.26 σ_m were present in welds whereas these stresses represented the function of one parameter, overstress p.

CONCLUSIONS

The tests of relaxation overstressing of seven huge spherical storage tanks for liquified hydrocarbons which could not be annealed due to their immense volume, were described in this paper. Three pressure cycles at temperature of vessel wall above crack arrest temperature were applied to relaxation. Maximum pressures evoked membrane stresses close to yield strength of material.

Strain gauge measurements proved that residual stress values measured on vessel surface decreased during prestressing and the dependences $\sigma = f(\varepsilon_t)$ and $p = f(\varepsilon_t)$ were practically linear after first prestressing cycle. Thus, it can be assumed that with pressure cycles the stresses induced by repair welding were decreased to a reasonable extent and at the same time it was assured that only elastic stresses would be present in vessel wall under operating pressures which attain 2/3 of pressure values of applied relaxation cycles. The storage tanks are in fluent operation again for one year.

REFERENCES

Holzmann, M. (1985). Dynamic fracture toughness of material for a spherical storage tank. ÚFM ČSAV Brno.
Hrivňák, I. (1985a). Int.J.Pres.Ves. and Piping, 20, 223-237.
Hrivňák, I. (1985b). Zváranie, 34, No. 1 and 2, 35.
Hrivňák, I. (1986). Heat treatment of welded joints in carbon and micro-alloy steels. Zváranie, 35, (in print).
Jaroš, P. (1986). Strain stress state in the vicinity of a weld in repair of spherical storage tank of liquified gas. Report No. 85 - 02126, SVÚSS, Prague.
Kálna, K. (1984). Assessment of brittle fracture resistance of materials for 2500 m^3 spherical storage tanks with unannealed

welds. Report of the Welding Research Institute.
Láncoš, J. and S. Vejvoda (1985b). Expertise Report No. 1312/85, ÚAM Vítkovice, Brno.
Séferian, D. (1962). Náuka o kovech ve svařování ocelí. SNTL, Prague.
Vejvoda, S., and I. Láncoš (1985a). Expertise Report No. 1293/85, ÚAM Vítkovice, Brno.
Vejvoda, S., and J. Láncoš (1985c). Discussion of results of strain gauge measurements of spherical storage tanks, ÚAM Vítkovice, Brno.
VÚZ (1985). Expertise Report No. Mg 96/21 from 25,7.

SESSION IV

ADVISABILITY AND OPTIMIZATION OF STRESS RELIEVING TREATMENT/OPPORTUNITE ET OPTIMISATION DES TRAITEMENTS DE RELAXATION DE CONTRAINTES

IV.1 Post weld heat treatment of a high integrity component of complex geometry
(C. M. White, W. P. Carter — United Kingdom/Royaume-Uni)   199

IV.2 Post heat treatment of welds on high strength steel nioval 47[1]
(M. Velikonja — Yugoslavia/Yougoslavie)   207

IV.3 Welding structures heat treatment application experience for stress relieving in heavy machine building industry
(L. P. Eregin et al. — USSR/URSS)   219

IV.4 Dependence of heat treatment parameters VS. Thermal conditions in welding and post-welded heterogeneity of properties
(F. A. Khromchenko — USSR/URSS)   221

IV.5 Comparison of the residual stress distibutions after stress-relief annealing of welded sheets of the high strength structural steel ST E 690 at different temperatures
(J. Heeschen, H. Wohlfahrt — FRG/RFA)   231

IV.6 Influence of the welding technology and the stress-relieving heat treatment on the corrosion cracking resistance of welded nitrogen-Alloyed stainless steel
(L. Kalev, V. Mihailov, A. Krustev — Bulgaria/Bulgarie)   239

IV.7 Relaxation of residual stresses during postweld heat treatment of submerged-arc welds in a C-Mn-Nb-Al steel
(R. H. Leggatt — United Kingdom/Royaume-Uni)   247

IV.8 Estimation methods for studying the degree of relaxation of residual welding stresses at appropriate heat treatment, as well as for evaluation of effect of non-relaxed stresses on the load-carrying capacity of structure members
(V. I. Makhnenko, E. A. Velikoivanenko, V. E. Pochinok — USSR/URRS)   257

IV.9 Qualification d'un traitement de relaxation
(A. Leclou — France)   269

IV.10 Electromagnetic monitoring of residual stress relaxation during heat treatment
(I. M. Zhdanov, V. V. Batyuk, A. A. Khriplivy, R. K. Gachik, K. B. Pastukhov, G. F. Kolot, A. V. Pulyayev — USSR/URSS)   277

IV.11 Postweld heat treatment of C-Mn and microalloyed steels: an evaluation on the basis of C.T.O.D. and wide plate tests
(W. Provost, A. Dhooge, A. Vinckier — Belgium/Belgique)   289

IV.12 Is high tempering always needed for low-carbon and low-alloyed steel structures?
(A. E. Asnis, G. A. Ivashchenko — USSR/URSS)   307

IV.13 Considerations of the post weld heat treatment of pressure parts
(I. G. Hamilton, A. R. G. Abbott — United Kingdom/Royaume-Uni)   313

IV.1

Post Weld Heat Treatment of a High Integrity Component of Complex Geometry

C. M. White and W. P. Carter

Whessoe Heavy Engineering Limited, Darlington, U.K.

ABSTRACT

This paper describes the successful post weld heat treatment of four large high integrity structures for the nuclear reactors, at Heysham II (Central Electricity Generating Board) and Torness (South of Scotland Electricity Board). The problems faced in conducting closely controlled heat treatments on such large and complex structures are reviewed and the solutions are described.

KEYWORDS

Heat treatment; structural integrity; process control; furnace; burners.

FUNCTIONS OF THE GAS BAFFLE ASSEMBLY

The gas baffle assembly of an Advanced Gas Cooled Reactor (AGR) is a high integrity steel fabrication at the heart of this type of nuclear power station. It performs a number of functions. Its outer envelope separates the primary coolant gas which is gaining heat from the core within it from that giving up heat to the boilers which surround it. This subjects it to the differential pressures needed to drive the pressurised CO_2 around the primary circuit. It provides structural support for the core and boilers, transmitting the deadweight and any seismic forces from the whole of the core and its shielding and part of the boilers through to the base slab of the prestressed concrete pressure vessel which houses the primary circuit. The top dome of the assembly carries a closely packed array of upstanding nozzles through which the hot gases from the fuel channels are ducted on their way to the boilers. These nozzles also provide location for the on-load refuelling of the reactor. At interstitial locations in this array are downstanding nozzles through which the reactor control rods operate. The assembly also provides support for various systems of instrumentation.

The fabrications need to be of the highest integrity to assure safety and to protect massive investments.

STRUCTURAL INTEGRITY

Preliminary design calculations for the Heysham II and Torness A.G.R's. established that a well proven fine grained carbon steel, chosen for its weldability, toughness and ductility could be used at conservative stress levels, and at section thicknesses well within existing manufacturing experience. Integrity was further enhanced by striving for simplicity and cleanliness of design with readily inspectable welds of simple geometry. Finally the decision was made to post weld heat treat the gas baffle assembly as a complete assembly.

By adopting these measures the critical defect sizes were demonstrated to be orders of magnitude larger than the sizes of defects which can be found reliably and consistently by well established non destructive examination techniques.

DESCRIPTION OF GAS BAFFLE ASSEMBLY

Material specifications were based upon B.S. 1501-224-28B (1964) with appropriate supplementary requirements. Each gas baffle assembly consists of five major sub-assemblies:-

1. The support skirt 13.85M inside diameter, 50mm thick containing eight reinforced openings 1.74M dia. for coolant gas circulation, and other connections for such purposes as man access. Cooling water pipes are welded to the skirt to control temperature gradients in operation.

2. The core support structure (or diagrid). This is a grillage structure 2M deep with plate thicknesses ranging from 38mm to 60mm. Loads are transmitted to the intersection points of the grillage through tubular stools with 91mm thick cap plates.

3. The gas baffle cylinder 13.85M inside diameter. General plate thickness 35mm, but with heavy forgings built into an 80mm thick strake to carry deadweight and seismic loadings from the boilers.

4. The restraint tank 13.55M outside diameter, 50mm thick concentric within the gas baffle cylinder. It carries an array of 450 fabricated brackets which serve as location points for the restraint mechanisms for the core and its shielding. Seismic loadings from the core and shielding would be transferred by the restraint tank to the periphery of the diagrid.

5. The torispherical dome 67mm thick containing 332 fuel nozzles, 120 interstitial nozzles and three large elliptical access openings. The set on fuel nozzles are of 330mm finished bore and of bi-metallic construction, generally austenitic, but with a narrow section of carbon steel welded to the dome. The transition welds are oblique, paralleling the local slope of the dome.

The five major sub assemblies were fabricated and brought together at the Dock Point (Middlesbrough) Facility of Whessoe Heavy Engineering Limited to form a structure some 14M diameter, 20.5M high weighing almost 1000 tonnes.

The problems of post weld heat treatment can now be appreciated. These were of performing a closely controlled operation on structures of these overall dimensions with section thicknesses varying from 7mm to about 250mm, taking into account such features as:-

The narrow annular gap between the restraint tank and the gas baffle cylinder.

The transition welds between carbon and austenitic steels in the fuel nozzles.

The potential restraining effects of the diagrid on the cylindrical components.

METHOD OF POST WELD HEAT TREATMENT

Support of the assembly during heat treatment did not pose any problems since advantage could be taken of the presence and massive strength of the diagrid. The assembly was supported by positioning a ring of roller supports under this structure to provide support and location whilst permitting radial expansion.

A detailed specification for the heat treatment cycle was written and approved. This was based on B.S. 3915-1965 but with supplementary requirements on heating and cooling rates and temperature gradients. Possible ways of carrying out the heat treatment within the limits of the specification were reviewed with a number of specialist contractors. The work was eventually entrusted to Kemwell Limited, supported by their consultants Thermal Hire Limited.

The assemblies were heat treated within the workshops, on their final build stations by enclosing each in turn in a lightweight externally framed cylindrical furnace shell. Rings of propane fired Appollo burners were fitted through the furnace wall at various levels. These were high velocity burners, some rated at 1 GJ/hr and others at 6 GJ/hr, with forced excess air enabling them to be operating at outlet temperatures down to $100^{\circ}C$. Some burners were arranged to fire into the annular space between the furnace wall and the assembly and some into the assembly. Most of the burner outlets were directed tangentially to generate and maintain swirl and turbulence of the furnace gases. The burners could be used in any desired combination. Inward leakage of cold air was prevented by operating the furnace at a small positive pressure by means of a damper in the exhaust gas duct.

The individual burners were computer controlled using a pre-programmable thermocouple measuring burner outlet temperature. Full advantage was taken of the operating characteristics of the burners to control the temperature of the assembly during the cooling phase of the cycle. The burners were controlled to operate marginally hotter than the assembly during heating, marginally cooler during cooling, and at the mean temperature of the assembly during soak.

Temperature of the assembly was monitored by a total of 252 thermocouples. These gave a continuous readout on chart recorders, and were also wired into a data logger to generate a printed readout at intervals - normally hourly, but additional sets of readings were taken when appropriate.

Nitrogen was circulated through the cooling pipes on the gas baffle skirt to reduce oxidation of their internal surfaces.

RESPONSIBILITIES FOR POST WELD TREATMENT

The heat treatment cycle and temperature limitations were determined by Whessoe Heavy Engineering Limited and approved by the National Nuclear Corporation. Detailed operational procedures and quality plans for the operation were prepared by Whessoe and Kemwell and approved by the National Nuclear Corporation.

The furnace, burners and instrumentation were provided by Kemwell Limited and their consultants, Thermal Hire Limited. Overall control of the heat treatment was the responsibility of a nominated Whessoe Engineer. The furnace and instrumentation were operated by a team of Kemwell/Thermal Hire Engineers. Changes to operating patterns or settings of the burners were agreed with the Whessoe Engineer in charge, and all operations logged.

The thermocouple readings were continuously monitored by a team of Whessoe Engineers to ensure that the various limitations were not exceeded. All operations were witnessed by the Independent Inspecting Authority and subject to surveillance by Engineers from the National Nuclear Corporation, the Generating Board concerned, and the Health and Safety Executive.

HEAT TREATMENT CYCLE

In addition to the requirements of B.S. 3915-1965 the following limitations were imposed:-

Rates of heating and cooling not to exceed $10^{\circ}C$ per hour for temperatures above $300^{\circ}C$ and $20^{\circ}C$ for temperatures below $300^{\circ}C$.

Maximum temperature differential on any two points on the assembly not to exceed $120^{\circ}C$ at mean temperatures up to $240^{\circ}C$, reducing progressively to less than $40^{\circ}C$ during the soak period.

Maximum surface temperature gradient during heating and cooling not to exceed $10^{\circ}C$ per metre. There shall be no significant temperature differences between the outer region of the diagrid and the adjoining regions of the gas baffle cylinder.

Maximum through thickness temperature difference not to exceed the larger of $10^{\circ}C$ or $1^{\circ}C$ per mm of thickness.

Cooling of the assembly to be controlled down to $150^{\circ}C$.

ACHIEVEMENT

All four gas baffle assemblies, two for Heysham II and two for Torness, were heat treated fully in accordance with requirements. The overall times to achieve the heat treatment cycles were 139, 180, 206, and 132 hours respectively. These times compare with a theoretical minimum cycle time of $85\frac{1}{2}$ hours.

During the soak periods the extreme range of thermocouple readings on each assembly was comfortably within the permitted $40^{\circ}C$. The extreme temperature range recorded at the end of the soak period varied from $25^{\circ}C$ to $30^{\circ}C$.

The thermocouples used for monitoring the temperature of the assembly proved extremely reliable. The thermocouple leads were attached directly to the assembly by capacitor discharge and only very small numbers of individual thermocouples had been 'lost' by the end of each of the extended heat treatments. At no time were we unsure of the condition of any part of the assemblies.

ADDITIONAL CONSIDERATIONS 1 - NOZZLES

There is a requirement to meet tight dimensional tolerances on the bore, position, tilt and heights of the array of nozzles in the dome. The measures adopted to meet this requirement were to weld to the dome nozzles having excess material in the bore and on length, and heat treat this major sub assembly to achieve dimensional stability before boring out and facing the nozzles to finished dimensions. This technique proved entirely satisfactory, particularly in the respect that subsequent movements of the nozzles during the final P.W.H.T. operations were negligible.

The stabilising heat treatments were carried out in accordance with procedures and quality plans similar to those for the final P.W.H.T. of the assembly, using the top section of the furnace and some of the burners. These heat treatments provided valuable experience and built up confidence prior to the post weld heat treatments of the full assemblies.

ADDITIONAL CONSIDERATIONS 2 - ANNULAR GAP

The narrow annular gap between the restraint tank and gas baffle cylinder would be difficult to clean effectively after P.W.H.T. It was therefore decided to clean the surfaces during final assembly and prevent undue deterioration during P.W.H.T.

Furnace tests in the laboratory established that a particular nuclear compatible coating applied to a shot blasted surface provided the necessary protection. The coating itself was degraded during the extended heat treatment cycle but remained tightly adherent and prevented oxidation of the steel surface.

This technique was adopted and immediately after the heat treatment had been completed the top and bottom of the annulus was sealed and nitrogen introduced to give long term protection to the surfaces.

CONCLUSIONS

The complex heat treatment operations on four gas baffle assemblies were completed successfully. Full dimensional re-surveys of the assemblies after P.W.H.T. showed that whilst some physical movements had taken place these were of a very minor nature and within the allowances assessed during tolerance build-up exercises in the design phase. All welds were accepted following full non-destructive examination after P.W.H.T.

ACKNOWLEDGEMENTS

The authors wish to express their gratitude to Whessoe Heavy Engineering Limited, the Central Electricity Generating Board, the South of Scotland

Electricity Board, and the National Nuclear Corporation for permission to publish this paper.

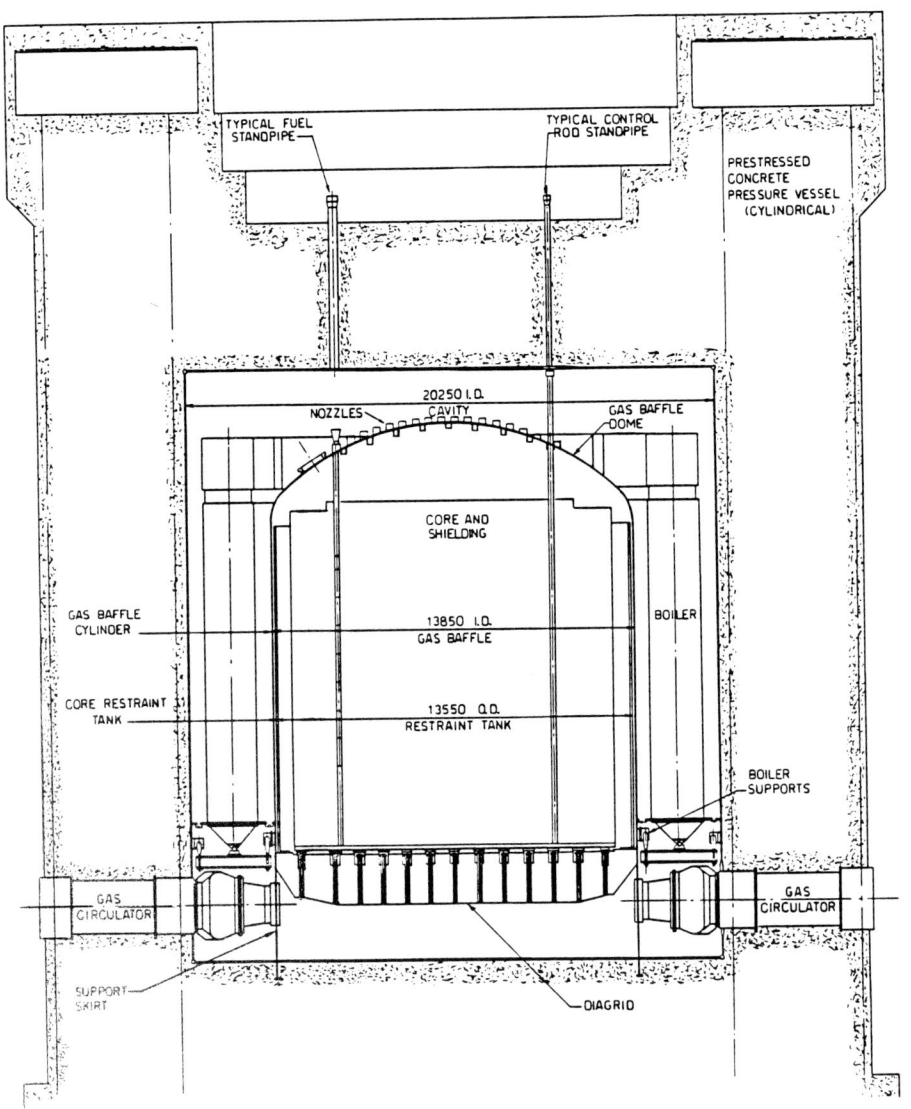

THE GAS BAFFLE ASSEMBLY OF AN ADVANCED GAS COOLED REACTOR

Preparing to heat treat the dome/nozzle assembly.

Final Assembly of the gas baffle.

Furnace for the PWHT operation.

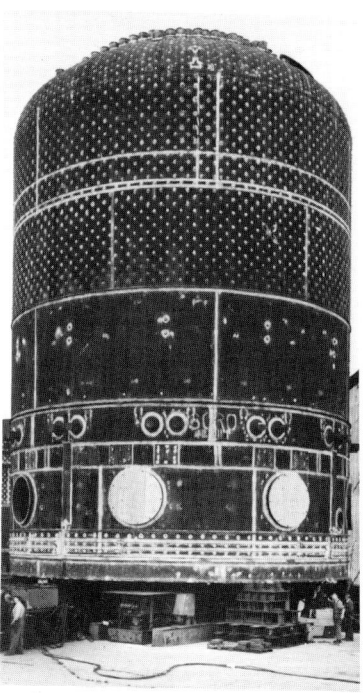

Gas baffle assembly after PWHT and final inspection.

Shipment to site of a gas baffle assembly.

IV.2

Post-heat Treatment of Welds on High Strength Steel Nioval 47[1]

M. Velikonja

Institute za varilstvo, Ljubljana, Yugoslavia

ABSTRACT

Welds on high strength structural steel Nioval 47 were examined as to their susceptibility to cracking during thermal residual stress relief.

Nioval 47 is fine-grained steel with yield strength guaranteed at 460 N/mm^2. Microalloyed additions are V and Nb.

Tests were carried out on test specimens experimentally welded and on test specimens with the structure of simulated heat affected zone. During continuous tensile test at constant annealing temperatures for stress relieving the reduction of stress or time to destruction were measured respectively. Test results showed that heat affected zone of the examined steel is bound to reheat cracking.

Annealing for reduction of internal stresses is not recommended. More concern should be devoted to the choice of technology and the execution of welding.

KEYWORDS

Post weld heat treatment, stress-relief annealing, high strength steel.

---

[1] This report is an abstract of research works: Study of heat treatment of welds II and III phase.

INTRODUCTION

In the present paper we would like to report upon our research on the susceptibility of high strength steel Nioval 47 to cracking during stress-relief annealing after welding.

The purpose of the investigation was to find out whether and under what conditions of subsequent heat treatment welded joints of tested steel become susceptible to cracking during post-heat treatment.

Research reports should help us as far as the choice of Nioval 47 steel for the fabrication of various different welded structures is concerned, where also requirements as to post-heat treatment are different.

In the 70's, among Yugoslav high strength steels for the production of pretentious welded structures the use of Nioval 47 steel expanded greatly. It is a fine-grained steel, microalloyed with vanadium and niobium. Orientational chemical composition and insured mechanical characteristics of some Yugoslav high strength steels are given in Tables 1 and 2.

TABLE 1 Chemical Composition of Some High Strength Yugoslav steels

Steel	max. % C	% Si	% Mn	% Cr	% Ni	% Mo	% V	% Nb
Niobal 43.	0,20	0,40	1,45	-	-	-	-	0,05
Nioval 47.	0,20	0,40	1,45	-	-	-	0,06	0,04
Nioval 50.	0,20	0,40	1,45	-	0,50	-	0,15	-
Nionicral 60 A	0,15	0,30	0,40	1,50	2,50	0,40	-	-
Niomol 490	0,10	0,35	1,20	-	1,40	0,40	-	-

All steels have also: min. 0,020 % Al and S and P under 0,030 %; Niomol 490 max. 0,020 S and P.

TABLE 2 Mechanical Characteristics of Some High Strength Steels

Steel	Yield Strength N/mm^2 for thickness 10 - 15 mm	Strength N/mm^2	Strain min. % $L_o = 5d$	Toughness ISO-V min (J)
Niobal 43	430	540-690	19	63 at +20°C 27 at -50°C
Nioval 47	460	540-740	18	47 at +20°C 27 at -50°C

Steel	Yield Strength N/mm^2 for thickness 10 - 15 mm	Strength N/mm^2	Strain min % $L_o = 5d$	Toughness ISO-V min (J)
Nioval 50	480	610-740	17	55 at +20°C 27 at -50°C
Nionical 60 A	590	690-830	18	118 at +20°C 118 at -60°C
Niomol 490	490	560-740	19	63 at +20°C 39 at -60°C

TESTING METHODS

Testing of the susceptibility of welded joints to cracking during thermal stress relief after welding was carried out in two ways:

Method I: Continuous tension test on test pieces with and without notch in the structure of the heat affected zone (HAZ) of a welded joint. The reduction of the initial stress in dependence of time to destruction at a constant annealing temperature has been measured. Test temperatures were: 580, 600, 625 and 650°C. Maximum time to destruction in clamped condition and at test temperature was 60 min. In critical cases, a premature destruction of test pieces was expected.

Method II: Slow hot tensile test on test pieces without notch. Contraction at the destruction of test pieces served as a criterium for the assessment of susceptibiliy to cracking during post-heat treatment. The shape of test pieces is shown in Fig. 1.

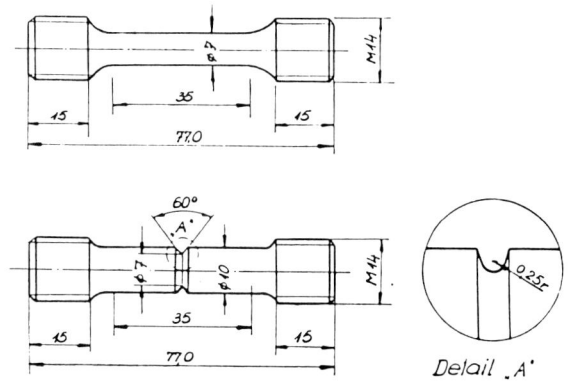

Fig. 1. Shape of test piece with and without notch

PREPARATION OF TEST PIECES

Test Pieces for Test Method I

- simulated structure of HAZ, achieved by quick heating up to the temperature of 1250 - 1280°C and by through hardening in oil. We draw near to the conditions of welding without preheating with small heat input.

- HAZ of weld, welded without preheating with heat input of 80 kJ/cm - weld no. 1.

- Heat affected structure of the weld, welded with preheating up to 150°C and with heat input of 80 kJ/cm - weld no. 2.

- heat affected structure of the weld, welded with preheating up to 150°C, heat input of 80 kJ/cm and subsequent building-up of the relieved run with heat input of 2,5 kJ/cm - weld no. 3.

The heat input stated exceeds the practical values. In case of single-run cored-wire MAG welding of 15 mm thick plate we did not manage to fill up single-V groove with lower heat input.

Test Pieces for Test Method II

Test pieces were prepared having simulated HAZ structure which was reached by quick inductive heating up to the temperature of 1250 - 1280°C and cooling:

       a) in oil
       b) in hot water.

TESTING RESULTS - METHOD I

Reduction of Initial Stress

In case of test pieces with simulated HAZ (Fig. 2.) it can be observed that the reduction of initial stress in flat test pieces at the annealing temperature of 600°C is even smaller than at that of 580°C. It can be concluded that the reduction of stress at the annealing temperature of 600°C was obstructed.

In case of test pieces with notch a premature destruction (after 6 and a half min. and 30 min. of annealing) occurred at the temperature of 580 - 600°C. Fig. 3. shows the structure of the flat test piece after stress-relief annealing at 600°C. Cracks along grain boundaries and signs of grain boundary separation are visible.

In case of the test pieces having HAZ of the weld no. 1, which was welded with heat input of 80 kJ/cm (Fig. 4.) it can be observed that the stress reduction is obstructed at the annealing temperature of 625°C with flat test pieces as well as those with notch.

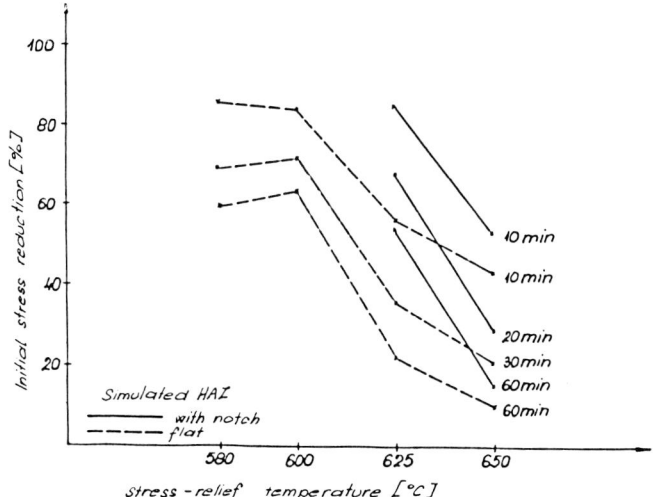

Fig. 2. Reduction of initial stress of simulated HAZ in dependence of time and annealing temperature.

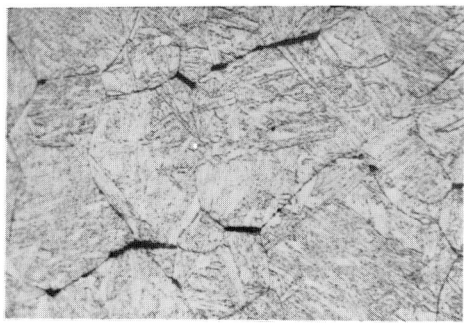

Fig. 3. Crack on grain boundaries on simulated test piece without notch after stress annealing at $600^{\circ}C$.
Enlarged 400 x.

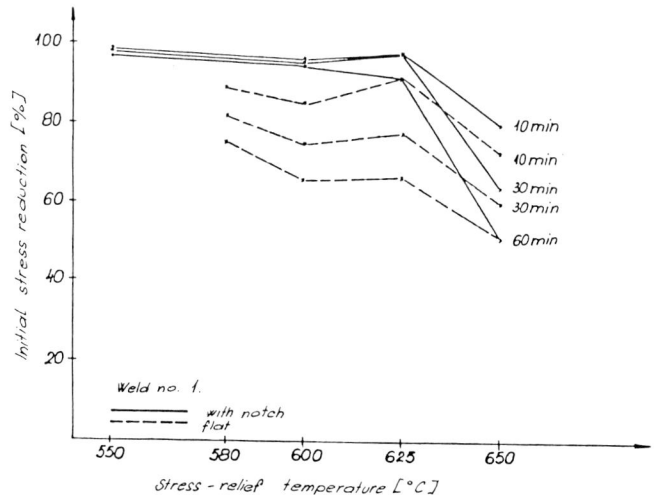

Fig. 4. Reduction of initial stress on test pieces of weld no. 1 in dependence of time and annealing temperature

In case of the test pieces having HAZ of the weld no. 2 which was welded with preheating and the heat input of 80 kJ/cm, the obstructed stress reduction of the flat test piece is observed at the annealing temperature of 600°C (10 min)(Fig. 5). In case of the test pieces with notch, a premature destruction occurred at the temperatures of 625, 650 and 675°C after 4, 34 and 45 minutes. It can be concluded that an exceeding heat input in welding increases the susceptibility of the HAZ to cracking at the annealing temperatures of 625°C and higher.

In case of the test pieces having HAZ of the weld no. 3 which was welded with preheating and the heat input of 80 kJ/cm and subsequent welding of the relieved run, it can be observed (Fig. 6) that the reduction of the initial stress is observable at the annealing temperature of 650°C. The premature destruction of the test pieces with a notch did not occur, which is to attribute to the "relieved run".

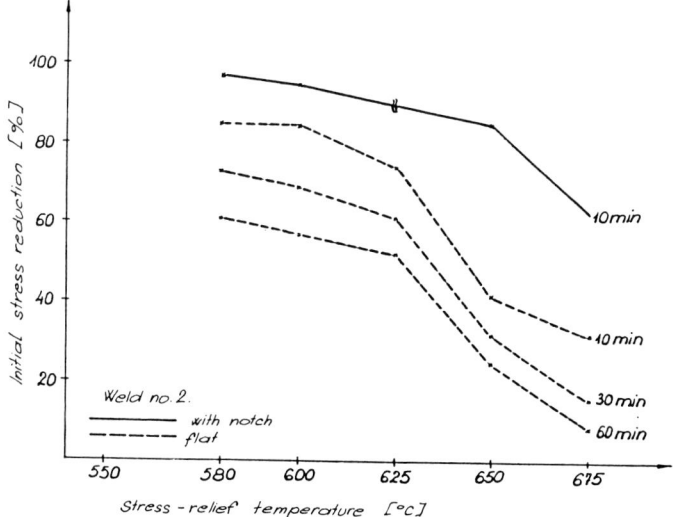

Fig. 5. Reduction of initial stress on test pieces of weld no. 2 in dependence of time and annealing temperature

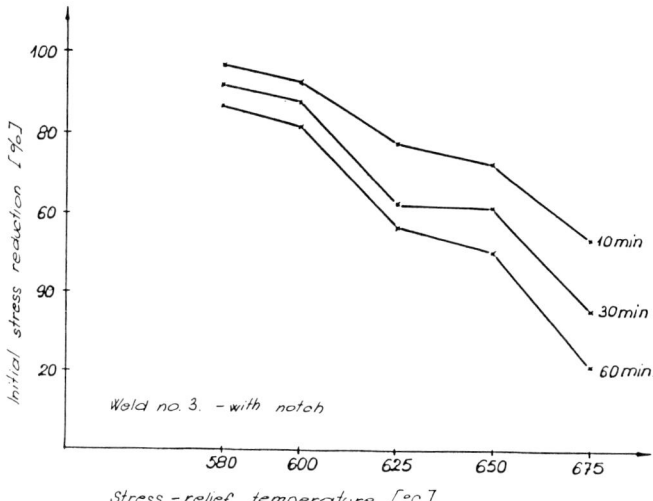

Fig. 6. Reduction of initial stress on test pieces of weld no. 3 with notch in dependence of time and annealing temperature

Discussion on the Course and Results of the Investigation

We are aware of the fact that neither the test pieces with simulated HAZ nor the test pieces of welded joints represented the HAZ which shows up at actual multi-pass welding in practice. Observations during operation and after stress-relief annealing called our attention to various problems, e.g.:

Flat test pieces with HAZ made by welding are not suitable, because at the process of stress relief, a complete welded joint is present, i.e. weld, HAZ and base material. Creeping occurs where the ductility of material is the biggest which, no doubt, is not the overheated HAZ.

Flat test pieces should be treated as welded joints in the structure, where there are no defects and no misalignements and where transverse residual internal stresses are those which have to be reduced by annealing. In such cases the reduction of internal stresses occurs because of creeping of more ductile base material and softer part of the HAZ. This is why the danger of cracking in such cases is at minimum.

Test pieces of welded joints with notch in HAZ represent welded joints with a defect, where stress raisers concentrate. They represent also circumferential welds with imposed creeping, necessary for stress relief in the HAZ, too, i.e. at the notch location. With these test pieces occurred the premature destruction in critical cases, too.

Test pieces with the structure of simulated HAZ and with a notch could be the most suitable. The problem lies in the execution of desired simulation of thermal cycle of welding, regarding that we do not have a simulator at our disposal.

TEST RESULTS - METHOD II

The criterium for the assessment of susceptibility has been the contraction at destruction during slow hot tensile test on simulated test pieces without a notch. In case the contraction at destruction is smaller than 20 %, it is considered that the HAZ of steel at test annealing temperature for the stress reduction is susceptible to cracking at reannealing.

Slow Hot Tensile Test

The results of slow hot tensile test are given in Table 3. For comparison also test results of Nioval 47 steel in as delivered state are given.

TABLE 3 Results of Hot Tensile Test

Variant of simulation	Test temperature °C	Breaking strength N/mm²	Breaking elongation %	Contraction %
a)	580	461	10,7	31,9
a)	600	343	15,3	52,2
b)	580	374	13,9	46,4
b)	600	361	16,5	53,7
b)	620	296	15,9	59,1
Nioval 47 in as delivered state	580	–	22,6	61,7
	600	303	23,2	66,7
	620	212	24,4	70,1

The test results show that in all cases the contraction is bigger than 20 %.

The metallographic examination showed that at simulation the corresponding heating occurred only on the surface, while the core of the test piece remained fine-grained. Not far away from the destruction a larger concentration of micro cracks in the layer with greater content of Mn-sulphide inclusions has been noticed. This area has been examined more precisely by electronic microanalyser.

The results of a linear microanalysis of Mn, Si, V and Nb content (Fig. 7, 8, 9) showed that the layer with a greater content of MnS inclusions contains also a greater concentration of Mn, Si and Nb. Individual concentrational peaks of Nb are noticed (Fig. 9).

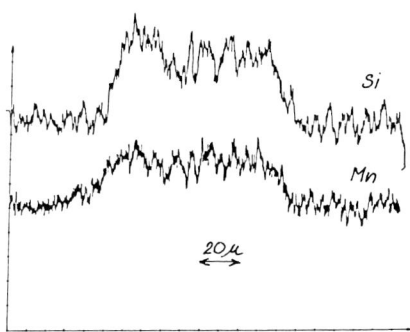

Fig. 7. Increased Mn and Si content in the layer with greater quantity of MnS inclusions.

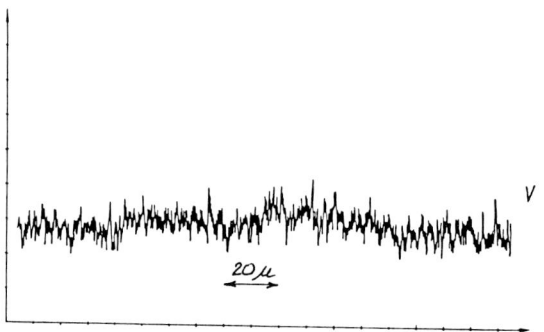

Fig. 8. Uniform V content in the layer with greater quantity of MnS inclusions

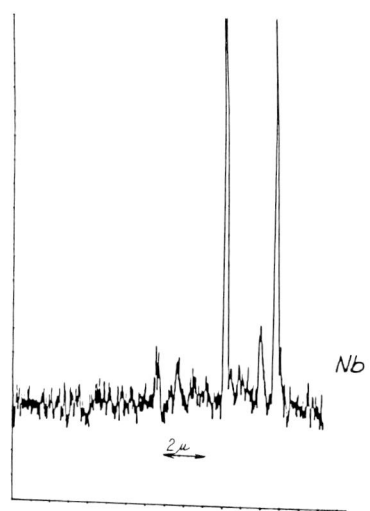

Fig. 9. Nb concentrational peaks in the layer with greater quantity of MnS inclusions

Regarding the higher concentration of cracks in this layer it is evident that the nonhomogeneities in material present additional points susceptible to cracking.

CONCLUSIONS

1. The results of the examination showed that in some cases welded joints on Nioval 47 steel are susceptible to cracking during stress-relief annealing, e.g.:

- during welding without preheating, when in HAZ of the weld a martensite or martensite-bainite structure can occur;

- during welding with excessive heat input, when the structure in HAZ is nor through hardened but coarse-grained, with carbon, vanadium and niobium in oversaturated solid solution and with impurities, concentrated on the boundaries of coarse grain.

2. For the reduction of crack appearance during stress-relief annealing of the workpiece of Nioval 47 steel it has been proposed:

- stress-relief annealing of those workpieces for which this is specially required by regulations, or not to use Nioval 47 steel for the workpieces which have to be stress-relief annealed after welding (pressure vessels).

- In any case of the production of workpieces of Nioval 47 steel, a special attention as well as working discipline should be paid to consistent execution of optimal welding parameters, specially as far as the preheating temperature and heat input are concerned. In this way we can avoid the occurrence of through hardened structures in HAZ, the residual internal stresses are lower and the post-heat treatment is not necessary.

- During steel fabrication, more attention should be paid to chemical composition, i.e. carbon and impurities content should be lowered, as well as the nonhomogeneity - stringers of the material.

3. It is reported about cracking on welded structures of Nioval 47 which are in use in oil industry. It is possible that the phenomenon of stress corrosion is in question.

In this regard a dilemma occurs about the influence of stress-relief annealing on the susceptibility of welded joints to cracking due to stress corrosion.

IV.3

Welding Structures Heat Treatment Application Experience for Stress Relieving in Heavy Machine Building Industry

L. P. Eregin *et al.*

USSR

The Publisher regrets that the manuscript for this contribution was unavailable at the time of going to press and apologises for the inconvenience caused to readers.

IV.4

Dependence of Heat Treatment Parameters vs. Thermal Conditions in Welding and Post-welded Heterogeneity of Properties

F. A. Khromchenko

VTI, Moscow, USSR

ABSTRACT

The effect of heat input conditions during welding on nonuniformity of mechanical properties of welded joints and their resistance to local failures in repeated heating is considered. Also, possibility of reducing high-temperature tempering time for heat-resistant Cr-Mo-V steel welded joints of steam pipes is shown.

The reliability and life of steam pipes of power plants depend mostly on quality and nonuniformity of properties of welded joints /1-3/. Nonuniformity of properties of zones of joints made of heat-resistant Cr-Mo-V steels is, in turn, a function of heat input, heat treatment conditions, composition of weld metal and service /2, 4, 5/.

Steam pipe service experience shows that failures of joints caused by brittle fracture /2, 6/ is because of brittle interlayers, both of high and low strength. The weld metal is characterized by about as high as 1.5-2.5 times level of short-term strength as compared to the base metal. One should suggest that in a general form sensitivity to brittle fracture increases with higher nonuniformity of properties (K^{HV}) because under binding loads, characteristic of steam pipe service, the near-weld zone, and the HAZ as a whole, are under the most infavourable conditions and here fractures occur most of all. The existing regulating codes (hereinafter referred to as RTM) /7/ do not account this factor. Therefore, determination of the effect of welding technology and postweld heat treatment on K^{HV} and its influence on the resistance to brittle fracture seem reasonable. Also, the permissible value of K^{HV}, as an additional criterion for estimating the workability of joints, is required. The pipe joints were welded by various methods as different heat inputs with design /8, 9/ values per welding length of 0.5-10 MJ/m with heating at 250-550°C and with no heating (see Table below). After welding, the joints were subjected to high-temperature tempering as follows: 650°C, 1 hr; 730°C, 1 hr; and 730°C, 3 hrs.

The hardness of the weld and base metals were tested and K^{HV} was evaluated. K^{HV} was used as a parameter, which is a ratio between the average values of the hardness (spread of $\pm$ 10%) of the weld and base metals. The results obtained are presented as a monogram in Fig. 1.

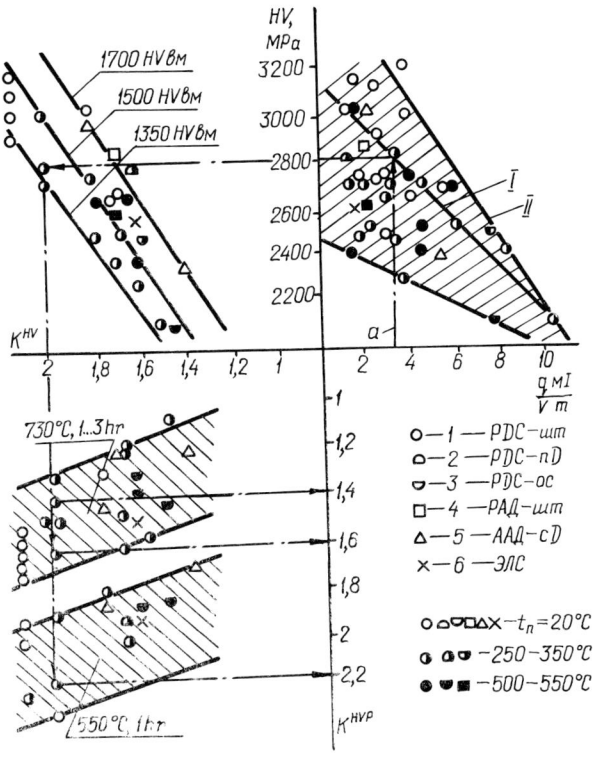

Fig. 1. Effect of heat input during welding (q/V, tn), base metal hardness and tempering (650-730°C) on K^{HV} for 20-60 mm thick wall pipe joints of Cr-Mo-V steels: hardness upper limits for welds with preheating (I) and without it (II).

One can see that with increase in welding heat input the weld metal hardness lower from 2800-3200 down 220-2400 MPa and, hence, K^{HV} in the initial state - from 1.8-2.2 up to 1.4-1.8 with hardness of the base metal of 1350-1700 MPa. Postweld high-temperature tempering may greatly affect K^{HV} (Fig.1,line a

Characteristic variation of the weld metal hardness and K^{HV} is shown in Fig. 2. The welding technology is of marked effect on

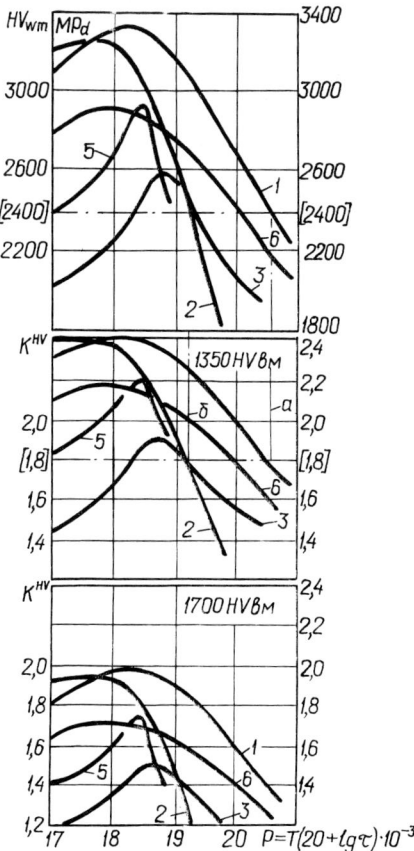

Fig. 2. Effect of welding technology (1-6 in Fig.1) and ageing (P) on weld metal hardness kinetics and K^{HV} with base metal hardness of 1350 and 1700 MPa.

the mechanism of metal dispersive solidification caused by various thermodeformation cycles of welding. In particular, for the joint with multilayer butt weld made with 2-4.5 MJ/m /7/ both sufficiently high metal weld hardness (up to 3400-3600 MPa) at the initial stage of repeated heating (P = 18.5, see Fig. 2), and subsequent reduction of up to 2200 MPa (P = 21) are observed. In this case the maximum value of K^{HV} reaches 2 and 2.6 with the hardness of the base metal of 1700 and 1350, respectively.

For single-pass joints with slit gaps the weld metal hardness at the initial stage of repeated heating (P = 18.5) increases up to 2600 MPa only and subsequently reduces down 2000 MPa ($\bar{P}$ = 20.5). This is due to specific effect of arc heat on the method, causing self-tempering /10/.

The maximum value of K^{HV} are 1.5 and 1.9 for base metal hardness of 1700 and 1350 MPa, respectively. Welded joints made by other welding methods are intermediate as regards K^{HV} (Fig.2).

The effect of K^{HV} on sensitivity to local brittle failure was estimated using the results of static bending tests (at 600°C and constant rate of $4.5 \cdot 10^{-5}$ m/min). This is a rather effective tool to solve the problem /2/. A 30-mm thick square specimen with a semi-circular notch (8 mm in radius) over the fusion zone (Fig. 3) was used. The choice of higher thick specimens was to reach real conditions as close as possible.

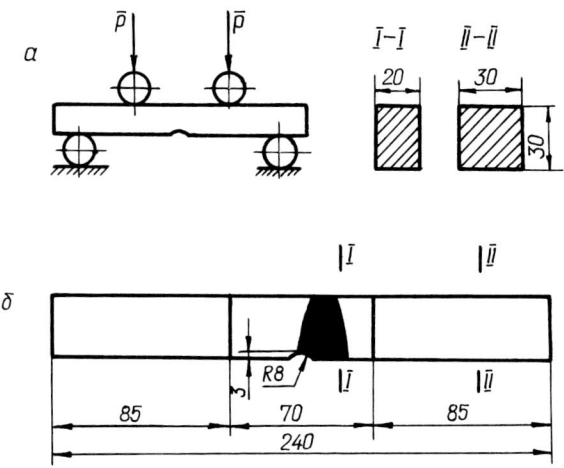

Fig. 3. Test scheme: (a) welded specimens; (b) static bending with constant strain rate at 600°C.

The notch was selected to meet the similarity of stress concentration for actual welded shaped components. The design stress concentration coefficient for the given notch of a bending specimen was 1.55-1.6 /11/. The 600°C temperature was taken to realize dispersive brittleness of joint zones as applied to long-term high-temperature service. As criteria to assess brittle failure use was made of values of specimen bending and relative elongation of the outer surface of the bending specimen until crack initiates. The tests under the selected loading conditions caused initiation of intercrystalline cracks in the near-weld zone (Fig. 4) and cracks along the mild interlayer in the HAZ, which is characteristic of failures of steam pipes during long service /6/. For comparative evaluation of resistance to brittle failure the joints welded as in /7/ and /10/ were tested. K^{HV} = 1.3-2.2 (points in Fig. 5) was observed.

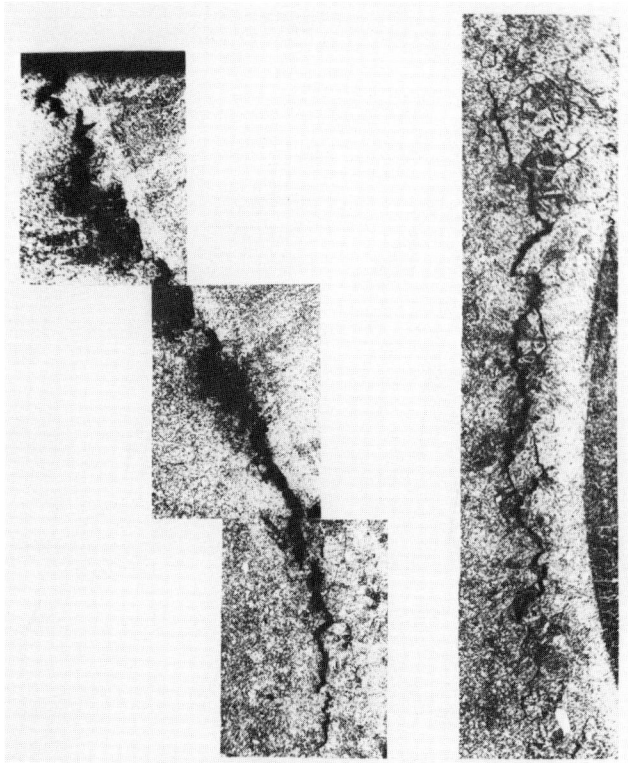

Fig. 4. Intercrystalline failure of the near-weld zone in testing welded specimens for static bending (see Fig. 3).

The test results have made one to conclude that the minimum strain resistance to crack initiation is observed at the initial stage of thermal ageing (P = 18.5) and falls within the maximum value of K^{HV} (Fig. 5). The curves of deformability of joints (δ_{50}, f) are as if "mirror" reflection of K^{HV} during ageing (Fig. 2). Minimum deformability and dispersive solidification of weld metal corresponds to maximum value of K^{HV}, and vice versa. From this it follows that reduction of K^{HV} adds to join workability. In this connection, the joints made by arc single-pass welding with slit gaps are of marked advantage.

Static bending test results presented as deformability vs. K^{HV} show that minimum values of f = 9 mm before crack initiation correspond to minimum level of K^{HV} = 1.8 (Fig. 6, line a) and it can be related to δ_{50} = 8% (line b in Fig. 6). K^{HV} = 1.8 relates to the joint made by the welding technology (curve 1 in Fig. 2) with the permissible weld metal hardness of

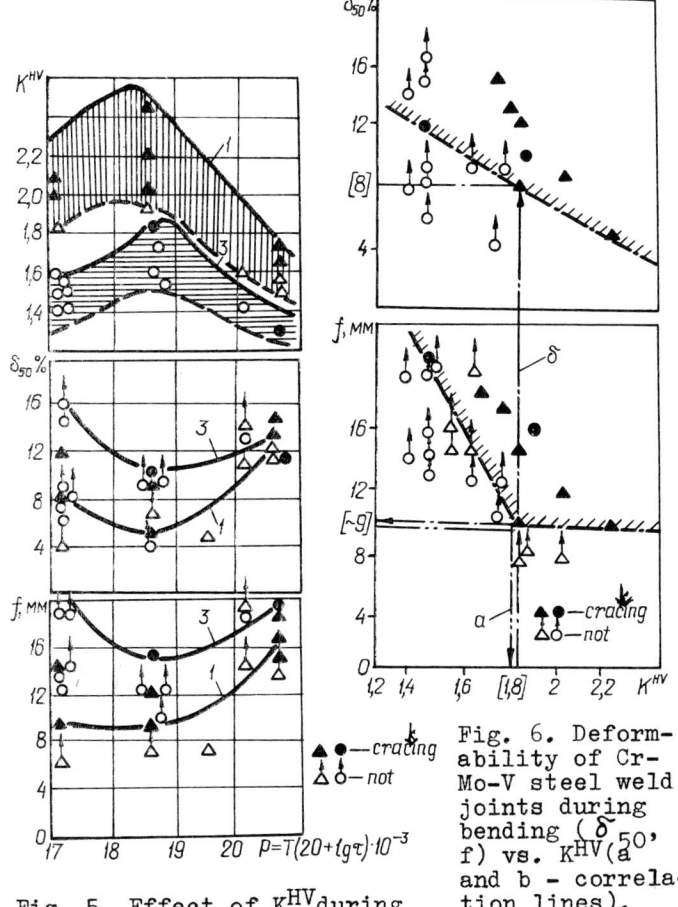

Fig. 5. Effect of K^{HV} during ageing (P) on sensitivity to local failure of welded joints made of 12ХIМФ steel under static bending: (a)-range of spread of K^{HV}; (b),(c)-ductility and deflection of specimens during bending; 1,3-lines of maximum ductility and deflection in initiation of joint cracks (for designations see Fig. 1).

Fig. 6. Deformability of Cr-Mo-V steel weld joints during bending (δ_{50}, f) vs. K^{HV} (a and b - correlation lines).

2400 MPa /7/.

Comparing welded joints with different values of K^{HV} (Fig. 2) one can see that K^{HV} is greatly influenced by both welding technology and heat treatment conditions and base metal hardness. Thus, optimal control of such factors enable the improved workability of welded joints. For example, pulsed arc welding methods used instead of those of /7/ allows shorter time for postweld heat treatment and ensures the required weld metal hardness and optimal value of K^{HV} (curves (a) and (b) in Fig. 2). Single-pass welding with slit gap /10/ seems even more promising in reduction the postweld heat treatment time due to self-tempering during welding. Lower K^{HV} can be also obtained with higher strength (hardness) of the base metal (see Fig. 2).

With this in view, reverse problems can also be solved, i.e. according to given K^{HV} optimal welding technology and heat treatment can be obtained.

Effect of welding method on reduction of tempering time for butt joints of 245-273 mm dia and 36-54 mm wall thick pipes of I2XIMФ steel made by various manual arc welding methods with heat input per welding length of 1.2-8 MJ/m, using 3-4 mm dia electrodes Э-09XIMФ and preheating at 300-500°C is examplified on the basis of dispersive solidification kinetic diagrams for ageing and crack initiation in joint zones (Table, Figs. 7 and 8).

TABLE

Welding method	Heat input per welding length, MJ/m	Pre-heat temperature, °C	Thermal cycles τ' (above 950°C)	τ'' (800-550°C)	Critical value of P	Required tempering time, hrs 700, °C	730, °C
РДС-ШТ	1.2-2	300	30-70	50-100	19.7	1.5	0.5
РДС-ШТ	1.2-1.5	500	60-80	800	19.1	0.4	0.1
РДС-ШТ	4.5-8	300	150-300	200-700	19.7	1.5	0.5
РДС-ОС	8-9	300	25-60	550-600	19.3	0.6	0.2
РДС-ПД	1.5	300	10-15	70-80	20.2	up to 3.5	up to 1.5

It has been found out that τ' = 10-80 and τ'' = 600-800 s are optimal and characterize welding thermal cycle as reaching to the ideal one.

In this case τ' up to 80 s eliminates impermissible grain growth in high-temperature heating zones and lowers the possibility of intense transfer of carbides to solid solution. With τ'' up to 600-800 s isothermal heating effect is observed with intense separation and coagulation of a dispersed phase. Most of all it is encountered in joints made by arc, single-pass welding and by multilayer welding with minimum heat input per welding length and 500°C preheat.

For joints made with manual arc welding to ensure the required values of HV and K^{HV} is in most cases tempering at 730°C during an hour is suffice (Fig. 7b and Table). This is proved by the results of crack resistance tests (Fig. 8) and heat-resis-

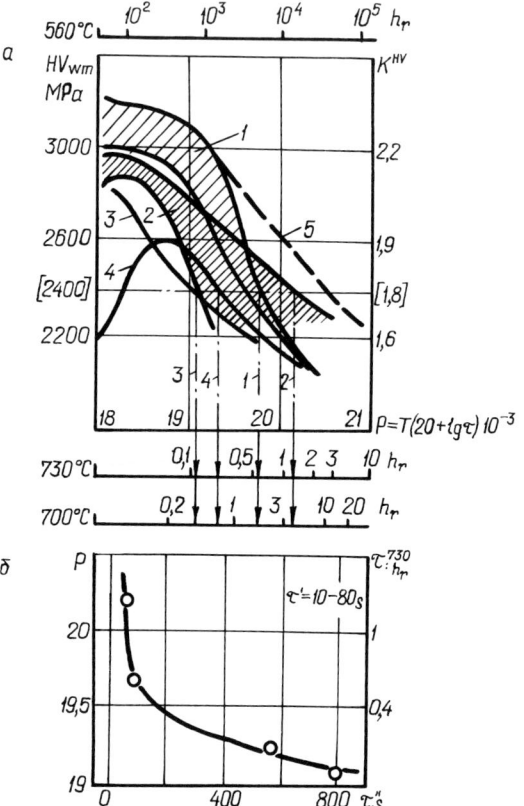

Fig. 7. Effect of welding technology on the required postweld tempering of 12ХIМФ steel joints with weld metal 09ХIМФ ensuring required weld hardness of 2400 MPa and permissible K^{HV}= 1.8: (a) - kinetics of hardness of weld metal 09ХIМФ (HV) and K^{HV} during ageing; (b) - tempering time (P, $\tau 730$) vs. duration of metal cooling during welding (τ'') after high-temperature heating (τ'=10-80 s): 1-for РДС-ШТ, 300°C; 2- РДС-ПД, 300°C; 3-РДС-ШТ,1.2-1.5MJ/m,500°C; 4-РДС-ОС,300°C; 5-РДС-ШТ.

tance of welded joints of steam pipes of 273X36 mm dia made of 12ХIМФ steel using manual arc welding РДС-ШТ with preheat temperature of 300°C and Э-09ХIМФ electrodes.

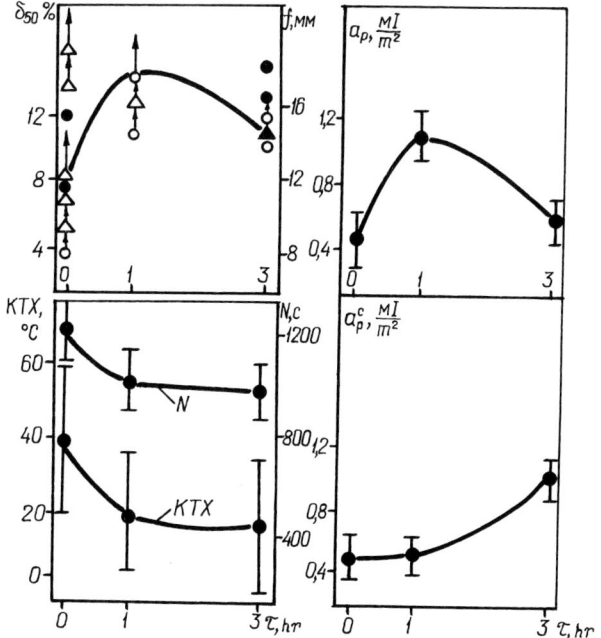

Fig. 8. Effect of tempering time at 730°C on resistance to failure of 12ХIМФ steel joint zones with weld metal 09ХIМФ :(a) - crack resistance (f,δ_{50}) of near-weld zone and HAZ during static bending at 600°C(Fig.3); РДС-ШТ, 4MJ/m, 300°C; РДС-ОС, 8MJ/m, 300°C; (b) - cold resistance brittleness critical temperature-KTX and low-cycle fatigue at 565°C weld metal; РДС-ШТ, 3.6MJ/m, 300°C; (a_p)-specific energy for development of failure of weld from the initial state (a_p) and after ageing at 565°C for $5 \cdot 10^3$ hrs (a_p^c) during static bending at 20°C; РДС-ШТ;: 3.6MJ/m, 300°C.

CONCLUSIONS

Fusion welding technology with postweld heat treatment substantially influences K^{HV} of 12ХIМФ steel joint as applied to steam pipes of 20-60 mm thick walls.

K^{HV} coefficient which is the ratio between the hardness of weld and base metals is considered as workability of welded joints. For steam pipe welded joints made of Cr-Mo-V steel the maximum recommended coefficient of K^{HV} is ≤ 1.8.

Reduction of KHV is obtained by using pulsed heat input welding instead of stationary arc welding, with the self-tempering effect in the former case. Single-pass welding with slit gaps is the most advantageous in this respect.

Reduced time of 730°C, 1 hr tempering is shown possible for joints of I2XIMΦ steel with 09XIMΦ weld metal.

REFERENCES

Stationary steam and water-tube boilers and steam and hot water pipes. Norms for strength calculations. OCT 108.031.02-75.TzKTI,1977, 107.

Zemzin V.N., and R.Z.Shron (1978). Heat Treatment and Properties of Welded Joints. Mashinostroyenie, 367.

Structural designing of welded components of power machines. PTM 24.940.08-74. TzKTI, 113.

Shron R.Z., Zemzin V.N., Mazel R.E., et al. Workability of steam pipe welded joints. Teploenergetika,1981, 11, 5-6.

Mazel R.F.(1966). The nature of loss of strength of thick-walled steam pipe welded joints made of I2XIMΦ steel. Teploenergetika, 4, 17-22.

The analysis of in-service failure of steam pipe welded joints made of heat-resistant pearlitic steels. Regulating instructions (PY) (1981). TzKTI, 45, 34.

Technical Regulating Instructions for Welding, Heat Treatment and Inspection of Boiler Systems and Pipes During Erection and Repair of Thermal Power Plant Equipment (PTM-IC-81)(1975). Energy, 272.

Rykalin N.N.(1951). Thermal calculations during welding.Mashgiz, 296.

Methodological instructions for using scales for structures and hardness in evaluating quality and workability of welded joints made of I2XIMΦ and I5XIMΦ steels (1972).STzNTI ORGRES, Energonot, 9.

Khromchenko F.A., V.D.Zalinova, B.M.Gruzdev, V.L.Melik-Shakhnazaryan (1984). Welding Production, 6, 13-15.

Petterson R.(1977). Coefficients of Stress Concentration. Mir Publishers, 302.

Khromchenko F.A., A.E.Anokhov, V.D.Zalinova, et al (1984). Structure and properties of welded I2XIMΦ steel joints with slit gaps. Welding Production, 8, 32-34.

IV.5

Comparison of the Residual Stress Distributions after Stress-relief Annealing of Welded Sheets of the High Strength Structural Steel ST E 690 at Different Temperatures

J. Heeschen* and H. Wohlfahrt**

*L.u.C. Steinmüller GmbH, Gummersbach, FRG
**Institute für Werkstofftechnik, Universität Gesamthochschule Kassel, FRG

ABSTRACT

Longitudinal and transverse residual stresses in sheets of the high strength structural steel StE 690 have been measured by means of X-ray diffraction after TIG-welding and after subsequent stress-relief annealing. Distributions of longitudinal residual stresses along a line perpendicular to the seam showed distinct maxima and minima. Characteristic differences of the remaining distributions of longitudinal residual stresses have been found after annealing at a relatively low temperature (400°C) and at a relatively high temperature (615°C). The reasons for these characteristic differences are discussed.

KEYWORDS

Residual stresses due to welding. Stress-relieving heat treatment. Optimisation of postweld annealing. Reduction of peak stresses. Measurement of residual stresses by means of X-rays.

INTRODUCTION

It has to be assumed that the reduction of residual stresses due to postweld annealing depends not only on the kind and the yield strength of the base and the filler material but also on the residual stress state, that is to say on the distribution and on the multiaxiality of the residual stresses. Already published experimental data on the relief of residual stresses due to post-weld annealing support this assumption, but nevertheless additional experimental results would be desirable.

After heating to 600°C within 20 hours Ueda and co-workers (1977a) have found zero residual stresses at the weld surface of a 200 mm thick plate of a Mn-Mo respectively a Cr-Mo steel which was submerged arc welded. Appreciable magnitudes of the residual stresses remained however in deeper layers, where residual stress peaks have been present before annealing. Gott (1977) has published a similar result. Residual stresses as high as the yield strength of the material have been measured in deeper layers of a multilayer welded plate (172 mm thick) of a reactor pressure vessel steel after a 5 hours annealing at 620°C. Contrasting with these results are the findings of Mang, Buczak and Steidl (1984) on butt welded rectangular tubes of the steel St52-3 with a wall thickness of 4 mm. Half an hour postweld annealing at 520°C was sufficient in this investigation to reduce peak residual stresses of +200 N/mm^2 and -250 N/mm^2 to values which were not different from zero within the boundaries of error.

Obviously more detailed information how residual stress distributions are changed due to postweld annealing should especially include better knowledge at which annealing temperatures and times the reduction of residual peak stresses becomes really pronounced. Such information is particularly necessary for structural steels with extreme ultimate tensile strengths as, on the one hand, residual stresses may have strong detrimental effects in high strength materials and, on the other hand, postweld annealing at distinct temperatures can result in an embrittlement of these steels. In the following paper experimental results are described on the change of complete residual stress distributions at the surface of high strength structural steels. The findings on annealing temperature and time necessary for a considerable peak stress reduction are discussed on the basis of the possible mechanisms for the thermal relief of residual stresses.

MATERIALS AND EXPERIMENTAL PROCEDURE

Plates of the steel StE 690 in the quenched and tempered condition with TIG dummy seams were used for the investigations. Table 1 shows the chemical composition of the steel, Table 2 indicates the strength and toughness values and in Table 3 values of the hot yield strength for this type of steel at different temperatures are listed, which have been taken from Degenkolbe (1968).

TABLE 1   Chemical Composition of the Investigated Steel (Weight-%)

C	Si	Mn	P	S	Ni	Cr	Mo	Cu	B	Ti	Al	Nb	Fe
.169	.71	.88	.011	.008	0.05	.73	.33	.09	.0007	.02	.045	.00	97

TABLE 2   Strength and Toughness Values of the Investigated Steel

Ultimate Strength N/mm^2	Yield Strength N/mm^2	Fracture Strain %	Notch Toughness at +20°C J	-50°C J
825	725	17	115	95

The dimensions of the plates were 110 x 90 x 10 mm^3. After the grinding of the surfaces all plates have been stress-relief annealed for 1 hour at 590°C. Pulsed arc TIG welding without a filler material has been used for the dummy seams on one surface of the plates. The direction of the seams was parallel to the longer side of the plates. The heat input was kept constant at 19,8 kJ/cm for all weldments.

TABLE 3 Hot Yield Strength of the Steel StE 690 from Degenkolbe (1968)

Temperature	20	100	200	250	300	400	500
Hot Yield Strength N/mm^2	722	700	680	670	660	590	500

After the welding operation the residual stress components parallel and transverse to the seam have been measured by means of X-rays along a line perpendicular to the seam at the surface of each plate. Then each plate was subjected to another annealing temperature or time. Table 4 represents the different annealing conditions. X-ray measurements of the residual stress components at the surface of each plate followed again the postweld heat treatments.

TABLE 4 Conditions of Stress-Relief Annealing

Annealing temperature in °C	400	400	520	520	520	615	615
Annealing time in minutes	5	90	5	90	270	5	90
Time for heating up in hours	2	2	2,5	2,5	2,5	3	3
Time for cooling in hours	4,5	4,5	5,5	5,5	5,5	7	7

The residual stress measurements by means of X-rays were performed with Cr-K$_\alpha$ radiation on a diffractometer with a ψ-set up. Irradiated areas of 1,5 x 6 mm^2 respectively 1,5 x 1,5 mm^2 were applied for the measurement of the transverse respectively the longitudinal residual stress components. According to the sin$^2 \psi$ - method the residual stresses have been evaluated from the shift of the diffraction lines of the {211} - lattice planes. The calculations were based on usual mechanical values of Youngs modulus and Poissons ratio.

EXPERIMENTAL RESULTS

Each of the following figures 1 to 6 compares the residual stress distributions after welding and after 90 minutes of annealing at one of the annealing temperatures 400°C, 520°C and 615°C. Fig. 2 and Fig. 5 additionally show residual stress distributions after an annealing time of 270 min at 520°C. Fig. 1 to 3 represent the longitudinal, Fig. 4 to 6 the transverse residual stresses at the surface. Residual stresses have been measured on both sides of the weld centre line, but only on one side measurements were also performed at bigger distances from the weld centre line. The stress distribution curves continue on the side with less measuring points as dashed lines under the assumption of symmetry to the weld centre line.

As described elsewhere (Wohlfahrt, 1986; Heeschen, Nitschke, Wohlfahrt,1987) the rather complicated residual stress patterns are a consequence of the combined effect of shrinkage and transformation stresses. In view of these complex processes which are connected with the development of the residual stresses it seems natural that their distributions after welding exhibit somewhat different maximum magnitudes in the different plates. Nevertheless the characteristic features of the residual stress patterns are the same in

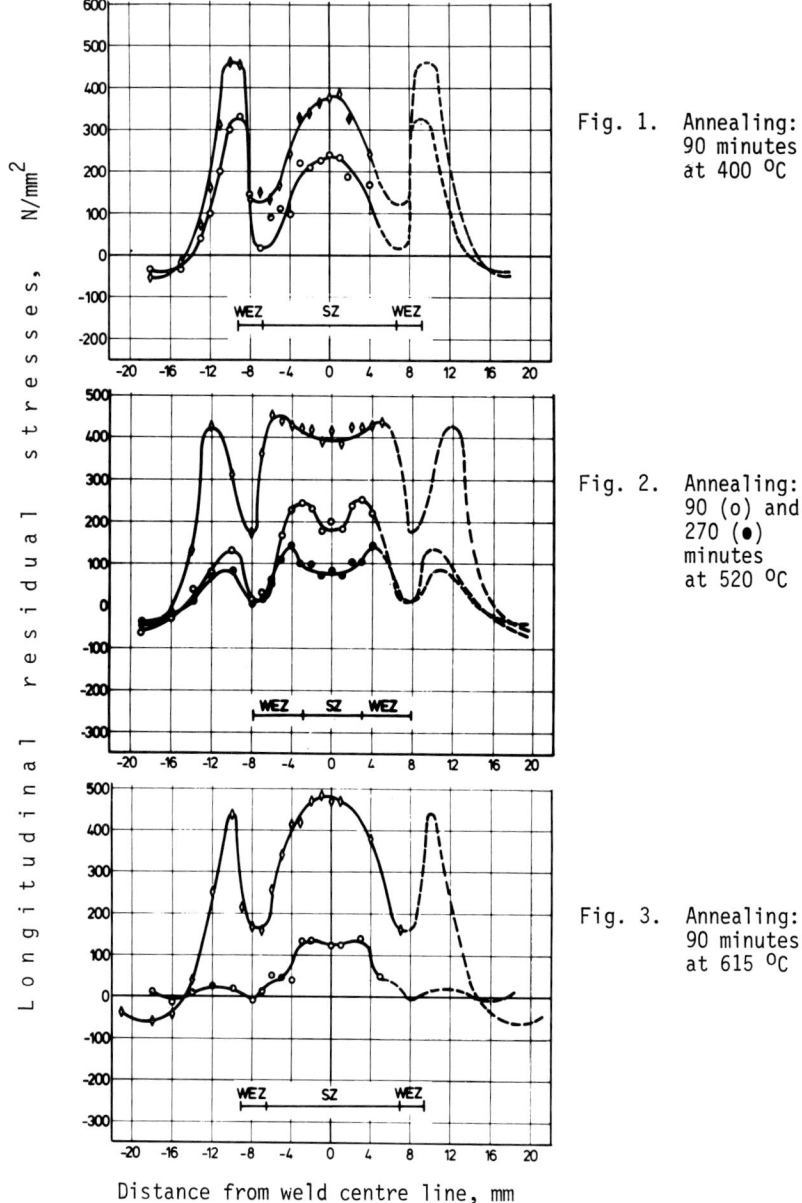

Fig. 1. Annealing: 90 minutes at 400 °C

Fig. 2. Annealing: 90 (o) and 270 (●) minutes at 520 °C

Fig. 3. Annealing: 90 minutes at 615 °C

Fig. 1 to 3. Distributions of longitudinal residual stresses versus distance from weld centre line.
◊ residual stresses after TIG welding.
o,● residual stresses after TIG welding and annealing at different temperatures.

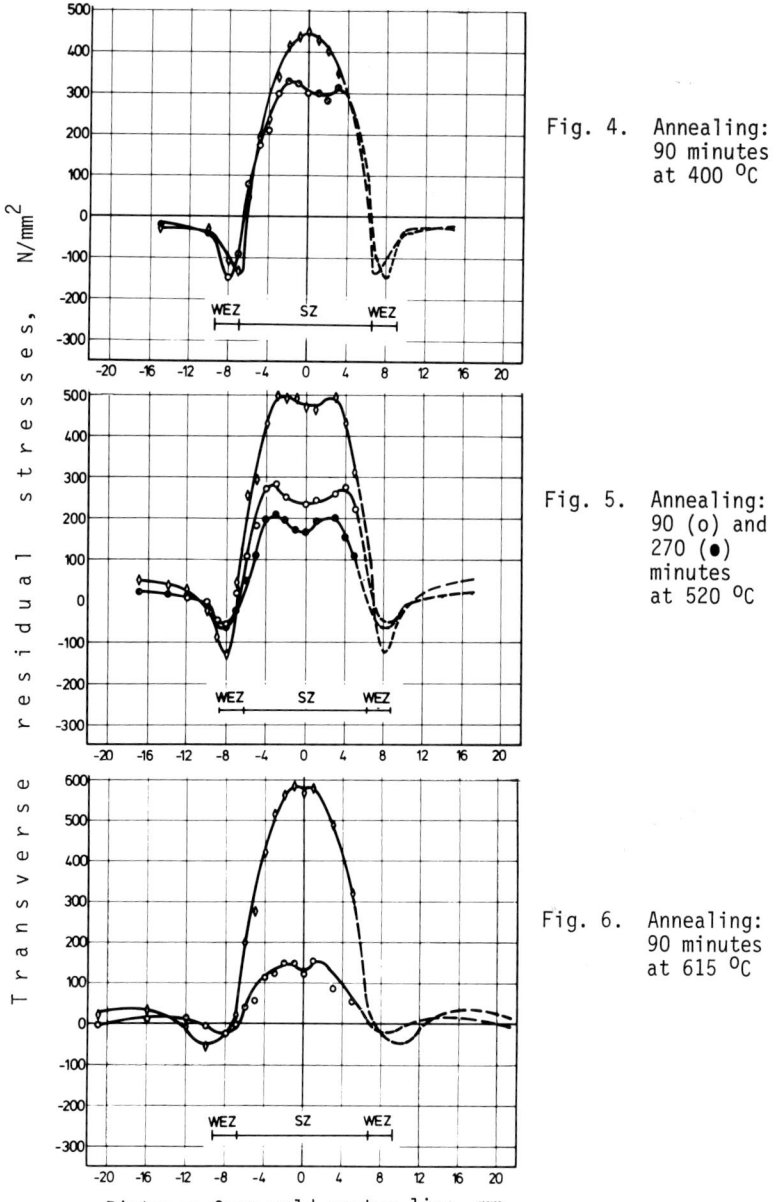

Fig. 4. Annealing: 90 minutes at 400 °C

Fig. 5. Annealing: 90 (o) and 270 (●) minutes at 520 °C

Fig. 6. Annealing: 90 minutes at 615 °C

Fig. 4 to 6. Distributions of transverse residual stresses versus distance from weld centre line.
◊ residual stresses after TIG welding.
o,● residual stresses after TIG welding and annealing at different temperatures.

each plate. The variations of the residual stress distributions and particularly the variations of their maxima with increasing annealing temperature shall be depicted in the following.

After annealing at 400°C all longitudinal residual stress values are reduced by nearly the same amount of 120 N/mm² to 150 N/mm² (Fig. 1). The original stress maxima are reduced only slightly more than the original stress minima. Thus the character of the residual stress distribution remains unchanged at this annealing temperature: the stress peak in the base material is still the highest one after annealing. The reduction of the transverse peak stresses (150 N/mm²; Fig. 4) is almost equal to the reduction of the longitudinal peak stresses, but it is difficult to determine the stress reductions due to annealing at 400°C in the steep flank of the distributions of transverse residual stresses.

At an annealing temperature of 520°C the maximum of longitudinal residual stresses in the base material is definitely reduced more than the residual stress peak in the weld seam (Fig. 2). In addition it should be mentioned that after an annealing time of only 5 minutes at 520°C the residual stress peak in the base material had already reached the same value as after an annealing time of 90 minutes. The transverse residual stresses which have been measured at the surface of the weld seam were higher than the longitudinal residual stresses after welding and remained a bit higher after annealing at 520°C.

At an annealing temperature of 615°C the residual stresses in the base material and also in the weld seam material are still more lowered. Whereas the residual stresses in the base material become nearly equal to zero, the weld seam shows still tensile residual stresses of more than 100 N/mm². Again it has to be added that already after an annealing time of 5 minutes the residual stress distribution was almost the same as after the annealing of 90 minutes at this temperature. It is worth to note that now after the annealing at 615°C the maximum of the transverse residual stresses is reduced to the same value as the maximum of the longitudinal residual stresses. At the other annealing temperatures the maxima of the transverse residual stresses have been higher than the comparable longitudinal residual stress maxima.

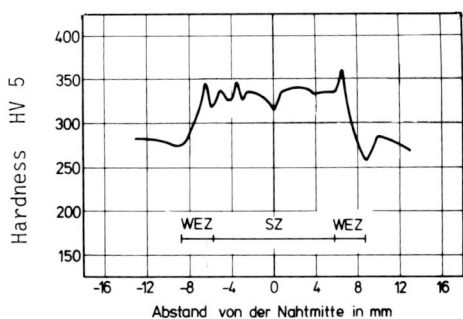

Fig. 7 Hardness distribution in the as-welded state

Fig. 7 represents the hardness distribution after welding. Obviously the hardness values - and consequently the strength values - are higher in the weld seam than in the base material.

DISCUSSION

As described the peak stresses in the weld seam and in the base material are lowered with increasing annealing temperature in a different way. This fact indicates that the relief of residual peak stresses during annealing cannot be based on an uniform mechanism. Literature offers information on two possible mechanisms as reasons for the thermal stress relief. Firstly the thermal relaxation of residual stresses has to be considered, which is a function of the time at elevated temperatures and of the original residual stress magnitude. Secondly a reduction of residual stresses can arise due to the decrease of the yield strength of the material with increasing annealing temperature. All those stress magnitudes must be reduced which are higher than the hot yield point at the annealing temperature. The stress relief due to this effect should occur instantaneously without any time dependence. A paper of Ueda and Fukuda (1977) distinguishes between these two mechanisms and older investigations of Lange (1967) indicate the efficiency of both mechanisms. The investigations of Lange deal with the relief of torsional load stresses during annealing. A pronounced stress reduction was observed already during the slow heating to the annealing temperature. This stress reduction was probably independent of the heating time and could be connected with the decrease of the hot yield point. The following gradual stress relief during annealing at a constant temperature consequently had to be attributed to the stress relaxation mechanism.

The experimental results of this paper can also be explained satisfactorily with the effect of the two stress relieving mechanisms and vice versa the experimental results offer a better understanding of the efficiency of both mechanisms.

At an annealing temperature of $400^\circ C$ the peaks of the residual stresses which were originally present, remain well below the hot yield point of the base or weld material. The reduction of the residual stresses at this temperature is solely a consequence of the stress relaxation mechanism. The amount of stress reduction increases with increasing annealing time (5 minutes and 90 minutes) and the higher original stress values are reduced a bit more than the lower ones. Thus - as already mentioned - the character of the residual stress distribution is not altered after annealing: the differences between maxima and minima remain nearly constant.

In order to explain the more pronounced reduction of the residual stress peak in the base material at an annealing temperature of $520^\circ C$ one has to assume that at this temperature the hot yield point of the base material is already lower than the magnitude of the stress peak. (The values of the hot yield point of a steel of the applied type in Table 4 neither support this assumption strongly nor disagree with it too much). Then, both stress relieving mechanisms - reduction of the hot yield point below the stress value and stress relaxation - are effective in the base material and the residual stresses can decrease appreciably more in the base material than in the weld. As the weld zone shows a higher hardness and hence probably a higher hot yield strength than the base material, it seems reasonable to assume, that the hot yield point of the weld material at $520^\circ C$ is still above the original peak stress in the weld and that only the relaxation mechanism is effective in the weld at 520°C. It has also to be taken into consideration that the state of the residual surface stress at a distance between 10 and 12 mm from the weld centre line is nearly uniaxial (compare Fig. 1 to 3 with Fig. 4 to 5). The residual stress state at the weld surface is biaxial. This difference may also contribute to the bigger amount of stress relief in the base material at 520°C annealing temperature. The fact that the stress reduction in the base material is almost the same after 5 minutes of annealing as after 90 minutes supports the given explanation strongly. The additional decrease of residual stresses after a prolonged annealing time of 270 minutes (Fig. 2 and Fig. 5) has to be attributed completely to the stress relaxation effect.

Finally one can assume that at an annealing temperature of 615°C the hot yield point of the weld material becomes lower than the peak stress in the weld and therefore the stress reduction in the weld zone is also promoted.

CONCLUSION

The experimental results of this investigation support the idea that during the postweld heat treatment of a high strength structural steel two stress-relieving mechanisms can be effective. The thermal stress relaxation occurs already at relatively low temperatures - for instance 400°C or even lower - and reduces high magnitudes of resiual stresses only somewhat more than low magnitudes. Therefore - with annealing times long enough - low residual stress values can decrease down to zero due to this stress relieving mechanism at relatively low annealing temperatures, but no homogenization of the distribution of residual stresses can be achieved within reasonable times, as the peak stresses are not reduced enough. At higher annealing temperatures - for instance 520°C and higher - the hot yield point of the material may fall below the peak stress values. Then the peak stresses decrease more drastically, that is to say for a pronounced reduction of peak stresses the higher annealing temperatures are recommendable.

REFERENCES

Degenkolbe, J. (1968). Schweizer Archiv Jan. 1968, 2 - 18.
Gott, K.E. (1977). Proc. Conf. on Residual Stresses in Welded Construction and Their Effects. The Welding Institute, Abington Hall, Cambridge 1977, 259-265.
Heeschen, J.,T. Nitschke and H. Wohlfahrt (1987). Proc. Int. Conf. on Residual Stresses. DGM-Informationsgesellschaft Verlag, Oberursel, to be published in 1987.
Lange, G. (1967). Schweißen u. Schneiden 19 (1967) 361-364.
Mang, F., Ö. Buczak and G. Steidl (1984). IIW-Doc. XIII-1122-84.
Ueda, Y. and K. Fukuda (1977). Trans. Japan Welding Soc. 8 (1977) 19-25.
Ueda, Y., K. Fukuda and K. Nakacho (1977). Proc. Conf. on Residual Stresses in Welded Construction and Their Effects. The Welding Institute, Abington Hall, Cambridge 1977, 27-37.
Wohlfahrt, H. (1986). Härterei-Techn. Mitt. 41 (1986) 248-257.

IV.6

Influence of the Welding Technology and the Stress-relieving Heat Treatment on the Corrosion Cracking Resistance of Welded Nitrogen-alloyed Stainless Steel

L. Kalev, V. Mihailov and A. Krustev

Institute of Metal Science and Technology, Bulgarian Academy of Science, Sofia, Bulgaria

ABSTRACT

Stress corrosion cracking tests were conducted on specimen with simulated HAZ, prepared from nitrogen-alloyed stainless steel, in order to study the effect of separate welding technology factors, including stress-relieving heat treatment, on the corrosion cracking resistance. The stresses in the welded joints were assessed by the finite elements method. As a result, a theoretical-experimental method is proposed for the assessment of the influence of the welding parameters, stress-relieving heat treatment and the level of the residual and working stresses upon the corrosion cracking resistance of welded joints from nitrgen-alloyed, nickel-free stainless steel.

KEYWORDS

Stainless steel; nitrogen; stress corrosion cracking; heat treatment; finite elements method; stress intensity.

INTRDUCTION

One of the most significant problems, encountered in the welding of stainless steels, is the preservation of their corrosion resistance in course of exploitation. Well known fact (Gooch, 1984) is that the corrosion cracking of welded stainless steels is due mainly to the residual tensile stresses and the precipitations at grain boundaries, occurring as a result of the thermal influence of the welding cycle. The practical use of austenitic nitrogen-alloyed, nickel-free stainless steels is complicated by the fact, that their structural state is unbalanced at room temperature and the thermal welding cycle can induce carbide-nitride precipitation along grain boundaries (Andreev, 1978). In the course of industrial exploitation of a welded construction, the stress, induced by an external load, combine with the welding stresses, thus activating the corrosion process. The possible ways to secure a good corrosion resistance are either reduction of the length of the thermal cycle, or reduction of the level of the tensile stresses by a proper heat treatment. The aim of this work is the comparative model investigation of the influence of separate technology parameters on the corrosion resistance of separate welded joints from nitrogen-alloyed austenitic stainless steel.

EXPERIMENTAL PROCEDURE

The inspection of corrosion cracking, observed in welded constructions from 4 mm nitrogen-alloyed stainless steel indicated, that cracking appears at a distance of 5-7 mm from the center of the bead. The joints are usually welded in two passes, without bevelling, the heat input varying from 350 to 516 KJ/m, depending on the electrode diameter and the welding gap. From a corrosion resistance viewpoint, it is of interest the parallel between the stress corrosion resistance of the base metal and the metal undergone a welding thermal cycle.

TABLE 1 Composition and properties of the used steel

Chemical composition				
C	Cr	Mn	Si	N
0.06	14.33	15.12	0.76	0.22

Thermal and mechanical properties											
Parameter	Temperature, °C										
	20	100	200	300	400	500	600	700	800	900	1000
λ	15.1	40.6	40.3	42.0	46.6	51.5	54.7	58.5	61.8	64.0	-
a	3.9	10.4	10.3	10.5	10.3	9.5	8.6	7.8	7.0	6.1	-
σ_{02}	470	460	430	390	340	280	205	200	150	95	62
E	20.7	20.2	19.3	18.3	17.5	16.8	16.1	14.9	13.7	-	-
ε^T	3.0	15.9	36.2	58.8	82.0	107.5	133.8	158.9	189.6	-	-
γ	0.3										

λ, W/(m.°C) - thermal conductivity coefficient
a x 10^6, m²/s - temperature conductivity coefficient
σ_{02}, MPa - yield strength
E x 10^{-4}, MPa - modulus of elasticity
ε^T x 10^4 - temperature deformation
γ - Poisson's coefficient

In this paper the authors compare the corrosion resistance of a zone, situated at a distance of 6 mm from the bead center with that of the base metal. The composition and the properties of the used steel are given in Table 1. As it was mentioned above, the thermal cycle is one of the main reasons for the loss of corrosion resistance. That's why the welding thermal cycles in the dangerous zone were registered with

the help of thermocouples Ni - Cr/Ni, 0.5 mm diameter, for the two
boundary values of the heat input.

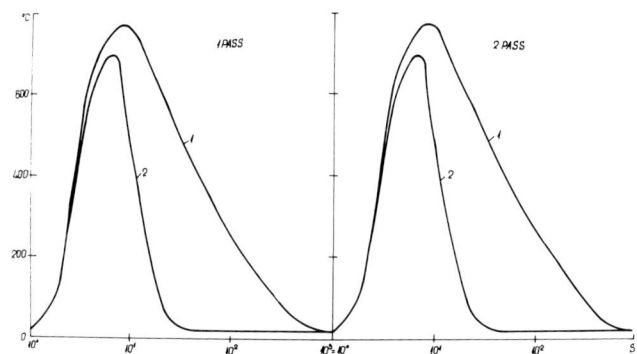

Fig. 1. Registered thermal welding cycles at a distance
of 6 mm from the weld bead center
1 - heat input 516 KJ/m
2 - heat input 350 KJ/m

The registered thermal cycles, Fig. 1, were simulated on test specimen
with the help of thermal cycle simulator, SMITWELD. Dimensions of the
test specimen are 150x15x4, and their chemical composition is given in
Table 1. The final shape and dimensions of the test specimen after a
mechanical preparation are given on Fig. 2.

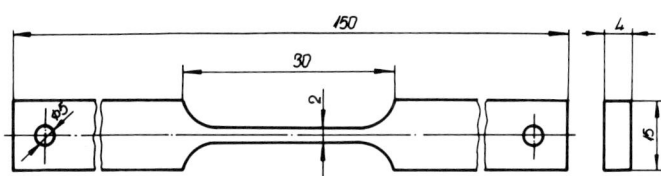

Fig. 2. Test speciman

The stress corrosion cracking test was conducted in a 45 % boiling
solutin ($154^{\circ}C$) of $MgCl_2$. The time for cracking at different stresses
was recorded.

THEORETICAL MODEL FOR THE STUDY OF THE STRESSED CONDITION

To assess the corrosion resistance of a real welded joint it is necessary to know the stressed condition. The most detailed picture of the thermo-mechanical process in the welding can be achieved by using the latest achievements of the calculating mathematics, combined with powerful computers. Several algorythms are known, solving the problem of the stressed and strained condition in the welded joints (Gatovskii, 1980; Mahnenko, 1976; Vershinskii, 1982).

In this work, the algorythm of Mahnenko (1976) was used, but adapted for the finite elements method (Gatovskii, 1980). It is based on the theory of the non-isothermic plastic flow. The transition of the ideal elastic-plastic material from elastic into plastic condition is carried out according the terms of constancy of stress intensity, i.e. Mises' term. In the welding, the influence of creep upon the stressed condition is insignificant, even in the area of the high temperatures. As the loading is active, the instant plastic strain is leading, compared with the diffusion plasticity. This is the usual reason to ignore the role of the relaxation processes. Due to lack of creep data, it is considered, that the low temperature stress-relieving heat treatment only leads to redistribution of the residual stresses.

Generally, the thermo-mechanical problem is related to the non-linear problems class. This is determined by the physical and geometrical non-linearities. The non-linearity of the connection between stress and strain is accomplished with the help of iteration process, based on the variable susceptibility method (Gatovskii, 1980; Pisarenko, 1981). The relatively low change of geometry is characteristic for the welding problems. That's why it can be adopted, that at each stage of the research, the geometric nonlinearity remains constant and is determined by the final condition of the previous stage.

The main features of the algorythm for the solving of the flat thermo-elastic-plastic problem by the finite elements method are:
1) The examined cross-section is approximated (manually or automatically) with the finite number two-dimensional symplex elements.
2) A non-linear, non-stationary thermal problem is solved, according to the method, described by Kalev (1986).
3) The thermo-elastic problem is solved by the shiftings method (Gatovskii, 1980; Segerlind, 1976).
4) The physical and geometrical non-linearities are considered with the help of the iteration procedure, described by Gatovskii (1980).
5) The components of the stresses and strains in the junctions of the elements are determined with the help of the coordinated results theory (Segerlind, 1976).
6) The stress and strains fields are formed and printed.
Thus stage by stage, beginning from the second point of the algorythm, a thermo-mechanical problem is solved, following the kinetics of temperature, stress and strain.

The described theoretical model was used to study the stressed condition in unspecified construction under the following terms: two plates, shown on Fig. 3, dimensions 200x100x4 mm, are welded both sides, with full penetration. The second bead is welded after the complete cooling of the plates. One plate only is considered, due to the symmetry. It is supposed, that the rest of the construction doesn't influence the stresses and strains field in the considered area.

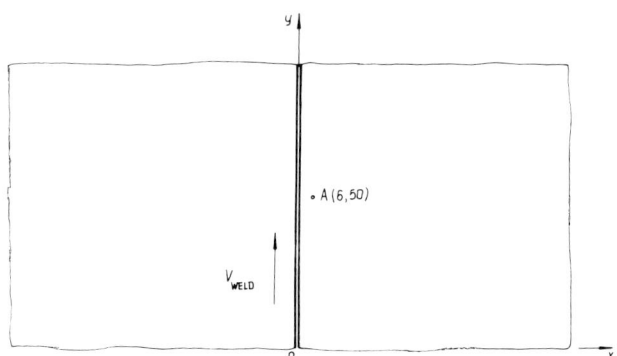

Fig. 3. Scheme of the welded joint

This means, that the deformation of the plate during welding is free and is carried out only under the influence of the thermal deformations, Fig. 4.1.

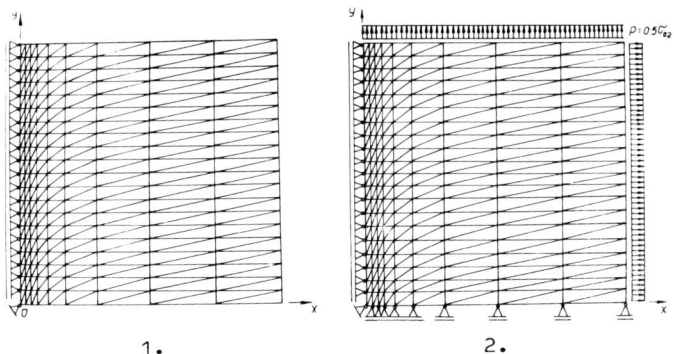

Fig. 4. 1. Scheme for calculating the welding stresses
2. Scheme for applying the working stresses

After welding, a low-temperature stress-relieving heat treatment is carried out. Imitation of the stressed condition in the course of exploitation is achieved by applying working stresses, equal to 1/2 of the yield strength, Fig. 4.2. It is supposed, that in the course of solving the thermo-elastic-plastic problem, the hypothesis of the stressed plane condition is valid.

RESULTS AND DISCUSSION

The experimental results from the stress corrosion cracking test are shown on Fig. 5. The influence of the welding cycle and the stress level upon the corrosion resistance of the test specimen is marked. The tendency towards corrosion cracking in the thermal treated specimen equalizes that of the base metal at stresses lower than 300 MPa.

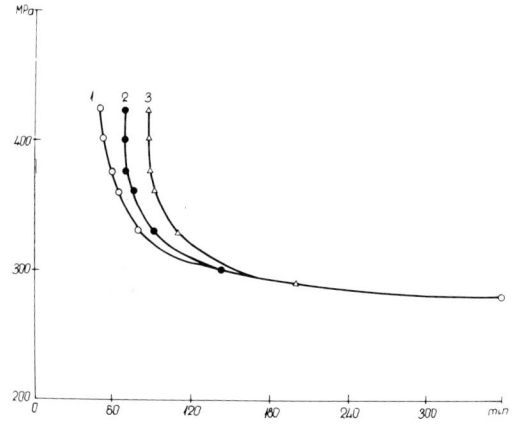

Fig. 5. Results from stress corrosion cracking test
1 - thermal cycle (1) from Fig. 1.
2 - thermal cycle (2) from Fig. 2.
3 - base metal

According to Pisarenko (1981), one of the often used criteria to compare the results of the one-, two- or three-dimensional tensile stress tests is the stress intensity. Figure 6 illustrates the theoretically determined stress intensities in point A, shown on Fig. 3., depending on the temperature of the heat treatment. The calculated values of the stress intensities for the two heat inputs differed in the boundaries of the assigned accuracy - 5 MPa.

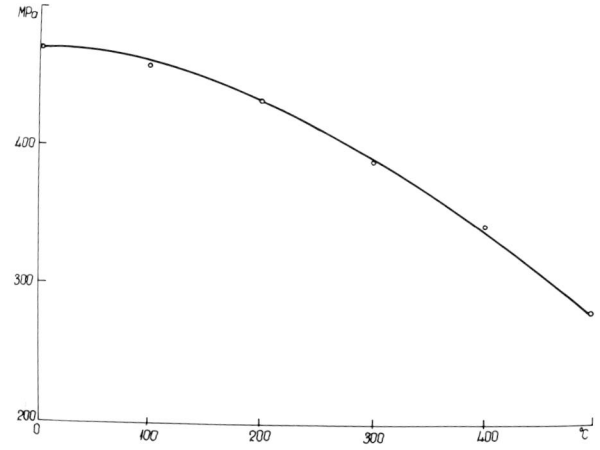

Fig. 6. Dependence of stress intensity in point A on the temperature of the stress relieving heat treatment

It is evident from Fig. 5 and Fig. 6 , that a stress relieving heat treatment at temperatures of 450 - 480 °C would secure a good resistance of the welded joint against stress corrosion cracking, as the stress intensity is lower than the critical, and according to Andreev (1978), at temperatures lower than 550 °C, the danger of precipitation at grain boundaries is minimum for this grade of steel.

CONCLUSION

Theoretical - experimental method is proposed for the assessment of the influence of the welding parameters, stress relieving heat treatment and the level of the residual and working stresses upon the corrosion cracking resistance of welded joints from nitrogen alloyed, nickel-free stainless steel.

REFERENCES

Andreev, C., R. Kovacheva, and L. Djambazova (1978). Otdeljane i mejdukristalna korozia pri izotermichno otgrjavane na hrom-manganova stomana s visoko azotno sadarjanie. Materials Science and Technology, 6, 35-40.
Gatovskii, K. M., and V. A. Karhin (1980). Teoria svarochnih deformacii i naprejenii. Leningrad.
Gooch, T. G. (1984). Stress corrosion cracking of welded austenitic stainless steel. Welding in the World, 3/4, 64-76.
Kalev, L., V. Mihailov, and V. Ushev (1986). Izpolzvane metoda na krainite elementi za presmjatane na temperaturni poleta pri mnogoslojno zavarjavane/navarjavane na stomani. Tehnicheska misal, 1, 107-112.
Mahnenko, B. I. (1976). Raschetnie metody issledovanja kinetiki naprejenii i deformacii. Naukova dumka, Kiev.
Pisarenko, G. S., and N.S. Mojarovskii (1981). Uravnenia i kraevie zadachi teorii plastichnosti i polzuchesti. Naukova dumka, Kiev.
Segerlind, L. J. (1976). Applied Finite Element Analysis. John Wiley and Sons, Inc., New York.
Vershinskii, A. V., and J. N. Kasymbekov (1982). Prilojenie metoda konechnih elementov pri raschete ostatochnih naprejenii i deformacii v svarnih kranovih metallokonstrukciah. Trudy MVTU, 371, 3-22.

IV.7

Relaxation of Residual Stresses During Postweld Heat Treatment of Submerged-arc Welds in a C-Mn-Nb-A1 Steel

R. H. Leggatt

The Welding Institute, Abington, Cambridge, UK

ABSTRACT

A study has been made of the effects of postweld heat treatment (PWHT) on residual stress levels in submerged-arc welds in a 50mm thick C-Mn-Nb-Al parent steel. Measurements were made of the surface and internal residual stresses in the welds after the application of PWHT at different temperatures and hold times, and for repeated PWHT cycles. Supporting measurements were made of the relaxation properties of all-weld metal tensile specimens using isothermal and anisothermal stress relaxation test procedures.

The stress relaxation tests showed that the weld metal was more resistant to stress relief and that its relaxation properties were more variable than for typical C-Mn parent steels. The maximum longitudinal residual stresses were up to $55 N/mm^2$ greater than the upper bound of the weld metal stress relaxation data for corresponding PWHT procedures, whilst the maximum transverse residual stresses were fairly low and were insensitive to the heat treatment conditions. The results suggest that residual stresses parallel to the welding direction after PWHT may be significantly greater than those predicted to occur using currently accepted methods.

KEYWORDS

Carbon manganese steels; stress relieving; residual stresses; weld metal; stress relaxation tests; post weld heat treatment

INTRODUCTION

Postweld heat treatment (PWHT) is usually applied to welded steel structures with section thickness in excess of 35 to 50mm (depending on the application) and sometimes in thinner section structures where a high degree of defect tolerance or resistance to stress corrosion cracking is required. The aims of PWHT are to relieve the high level of residual stresses induced by the welding process, and to modify the microstructure and mechanical properties in the weld metal and heat affected zone (HAZ).

Suitable PWHT conditions are specified in design codes such as BS5500 and the ASME Boiler and Pressure Vessel Code for a range of steels used in pressure vessel construction. These codes do not give any indication of the degree of stress relief which can be expected. The usual method for quantifying the level of residual stresses associated with a specific PWHT procedure is on the basis of stress relaxation tests. These are tests in which the stress changes in a uniaxial tensile specimen held at constant strain at elevated temperature are monitored over a period of time. Stress relaxation data for a range of pressure vessel steels have been published, although very few data are available for weld metals. The maximum residual stress in a weldment after PWHT is assumed to be equal to that observed in relaxation tests on the relevant materials after the specified hold time and temperature (Barr and Moir, 1970).

This assumption does not appear ever to have been systematically investigated. A small number of papers have been published describing investigations in which residual stresses in weldments after PWHT have been measured in materials for which weld or parent metal stress relaxation data are available (Ferril, Juhl and Miller, 1966; Grotke, Bush and Kromer, 1973; Ueda and co-workers, 1977; Fidler, 1978; Gott, 1979; Larsson and Sandstrom, 1982). Examination of these data shows that the peak residual stresses were usually greater than the expected value based on relaxation data, and that residual stresses up to $75N/mm^2$ greater than the expected value have been recorded. This observation suggests that current methods for estimating the level of residual stresses in weldments subject to PWHT may be significantly unconservative.

The objective of the present study was to conduct a systematic investigation of the relaxation properties of, and residual stresses in, submerged-arc welds deposited in a C-Mn-Nb-Al parent steel and subject to a range of PWHT conditions. The project is described in greater detail in two reports available to research members of The Welding Institute (Leggatt, 1986; Threadgill and Leggatt, 1984).

TEST PANEL MANUFACTURE AND HEAT TREATMENT

The tests described in this paper were carried out on a continuously cast and normalised 50mm thick BS1501-225-490B-LT50 steel. The steel is a C-Mn grade which has been Si-killed and Al and Nb-treated. The welds were deposited in a 2/3:1/3 K-preparation using the submerged arc process. The chemical compositions and mechanical properties of the materials are given in Table 1. The parent plate satisfied the chemical and mechanical requirements of both BS1501-225 and BS4360 Grade 50E.

Seven nominally identical test panels were manufactured. Each was 1500mm long, 485mm wide, and 50mm thick. Four strongbacks were attached to each panel to provide restraint against transverse shrinkage and rotation during welding, and were left on the panels during PWHT and residual stress measurement.

After welding, a 100mm length of material was cut off each panel for the provision of samples for stress relaxation testing. The panels were then subjected to heat treatment in accordance with the schedule given in Table 2. The severity of the heat treatments is expressed in Table 2 in terms of the effective hold time, τ, and the Holloman-Jaffe parameter, H_p (Holloman and Jaffe, 1945). These are defined as follows

TABLE 1 Chemical Composition and Mechanical Properties

Sample	Element weight, %											
	C	S	P	Si	Mn	Ni	Cr	Mo	V	Cu	Nb	Al
Plate	0.14	0.007	0.015	0.39	1.32	0.12	0.03	0.01	<0.002	0.16	0.030	0.024
Wire	0.09	0.013	0.015	0.24	1.69	0.05	0.05	0.01	0.005	0.22	<0.002	<0.003
Weld metal	0.08	0.009	0.016	0.28	1.31	0.11	0.06	0.02	<0.002	0.23	0.007	0.016

Sample	Data source	Yield strength N/mm^2	Tensile strength N/mm^2	Elongation %	Reduction of area, %
Plate	Manufacturer	353	512	29	-
Weld metal, as-welded	Manufacturer	448	520	30	74
Weld metal, after PWHT,					
min	Welding Institute	362	509	33	69
max	Welding Institute	440	558	41	76

$$\tau = \left(t + \frac{T}{2.3K_1 (20-\log_{10} K_1)} + \frac{T}{2.3K_2 (20-\log_{10} K_2)}\right) N \quad [1]$$

$$H_p = T (20 + \log_{10} \tau) \times 10^{-3} \quad [2]$$

where T is the hold temperature (degrees K), t is the hold time (hours), N is the number of heat treatment cycles (if applicable), and K_1 and K_2 are the heating and cooling rates (°C/hour). The effective hold time (Gulvin and co-workers, 1972-3) includes an allowance for time spent at elevated temperature during heating and cooling (if applicable). The Holloman-Jaffe parameter is a function of both time and temperature, and is intended to allow results from specimens subjected to different heat treatment conditions to be plotted on a single graph.

TABLE 2 Postweld Heat Treatment Conditions

Panel	Hold temperature		Hold time, hr	Cycles	Effective time, τ, hr	Holloman-Jaffe, H_p
	°C	K				
W1	600	873	2	X1	2.68	17.83
W2	600	873	2	X2	5.37	18.10
W3	600	873	2	X3	8.05	18.25
W4	550	823	2	X1	2.64	16.81
W5	600	873	1	X1	1.68	17.66
W6	600	873	4	X1	4.68	18.05
W7	650	923	2	X1	2.72	18.86

Heating rate: 30°C/hr
Cooling rate: air cool

ISOTHERMAL STRESS RELAXATION TESTS

Six all-weld metal tensile specimen blanks were supplied to ERA Technology Limited, who carried out the machining of the specimens and isothermal stress relaxation testing to BS3500:Part 6. The specimen dimensions were 9mm diameter and 25mm gauge length. The heating and soaking times are given in Table 3, together with the test results.

TABLE 3 Results of Isothermal Stress Relaxation Tests

Specimen identity	Test temperature °C	Heating time, hr	Soaking time (unloaded), hr	Initial stress, N/mm^2	Total strain %	Hold time, hr	Final residual stress, N/mm^2
W5/2	525	2.5	1.0	318	0.362	21.0	56
W5/4	550	2.0	1.25	273	0.350	21.0	64
W6/2*	575	16.0	1.0	254	0.372	21.0	32
W6/4	600	2.0	1.0	181	0.321	21.0	13
W7/2**	625	2.0	1.5	132	0.305	9.5	0
W7/4**	650	2.0	1.0	115	0.286	11.5	0

* Heating cycle interrupted on W6/2. Time above 500°C was 4 hr only
** Tests ended prematurely on W7/2 and W7/4 because loads fell below stable control level

The sequence of operations was as follows:

1. Load specimen in furnace in 160kN servohydraulic test machine.
2. Raise to test temperature and 'soak' for approximately 1 hour to allow temperature to equalise.
3. Load at 0.2% strain per minute up to 0.2% proof stress.
4. Hold total strain constant by automatic servohydraulic control and monitor load reduction continuously from load cell output of machine.
5. After 21 hours at constant strain (or sooner in two specimens in which the stress had reduced to a very small value at which level strain control was unreliable), unload.
6. Cool to ambient.

ANISOTHERMAL STRESS RELAXATION TESTS

Six all-weld metal tensile specimen blanks were supplied to Alsthom-Atlantique, who carried out the machining of the test specimens and anisothermal stress relaxation tests according to procedures described by Leymonie (1980). The test parameters are given in Table 4. The sequence of operations was as follows:

1. Locate test specimen and unloaded control specimen in furnace in testing machine.
2. Load test specimen to 0.2% proof stress at ambient temperature.
3. Link strain control on test specimen to thermal strain in unloaded specimen.
4. Raise both specimens to PWHT temperature at 30°C/hr. Record stresses at half hour intervals.

5. Hold temperature for 2 hours. Record stresses at start of hold time, half hour later, and end of hold time.

TABLE 4   Results of Anisothermal Stress Relaxation Tests

Test specimen	Control specimen	Heating rate, °C/hr	Test temperature, °C	Hold time, hr	Initial stress N/mm^2	Relaxed stress after hold time, N/mm^2
W5/1	W5/3	30	550	2	400	135
W6/1	W6/3	30	600	2	420	111
W7/1	W7/3	30	650	2	435	11

STRESS RELAXATION TEST RESULTS

The stress relaxation test results are plotted on Figs 1 and 2, and are replotted on Fig.3 as a function of the Holloman-Jaffe parameter, together with previous results by Gulvin and co-workers (1972-3) for a similar parent steel, BS1501-213.

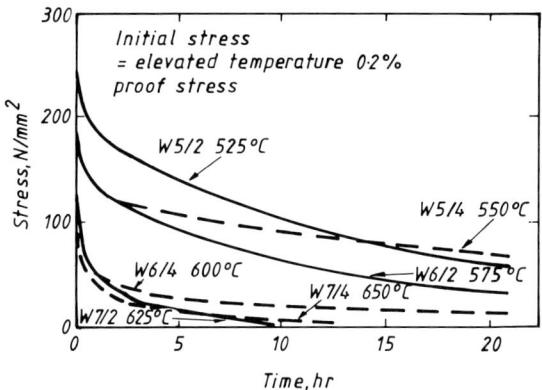

Fig. 1. Isothermal stress relaxation test results

It can be seen that the weld metal was much more resistant to relaxation than the parent steel, and that its relaxation properties were subject to more scatter. However, there were no significant differences between the results from the two test procedures, nor between specimens extracted from root or fill regions of the weld.

RESIDUAL STRESS MEASUREMENTS

Surface residual stresses in weld metal and at the fusion boundaries were determined using the centre-hole rosette gauge technique, using 2mm diameter holes. Between 6 and 14 measurements were made on each panel. The measured surface residual stresses after PWHT are plotted as functions of heat treatment temperature and effective hold time at 600°C in Fig. 4.

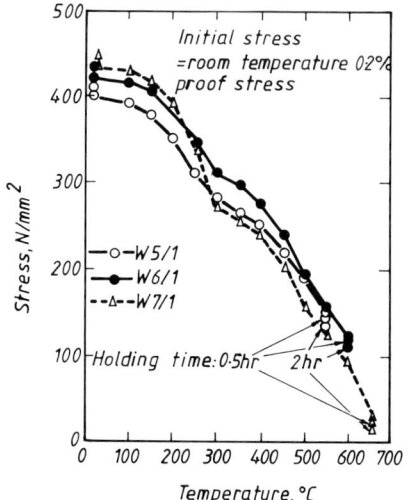

Fig. 2. Anisothermal stress relaxation test results

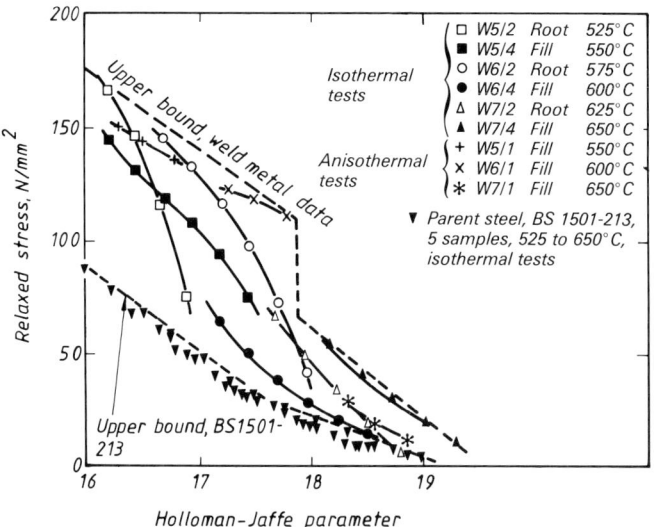

Fig. 3. Relaxed stress v. Holloman–Jaffe parameter

The through-thickness distributions of internal residual stresses in weld metal in the central region of each panel after PWHT was measured by a two-stage sectioning procedure known as block removal and layering. The reaction stresses (i.e. the sum of the membrane and bending components of the through-thickness distribution of residual stresses in the weld) were measured by removing strain gauged blocks of material from the panels. The gauges were attached to the panels whilst the strongbacks were still in place, such that the measured stresses included the reaction stresses acting on the weld due to the presence of the strongbacks. After removal of the relaxation blocks, the remaining non-linear distribution of residual stresses was determined from measurements of the strain changes on the bottom of the blocks when 3mm thick layers of material were removed from the opposite face using a vertical milling machine. The through-thickness distribution of internal residual stresses in Panels W1, W4 and W7 are plotted in Fig. 5.

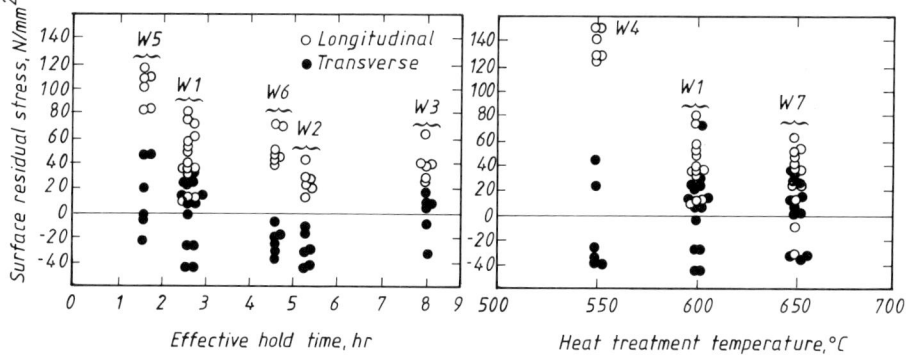

Fig. 4. Variation of measured surface residual stress with heat treatment time and temperature

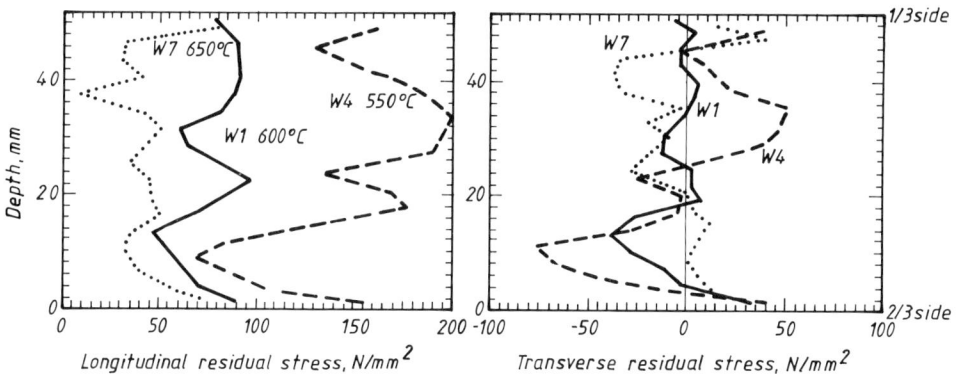

Fig. 5. Internal residual stresses after 2 hours PWHT

DISCUSSION

The maximum residual stresses measured in the directions parallel and transverse to the weld in the welded panels after PWHT are plotted in Fig. 6 as a function of the Holloman-Jaffe parameter. Three different maximum stresses are plotted for each stress direction, namely the maximum value of the measured internal residual stresses, the maximum value of the measured surface residual stresses, and the maximum reaction stress. The latter quantity is a measure of the bulk stresses occurring in the weld ignoring localised peak values. It is equal to the sum of the membrane and bending components of the residual stresses at the weld. The upper bound line for the stress relaxation data shown in Fig. 3 is also plotted in Fig. 6. It can be seen that the maximum measured longitudinal residual stresses were up to 55N/mm² above the line corresponding to the upper bound of the relaxed stresses as measured during the temperature hold period.

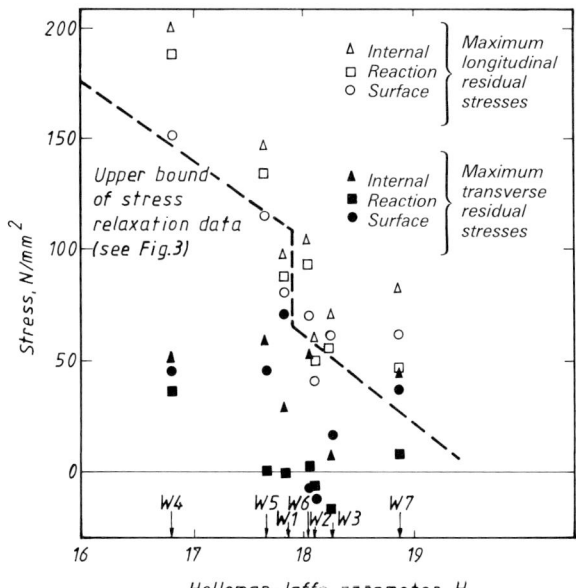

Fig. 6. Comparison of maximum measured residual stresses and upper bound of stress relaxation data

This finding is in line with the previous observations discussed in the Introduction to this paper. Leymonie (1980) has measured stresses in anisothermal test specimens during the cool-down period, and found that they increased. This was presumably due to the increase of Young's modulus with falling temperature, and could account for the discrepancy between measured residual stress and stress relaxation data. Unfortunately, the stress recovery during cool-down was not measured in the present project. It was also observed in the present project that the transverse stresses after PWHT were significantly smaller than the values indicated by stress relaxation testing. This was presumably due to the different restraint conditions acting in the transverse direction, which would be difficult to model in a stress relaxation test. Hence, it may be concluded that direct residual stress measurement is the only suitable experimental method for determining the transverse residual stresses in a weld after PWHT. Ideally, internal residual stress measurements would be carried out in order to determine the peak stresses at interior locations.

In cases where no stress relaxation data are available, it is often assumed that the maximum residual stresses after PWHT may conservatively be assumed to be less than or equal to 20% of the room temperature yield stress of the weld metal. The heat treatment conditions specified in BS5500:1976 for 50mm thick weldments in C-Mn steels are a median hold temperature of 600°C, a hold time of 2.1 hours, and a maximum heating and cooling rate of 137.5°C/hr. The corresponding effective hold time is τ = 2.4hr, and the Holloman-Jaffe parameter is 17.8. Based on the observed maximum stresses plotted in Fig. 6 for the five panels which were heat treated at 600°C, the expected maximum residual stress at H_p = 17.8 is about 130N/mm^2. This represents 29% of the typical weld metal yield stress as-welded. Hence, it appears that the '20% rule' is not conservative. It is believed that this lack of conservatism has arisen because most assessments have been based on the results of stress relaxation tests (and in some cases on parent metal stress relaxation tests), which, as illustrated in the present project, do not give an accurate guide to the maximum residual stresses.

The width of the scatterband of the stress relaxation data in Fig. 3 was between 30 and 80N/mm^2. Similar ranges of data values were observed in the sets of longitudinal and transverse surface residual stress measurements shown in Fig. 4 and in the internal stress distributions shown in Fig. 5. It appears that variability in the relaxation properties of the weld metal contributes to the scatter of residual stresses in welds after PWHT.

There was some evidence that the specimens subjected to repeated PWHT cycles (W2 and W3) had lower maximum residual stresses than the general run of data in Fig. 6. This suggests that there may be some additional effect associated with repeated PWHT, or that the allowance for heating and cooling time in the Holloman-Jaffe parameter (see Eq. [1]), which is more significant in the case of repeated cycles, is inadequate. However, it was felt that the present data was too scattered to provide a reasonable basis for proposing any modification to the Holloman-Jaffe formulation.

CONCLUSIONS

1. The measured maximum residual stresses in the test panels and the measured relaxed stresses in the tensile specimens both showed the expected overall decrease in stress level with increasing heat treatment temperature and hold time.
2. The maximum residual stresses in the welds were always those in the longitudinal direction parallel to the weld length. The measured transverse residual stresses lay mainly in the range ±50N/mm^2, and were insensitive to heat treatment conditions.
3. The maximum measured residual stresses in the test panels were up to 55N/mm^2 greater than the upper bound of the stress relaxation data for corresponding PWHT temperatures and hold times.
4. Based on the present results, the expected maximum residual stress in the weldments under investigation after heat treatment in accordance with BS5500:1976 is 130N/mm^2. This represents 30% of the typical room temperature yield stress of the weld metal in the as-welded condition.
5. Conclusions 3 and 4 above indicate that currently accepted methods for predicting the maximum residual stresses in weldments after PWHT would have produced values which were significantly less than those measured in the present investigation.
6. The relaxed stresses in the all-weld metal test specimens were between two and four times greater than those previously observed in C-Mn parent steels for corresponding PWHT conditions. The weld metal stress relaxation data showed more variability than parent metal data.
7. Within the scatter of stress relaxation test results, similar relaxed stresses were obtained for corresponding hold times and temperatures using either isothermal or anisothermal test procedures. The anisothermal stress relaxation test is also capable of indicating the stress recovery that occurs during cooling after PWHT, though this was not done in the present project but would be desirable.

8. In view of the variability of relaxation properties and their inability to predict the levels of transverse residual stresses, consideration should be given to using direct residual stress measurement as an alternative to stress relaxation testing for determining the residual stresses in weldments after PWHT.

ACKNOWLEDGEMENTS

The work was jointly funded by Research Members of The Welding Institute and the Minerals and Metals Division of the UK Department of Trade and Industry. Thanks are due to Dr. Leymonie, Mme Lecocq and M. Chalet of Alsthom-Atlantique who kindly carried out the anisothermal tests as a contribution to the project, and to Messrs How and Meecham of ERA Technology Limited, who carried out the isothermal tests.

REFERENCES

Barr, R.R. and P.E. Moir, (1970) 'The short time uniaxial stress relaxation characteristics of some carbon-manganese and low alloy weldable steels'. Proc. The Welding Institute Conference on 'Welding Creep Resistant Steels'.

Ferril, D.A., P.B. Juhl and D.R. Miller, (1966) 'Measurement of residual stresses in a heavy weldment'. Welding Research, Vol.31, No.11, pp.504s-514s.

Fidler, R. (1978) 'A finite element analysis for the stress relief of a CrMoV-2CrMo main steam pipe weld'. CEGB Report R/M/R270, TC375, October.

Gott, K. 'Measurement of residual stresses in the PVRC nozzle specimen 204'. (3. Final Report), STUDSVIK/E1 - 79/99.

Grotke, G.E., A.J. Bush and F.J. Kromer, (1973) 'Residual welding stresses in an 11in. thick pressure vessel'. Westinghouse Research Report 73-1D4-WELST-R2.

Gulvin, T.F., D. Scott., D.M. Haddril and J. Glen, (1972-73) 'The influence of stress relief on the properties of C and C-Mn pressure-vessel plate steels'. Journal of the West of Scotland Iron and Steel Institute, Vol.80.

Holloman, J.H. and L.D. Jaffe, (1945) 'Time temperature relations in tempering steel', Trans. Am. Inst. Min. Met. Engrs, Vol.162, p.223.

Larsson, L.E. and R. Sandstrom, (1982) 'Method for evaluation of the 3-D residual stress field from X-ray diffraction measurements in heavy weldments'. Report IM-1666, Institutet for Metallforsking.

Leggatt, R.H. (1985), 'Relaxation of residual stresses during postweld heat treatment of submerged-arc welds in a C-Mn-Nb-Al steel', The Weld. Inst. Res. Report 288/1985.

Leymonie, C. (1980) 'Utilisation of anisothermal stress relaxation tests'. In Proc. Int. Conf. on Engineering Aspects of Creep, Vol.1, pp.115-120, Sheffield.

Threadgill, P.L. and R.H. Leggatt, (1984) 'Effects of postweld heat treatment on mechanical properties and residual stress levels of submerged-arc welds in a C-Mn-Nb-Al steel', Weld. Inst. Res. Report 253/1984.

Ueda, Y., K. Fukunda., K. Nakacho., E. Takahashi and K. Sakamoto, (1977) 'Transient and residual stresses from multipass weld in very thick plates and their reduction from stress relief annealing'. Third Int. Conf. on Pressure Vessel Technology, ASME.

IV.8
Estimation Methods for Studying the Degree of Relaxation of Residual Welding Stresses at Appropriate Heat-treatment, as well as for Evaluation of Effect on Non-relaxed Stresses on the Load-carrying Capacity of Structure Members

V. I. Makhnenko, E. A. Velikoivanenko and V. E. Pochinok

E. O. Paton Electric Welding Institute of the Ukrainian SSR Academy of Sciences, Kiev, USSR

The opportunity to obtain data about the degree of relaxation of residual welding stresses depending upon the nature of the stressed state and main parameters of the tempering conditions for different materials by using the estimation methods are of a great practical interest. In this direction the work /1/ is the most known. However, some inaccuracies, made in mathematical description of the process of stress relaxation, impedes the application of estimation algorithms suggested in this work. The more strong approach, based on the theory of a non-isothermal yielding of the elastic-viscous plastic medium, is suggested in work /2/. It allows to consider separately two known mechanisms of relaxation of residual stresses in heat-treatment. First one is due to "instantaneous" plastic deformations, caused mainly by the reduction of material yield strength in heating and the second one - due to "diffusion" ductility (creep) of the material during a certain time. In accordance with this approach the increment of strain tensor $d\varepsilon_{ij}$ in any point of the element considered at an arbitrary moment of time t is determined by the sum

$$d\varepsilon_{ij} = d\varepsilon_{ij}^{e} + d\varepsilon_{ij}^{p} + d\varepsilon_{ij}^{c} , \qquad (1)$$

where ε_{ij}^{e} is the tensor of elastic strains; ε_{ij}^{p} is the tensor of strains of an "instantaneous" ductility and ε_{ij}^{c} is the tensor of strains of "diffusion" ductility (creep).

The tensor ε_{ij}^{e} is associated with the stress tensor σ_{ij} by Hooke's law, i.e.

$$d\varepsilon_{ij}^{e} = d\left[\frac{\sigma_{ij} - \delta_{ij}\sigma}{2G} + \tilde{\delta}_{ij}(K\sigma + \varphi)\right] , \qquad (2)$$

where G is the shear modulus, K is the coefficient of uniform compression; $\tilde{\delta}_{ij}$ is the single tensor, σ is the mean value of normal components of tensor σ_{ij}, φ is the function of free volumetric changes caused by the temperature elongation and structural transformations.

Tensor ε_{ij}^{p}, in accordance with the theory of yielding, associated with the Mises condition of yielding is related to the σ_{ij} by relationship /2/

$$d\varepsilon_{ij}^{p} = d\lambda(\sigma_{ij} - \delta_{ij}\sigma), \qquad (3)$$

where the scalar function $d\lambda$ is determined by a stress intensity σ_i and material yield strength σ_T at a given temperature. Tensor ε_{ij}^c according to /2/ is, respectively, connected with stresses σ_{ij} by the hypothesis of yielding, i.e.

$$d\varepsilon_{ij}^{c} = \dot{\Phi} dt (\sigma_{ij} - \delta_{ij}\sigma), \qquad (4)$$

where $\dot{\Phi} = \dot{\Phi}(\sigma_i, T)$ is the function of material creep, determined experimentally by simple experiments. In particular, when using the curves of relaxation of uniaxial stresses σ_1, in specimens at isothermal condition

$$\dot{\Phi} = \frac{3}{2\sigma_1} \frac{d}{dt}\left(\frac{\sigma_1}{E}\right), \qquad (5)$$

where E in the elasticity modulus. Fig. 1 presents the typical curves of values of $\dot{\Phi}(\sigma_i, T)$ functions for steel 35 and aluminium alloy D16AT.

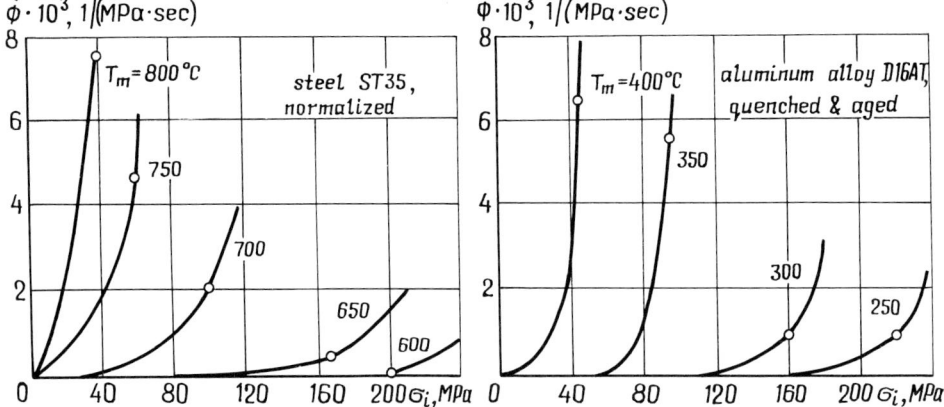

Fig. 1. General view of creep function for St35 steel and D16AT aluminium alloy.

The curves were obtained by a proper processing of experimental results from works /3, 4/. It follows from these data that the reep function $\dot{\Phi}(\sigma_i, T)$ largely depends on temperature and stress intensity. The rate of relaxation of residual stresses at the temperature of heat treatment of welded structures is determined both by the temperature of heating and the value of non-relaxed stresses at the given moment.

The presence of an active zone, where as a result of welding the plastic deformations occurred, and a passive zone (the rest part of the structure), the stresses of which compensate the active zone stresses, is typical for the residual welding stresses in the welded structures. Since the stresses in the active zone have, as a rule, higher σ_i than those in the passive zone then the rate of relaxation of the residual stresses in the welded structure during the isothermal heating are determined by the rate of stress relaxation in the active zones. In this connection in heat-treatment of a homogeneous welded structure in the conditions that provide the uniform heating and cooling it is possible to evaluate the degree of relaxation of the residual stresses with an accuracy sufficient for the engineering purposes on the basis of considering the stress relaxation only in the active zones depending on the preset temperature cycle of heat-treatment $T(t)$ (Fig. 2) by using the relations

$$\frac{d}{dt}\left(\frac{\sigma_i}{E}\right) + \frac{2}{3}\left[\dot{\phi}(\sigma_i,T) + \frac{d\lambda}{dt}\right]\sigma_i \approx 0 \qquad 0 < t < t_{o\delta} \qquad (6)$$

$$t = 0 : \quad \sigma_i = \sigma_i^{ocr}, \quad \sigma_i \leqslant \sigma_T(T_m), \qquad (7)$$

where $t_{o\delta}$ is the time of treatment (Fig. 2).

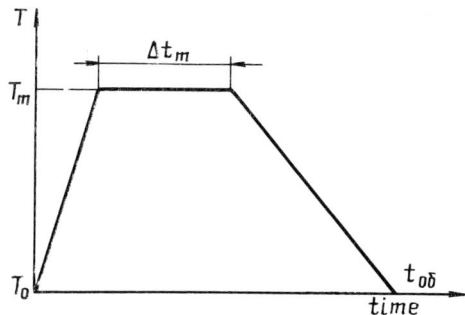

Fig. 2. Diagram of temperature cycle in heat treatment.

If to neglect the stages of heating and cooling (Fig. 2) then it is possible to obtain from (6) and (7) the relationship for the duration Δt_m at the temperature of heat-treatment (tempering) T_m

$$\Delta t_m = \frac{3}{2}\int_{\sigma_i^{Hp}}^{\sigma_i^0} \frac{d(\sigma_i/E)}{\dot{\phi}(\sigma_i,T)\sigma_i}, \qquad (8)$$

where $\sigma_i^0 = \sigma_i^{ocr}$, if $\sigma_i^{ocr} \leqslant \sigma_T(T_m)$ or

$\sigma_i^0 = \sigma_T(T_m)$, if $\sigma_i^{ocr} > \sigma_T(T_m)$;

σ_i^{Hp} are the non-relaxed stresses. Fig. 3 gives the curves of stress relaxation for two typical alloys at different temperatures T_m and the condition that $\sigma_i^{ocr} = \sigma_T(T_m)$ by using the relationship (8) and data of Fig. 1. It is well seen from these data that if the residual stresses in the active zone are higher than $\sigma_T(T_m)$, then the stress relaxation below the level $\sigma_T(T_m)$ at the initial period of relaxation is rather active by the diffusion mechanism at temperatures $T_m \geqslant T_*$, where $T_* = 0.5 \ (T_{melt} - 273°C)$, T_{melt} is the material melting tempera-

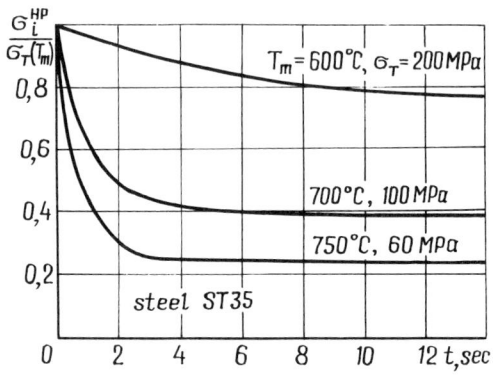

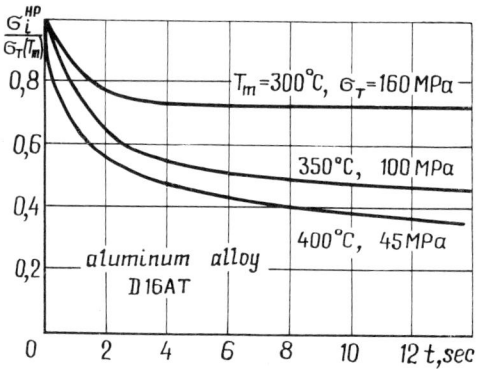

Fig. 3. Estimated curves of stress relaxation in homogeneous welded joints of ST35 steel and D16AT aluminium alloy.

ture in °C; then at large t the relaxation rate is sharply decreased. It is possible to approximately evaluate the components of non-relaxed stress tensor in the active zone by value σ_i^{HP}. In particular, if $\sigma_i = \sigma_1 \chi$, where $\sigma_1 > \sigma_2 > \sigma_3$ are the main stresses and $\sqrt{2} \cdot \chi = \sqrt{(1-\sigma_2/\sigma_1)^2 + (1-\sigma_3/\sigma_1)^2 + (\sigma_2/\sigma_1 - \sigma_3/\sigma_1)^2}$, then assuming that χ value remains constant at a simple stress relaxation, we obtain

$$\sigma_1^{HP} = \frac{\sigma_i^{HP}}{\chi}$$

For non-homogeneous structures the heating and cooling in heat treatment, relaxing to a certain degree the residual welding stresses, contribute to a creation of a new stressed state that in some cases can make the rationality of application of the appropriate heat treatment questionable.

As an example the results of an estimated study of stress redistribution in the dissimilar two layer plate are given: - base metal of plate is 20ГСЛ steel of δ_0 thickness, deposited layer is 30ХIOГIO cavitation steel. The required data about the creep function Φ were obtained experimentally and processed in the form of

$$\dot{\phi} = K_1 \exp(\beta \sigma_i + \mu T + C) \qquad (9)$$

where K_1, β, μ, C are constants (see Table 1). The stressed state was studied by estimation in the distance from plate edges at a complete deposition of surface that permitted to apply the assumption that $d\varepsilon_1 = d\varepsilon_2(z)$, $\sigma_1 = \sigma_2(z)$, $\sigma_3 = 0$, where 1 and 2 are the main directions in the plate plane, z is the space coordinate by thickness (Fig. 4a). Fig. 4b presents the estimated data on kinetics of changing the stress distribution $\sigma_1(z)$ depending upon the temperature cycle of heating and cooling according to a corresponding curve in Fig. 6a. In this case it was assumed that the coefficient α of the linear temperature elongation of the material is the function of temperature by Table 2, as well as the values $\sigma_T(T)$ and $E(T)$ for surfacing, base metal and heat-affected zone (Fig. 5).

TABLE 1   EXPERIMENTAL DATA FOR PARAMETERS OF EXPRESSION (9)

Steel	K_1, (MPa·s)$^{-1}$	β, MPa^{-1}	μ, 1/°C	C
20ГСЛ	10^{-7}	0.00030	0.023	-14.3
30ХIОГIО	10^{-7}	0.00055	0.067	-41.8

TABLE 2   RELATIONSHIP BETWEEN THE COEFFICIENT α OF TEMPERATURE ELONGATION AND TEMPERATURE FOR STEELS 20ГСЛ AND 30ХIОГIО

Steel	$\alpha \cdot 10^5$, deg.$^{-1}$ for T, deg.								
	100	200	300	400	550	650	700	750	800
20ГСЛ	1.2	1.25	1.28	1.30	1.37	1.40	1.41	1.83	1.83
30ХIОГIО	1.66	1.71	1.75	1.83	1.93	1.95	1.99	2.01	2.05

Fig. 6 gives the kinetics of changing the stresses in two typical points at the interface both from the side of base metal and from the side of surfacing for three concrete cycles of heat-treatment of such deposited parts. The analysis of these estimated data show that the initial stresses occurred during deposition and at the stage of heating have time to be relaxed practically completely after 5 hour soaking above 550°C.

However, the subsequent cooling leads to the formation of the new stressed state specified by different coefficients of the temperature elongation, as well as by a high relaxation resistance below 300-400°C, the level of maximum stresses after heat-treatment being not decreased.

It follows from /1/ and the present work that achievement of the significant decrease of residual stresses in welded (deposited) structures by heat treatment is hardly possible due to different reasons (physical, technological, economic).

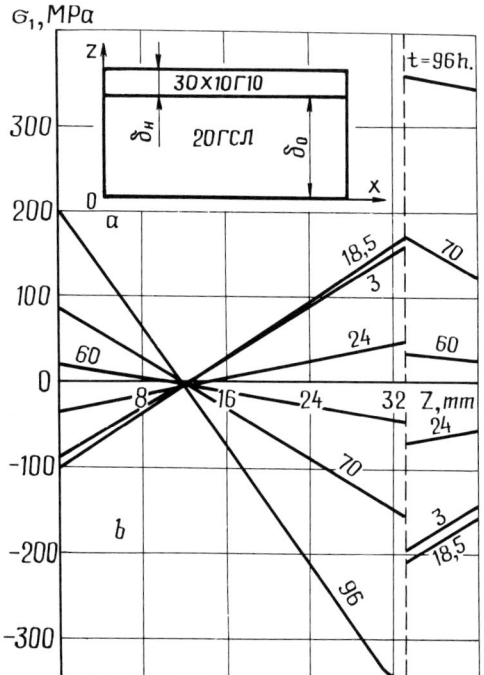

Fig. 4. Sketch of deposited plate (a) and distribution of stresses $б_1$, across the thickness of non-homogeneous joints of steels 20ГСЛ ($δ_0$ = 33 mm) and 30ХIОГIО ($δ_H$ = 7 mm) at heat treatment in accordance with curve 1 in Fig. 6a.

In this connection the possibility to evaluate the degree of effect of non-relaxed residual stresses on various technological or service characteristics, including the load-carrying capacity of corresponding elements of the welded structures by estimation is of a great practical interest. This problem is paid a great attention in different investigations, especially in those devoted to brittle and fatigue fractures of welded structure members. As a rule, the degree of effect of residual non-relaxed stresses in these investigations were experimentally evaluated that is caused by large difficulties for theoretical descriptions of effectiveness of this effect especially for sufficiently general cases of loading. Below the estimation approach is considered for the evaluation of effect of residual non-relaxed transverse stresses on the load-carrying capacity of butt joints with preset lacks of penetration in plate under loading with static and alternate transverse loads $б_{yy}^{\infty}$ (Fig. 7).

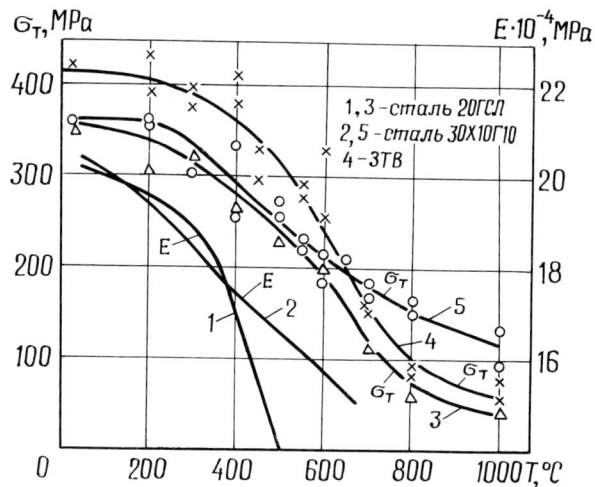

Fig. 5. Temperature relationship between the elasticity modulus E and yield strength σ_T of steels 20ГСЛ and 30ХІОГІС and HAZ of their joint.

The two-parameter criterion of limiting state at static loading from /5/ taking into account the generalizations /6/ was used

$$\left(\frac{\sigma_i^{max}}{\sigma_B}\right)^{q+1} + \left(\frac{K_I^{max} + K_{Ir}}{\eta K_{IC}}\right)^2 = 1.0 , \quad (10)$$

where σ_i^{max} is effective stress in the net section caused by external loads; σ_B is ultimate strength of material; K_I^{max} is stress intensity factor at the partial penetration tip due to external loads; K_{Ir} is that due to unrelaxed part of welding residual stresses; K_{IC} is plane strain fracture toughness of the weld metal/heat affected zone; η is the correction factor accounting for the deviation of strain conditions at the uncomplete penetration tip from that of plane strain at crack tip; following relationship was used /6/ $\eta = \eta_1(r) \cdot \eta_2(h)$, (11)

where
$\eta_1(r) = 1.0$, if $r < r_c \approx 0.1$ mm ,
$\eta_1(r) = \sqrt{r/r_c}$, if $r > r_c$;
$\eta_2(h) = 1.0$, if $h > m(K_{IC}/\sigma_T)^2 = h_m$
$\eta_2(h) = \sqrt{h_m/h}$, if $h < h_m$

where $m = 1 - 2.5$, $q \geq 1.0$.

For the case of loading considered in Fig. 7

$$\varepsilon_i^{max} = \frac{(\sigma_{yy}^{\infty})^{max}}{1 - a/S} , \quad K_I^{max} = (\sigma_{yy}^{\infty})^{max}\sqrt{\pi a}\, F\left(\frac{a}{S}\right) , \quad (12)$$

$$K_{Ir} \approx \sigma_{yy}^{HP}\sqrt{\pi a}\, F\left(\frac{a}{S}\right) .$$

The effect of lack of penetration in combination with residual stresses σ_{yy}^{HP} can be presented by a value of additional safety factor n_{xr} to the condition of ductile failure $\sigma_i \approx \sigma_B/n_{xr}$,

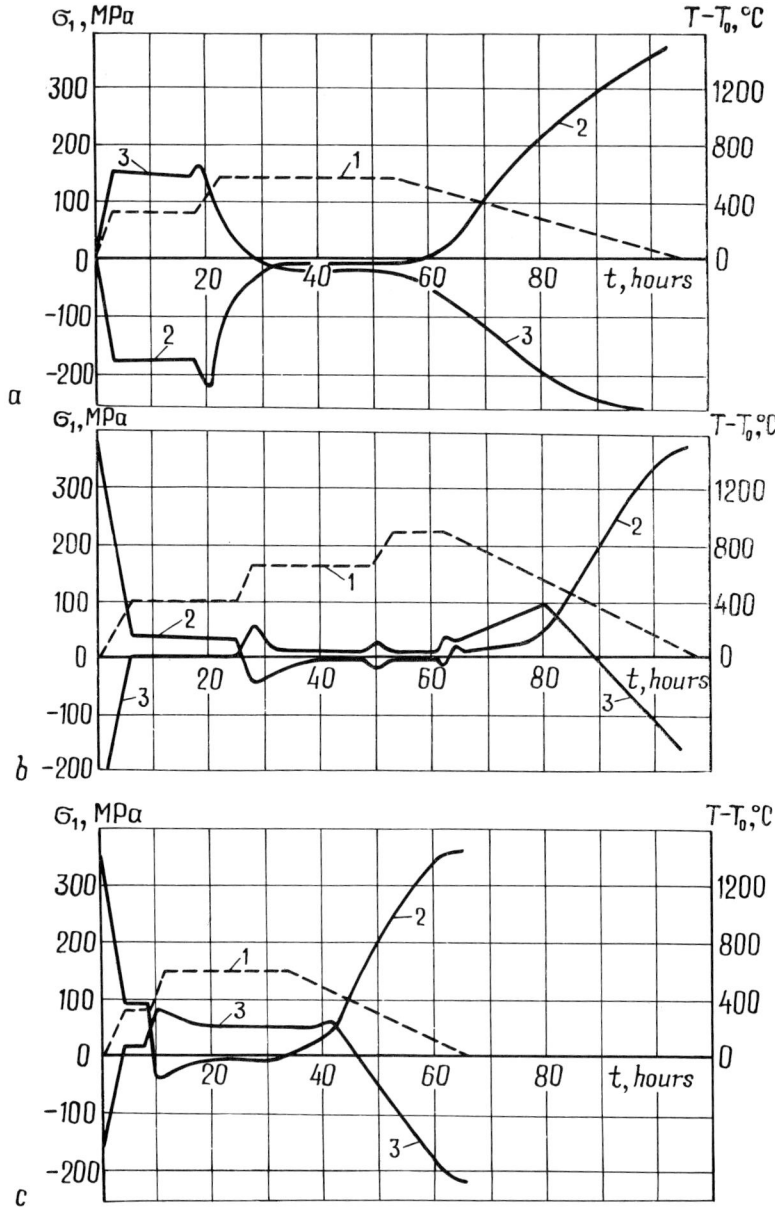

Fig. 6. Kinetics of changing the stresses σ_1, at the interface of layers at a subsequent heat treatment according to corresponding cycle a-c; 1- temperature cycle, 2 - stresses in deposited layer (z = 34 mm); 3 - stresses in base metal (z = 32 mm).

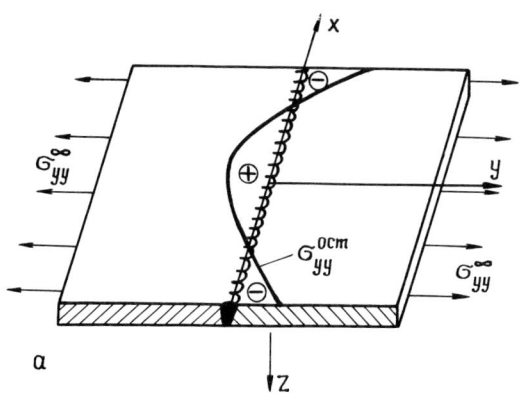

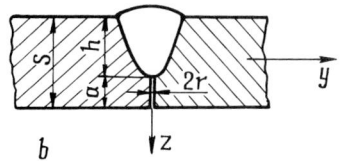 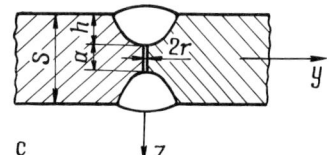

Fig. 7. Sketch of butt joint with σ_{yy}^{ocm} transverse residual stresses (a) and lack of penetration in weld root (b) and in centre (c).

where
$$n_{xr} = \frac{A\bar{\sigma}_{yy}^{HP} + \sqrt{1+A-A\bar{\sigma}_{yy}^{HP\,2}}}{1-A\bar{\sigma}_{yy}^{HP\,2}} \quad , \quad \bar{\sigma}_{yy}^{HP} = \sigma_{yy}^{HP}/\sigma_{B} \; , \quad A = \left[\frac{\sigma_{B}\,F(\frac{a}{\delta})}{\eta K_{IC}}\right]^{2} \pi a$$

Table 3 gives the values n_{xr} depending upon the value $\bar{\sigma}_{yy}^{HP}$ and A. It is seen that the role of the non-relaxed residual stresses abruptly rises with the growth of value A that contains the properties of material and geometry of "crack-like defect"(lack of penetration). At alternate (cyclic) loading of joints of the type given in Fig. 7 the non-relaxed residual stresses can decrease the fatigue life due to: 1) decrease in threshold value of range $\Delta\sigma_{yy}^{\infty}$ of stresses σ_{yy}^{∞} at which the fatigue crack at the apex of lack-of-penetration begins to grow. 2) increase in rate of fatigue crack growth dl/dN; 3) decrease in critical length l_{Kp} of a fatigue crack, at which the joint fracture occurs.

For the estimated evaluation of these effects the relation (10) in combination with a known Paris-Erdogan relationship can be used which according to /6/ is convenient to approximately present in the form of
$$\frac{dl}{dN} = \frac{\bar{C}}{(1-r)^{m}} \Delta K_{I}^{n} \left(1 + \frac{K_{Ir}}{K_{I}^{max}}\right)^{m} , \qquad (13)$$

TABLE 3 Effect of $\bar{\sigma}_{yy}^{HP}$ and A on Value η_{xr}

$\bar{\sigma}_{yy}^{H\rho}$ \ A	0.1	0.5	1.0	2.0	4.0	8.0	16.0
0	1.05	1.22	1.41	1.73	2.23	3.00	4.12
0.1	1.06	1.28	1.52	1.97	2.74	4.12	6.70
0.2	1.07	1.34	1.67	2.29	3.57	6.68	20.1
0.3	1.08	1.42	1.85	2.78	5.14	19.1	-

where $\bar{C} = 0$, if $\Delta K_I < \Delta K_{th}(R)$;
$\bar{C} = C_0$, if $R = \dfrac{K_I^{min} + K_{Ir}}{K_I^{max} + K_{Ir}} > 0$ and $K_I^{min} + K_{Ir} > 0$, $r = \dfrac{K_I^{min}}{K_I^{max}}$

$\bar{C} = \dfrac{C_0}{(1-R)^{n-m}}$, if $R < 0$ and $K_I^{min} + K_{Ir} < 0$
$\bar{C} = 0$, if $R > 0$ and $K_I^{min} + K_{Ir} < 0$,

where C_0, n, m, $\Delta K_{th}(R)$ are the experimental characteristics of the material.

From (13) it follows that the effect of decreasing the threshold range $\Delta \bar{\sigma}_{yy}^{HOP}$ of external stresses from the conditions of initiation of effective growth of fatigue crack is determined by type of relationship $\Delta K_{th}(R)$. If this relationship by /7/ to take in the form of

$$\Delta K_{th}(R) = \Delta K_{th}^0 (1 - 0.85 R), \quad (14)$$

then taking into account (12) we obtain

$$(\Delta \bar{\sigma}_{yy}^{\infty})^{\text{nop}} = \dfrac{\Delta K_{th}^0}{\sqrt{\pi a}\, F(a/S)} \cdot \xi(r,\bar{\sigma}_r), \quad (15)$$

where $r = \dfrac{(\sigma_{yy}^{\infty})^{min}}{(\sigma_{yy}^{\infty})^{max}}$, $\bar{\sigma}_r = \dfrac{\sigma_{yy}^{HP}}{(\sigma_{yy}^{\infty})^{max}}$,

$$\xi(r,\bar{\sigma}_r) = \dfrac{1 - 0.85 r + 0.15 \bar{\sigma}_r}{1 + \bar{\sigma}_r} \quad (16)$$

Thus the value ξ according to (16) determines the degree of effect of the non-relaxed residual stresses by value $(\Delta\bar{\sigma}_{yy}^{\infty})^{nop}$. Table 4 gives the values $\xi(r,\bar{\sigma}_r)$ for r and $\bar{\sigma}_r$, that illustrate the extent of effect $\bar{\sigma}_r$ at various r. The effect of the non-relaxed stresses on the rate of fatigue crack growth dl/dN at similar parameters of the external loading K_I^{max} and r can be expressed in (13) by a factor λ to the rate of growth dl/dN at $K_{Ir} \neq 0$. Respectively, we obtain

$$\lambda = \left(1 + \dfrac{K_{Ir}}{K_I^{max}}\right)^m \quad \text{at } R > 0,\ K_I^{min} + K_{Ir} > 0 \quad (17)$$

$$\lambda = \left(1 + \dfrac{K_{Ir}}{K_I^{max}}\right)^n \quad \text{at } R < 0,\ K_I^{min} + K_{Ir} < 0$$

As the values m are within the ranges $0.25-1.0$ and $n = 2-5$, then from (17) it follows that at $r = 0$, the compressive non-relaxed stresses $\bar{\sigma}_r$ more intensively decrease the value λ while at the tensile stresses $\bar{\sigma}_r$ it is increased, that is good correlated with the experiment.

TABLE 4   Values $\xi(r,\bar{\sigma}_r)$

$\bar{\sigma}_r$ \ r	-0.1	-0.5	0	0.5
-0.5	3.55	2.70	2.15	1.0
0	1.85	1.425	1.0	0.575
0.5	1.28	1.0	0.717	0.433
1.0	1.0	0.788	0.575	0.362

REFERENCES

Vinokurov V.A. (1973). Tempering of welded structures for stress relieving. Mashinostrojenije, 215.
Makhnenko V.I. (1976). Estimated methods of study of welding stress and strain kinetics. Kiev, Naukova Dumka, 320.
Rabotnov Yu.N., and S.T.Milejko (1970). Short-time creep. Nauka.
Bezukhov N.I., et al (1965). Calculations of strength under the high-temperature conditions. Mashinostrojenije.
Vasilchenko G.S., Morozov E.M., and D.M.Shur (1979). Evaluation of the bearing capacity of cracked components of welded structures. In: IIW Colloquium on Pract.Appls.Fracture Mechs., Bratislava, 52-59.
Makhnenko V.I., Pochinok V.E. (1984). Cyclic load resistance of welded joints having welds with uncomplete penetration. Avtomaticheskaya svarka, 10, 33-40.
Rolfe S.T., and J.M.Barsom (1977). Fracture and fatigue control in structures. Applications of fracture mechanics. - Prentice-Hall, Inc., Englewood Cliffs, New Jersey, 562.

IV. 9

Qualification D'un Traitement de Relaxation

A. Leclou

Commissariat à L'Energie Atomique, Saclay, France

RESUME

L'efficacité d'un traitement de relaxation doit être évalué selon l'application concernée. Nous traitons ici plus particulièrement des aspects niveau moyen de contraintes résiduelles et stabilité dimensionnelle. Nous présentons des assemblages d'essai permettant d'une part (assemblage plan) de déterminer un taux de relaxation et d'autre part (assemblage flexion) de classer les traitements de relaxation à partir de mesures de flèches correspondant au retour élastique de pièces traitées. Ces techniques d'essai ne nécessitent que du matériel léger : extensométrie mécanique par exemple et métrologie ; elles ne nécessitent pas la connaissance des caractéristiques réelles à 20° C et à la température de traitement.

L'efficacité relative des traitements d'assemblages en aciers austénitiques est chiffrée. Nous n'avons pas constaté de relaxation résultant du traitement par vibration.

MOTS CLES

Traitement de relaxation ; contraintes résiduelles ; déformations de soudage ; assemblage d'essai.

PREAMBULE

Certaines réalisations mécano-soudées nécessitent un examen attentif des contraintes résiduelles et des déformations qu'elles risquent de causer en cours de fabrication ainsi que de la stabilité dimensionnelle ultérieure.
Pour cela, nous utilisons deux types d'assemblages :
- assemblages plan (ou "assemblage en H") :
    il tend à la détermination d'un taux de relaxation des contraintes.
- assemblages flexion :
    ils permettent de classer l'effet des traitements de relaxation et des procédés sur l'aptitude d'un assemblage à se déformer par la suite sous l'effet de l'usinage par exemple.

RAPPEL RELATIF AUX CONTRAINTES RESIDUELLES ET DEFORMATIONS.

La réalisation d'un appareil par soudage ne peut pas se faire sans contraintes "résiduelles" qui résultent des dilatations et contractions inhérentes à ce procédé.
Selon la catégorie de conséquences que l'on examine, les aspects les plus déterminants pourraient être classés ainsi :
- niveau de contraintes,
- localisation des contraintes,
- nature et répartition des contraintes.

Niveau des contraintes

Nous portons ici l'attention plus particulièrement sur le chargement d'une structure par le retrait transversal.
Le retrait transversal charge la partie courante de la structure en fonction de la valeur du retrait et de la rigidité (sens travers) qui est mise en cause. Ce mode de chargement (et le niveau des contraintes) a des conséquences, entre autres, sur les phénomènes suivants :
- la fissuration au retrait,
- la fissuration à froid (aciers faiblements alliés),
- la fissuration en relaxation (et son cas particulier, la fissuration au chauffage),
- la corrosion sous tension,
- la rupture fragile,
- un chargement non prévu par le bureau d'études,
- les déformations en cours de traitement, usinage, service...
    Evidemment, une structure sera d'autant moins affectée par ce phénomène qu'elle sera bien conçue et réalisée selon des procédés convenables.
L'assemblage plan que nous examinerons plus en détail permet d'évaluer cet effet et d'annoncer un taux de relaxation de la contrainte moyenne en traction sens travers à la soudure.

Localisation des contraintes

Selon le sens transversal par exemple le niveau des contraintes atteindra des valeurs élevées au voisinage de la zone fondue, pour s'atténuer rapidement et s'annuler à une distance relativement courte si la pièce n'est pas bridée.
L'effet sera plus ou moins localisé selon, par exemple, l'énergie spécifique mise en jeu. Nous ne traiterons pas de cet aspect par ce que, pratiquement, il peut être réglé par un traitement de relaxation dont l'efficacité aura été vérifiée. D'autres techniques, mais elles sont difficiles à évaluer et à contrôler , peuvent améliorer la situation :
- passe TIG (nécessairement automatique),
- grenaillage,
- marteau à aiguilles,
- modification de la charge par flamme, etc.

Nature et répartition des contraintes

Nous prenons en exemple deux types de contraintes :
- Contraintes de flexion résultant du retrait transversal (contraintes résultant de la section variable du bain de fusion selon l'épaisseur, de la répartition de l'énergie, de l'effet de charnière...).
il est évident que les séquences convenables et que la nature du procédé ont un effet direct sur ce phénomène.
- Contraintes réparties dans le produit : Ce sont les contraintes qui maintiennent le produit dans une forme donnée :
 . Tôles à l'état de livraison : contraintes résiduelles de laminage et de planage.
 . Tôles cintrées : contraintes résiduelles de flexion qui maintiennent le produit à l'équilibre à un rayon de courbure défini.
Dans ces deux exemple ci-dessus, l'enlèvement de matière rompt l'équilibre des contraintes. Même si le niveau le niveau de ces contraintes est relativement faible, il en résulte une déformation qui peut être redhibitoire.
Le traitement thermique global de l'ensemble améliore la situation. Cependant les contraintes résiduelles restent souvent d'un niveau tel que la rigueur géométrique ne peut pas être garantie.
Il faut donc, aussi, prendre d'autres précautions, telles que : emploi de séquences d'usinage et de soudage judicieuses, mise en oeuvre de procédés amenant le minimum de déformation.

Les "assemblages flexion" que nous examinerons permettent de classer les effets évoqués ci-dessus et les traitements qui tendent à les réduire.

ASSEMBLAGE PLAN

Ce type d'assemblage (fig. 1) a été utilisé fréquemment pour des essais de soudabilité et a déjà fait l'objet d'études sur les contraintes et le bridage [1] [2].

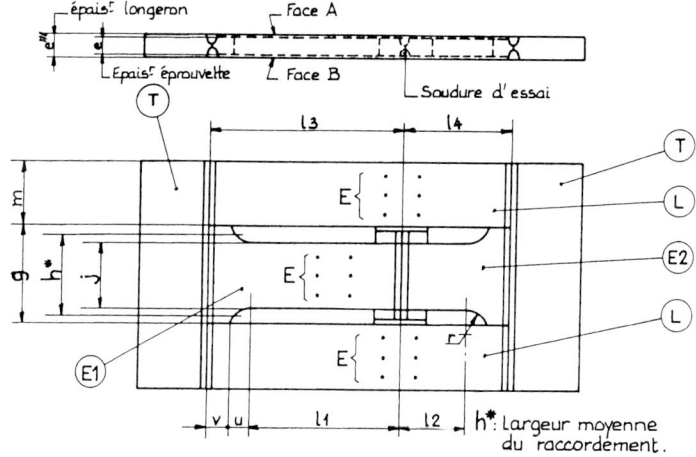

FIG.1 _ ASSEMBLAGE PLAN _ CAS GÉNÉRAL .

Principe

Les principes essentiels de l'expérimentation sont les suivants :
- on évalue l'efficacité d'un traitement de relaxation à partir de la mesure de la contrainte résiduelle moyenne selon le sens transversal à la soudure, en partie courante d'assemblage.
- l'assemblage est auto-contraint. On utilise la propriété des déformations selon le sens transversal d'être d'un niveau élevé et bien reproductible pour mettre en charge l'assemblage.
- les déformations selon le sens transversal ont aussi la propriété d'être particulièrement caractéristiques d'un "procédé" donné, (conditions opératoires), ceci permet donc d'introduire éventuellement le paramètre "procédé" pour une expérimentation incluant l'ensemble des conditions de réalisation.
- Les dispositions adoptées dans cette technique d'essai rendent faciles la pratique et l'interprétation de l'extensométrie, puisque l'on mesure des contraintes moyennes en partie courante d'assemblage. Nous utilisons l'extensométrie mécanique, qui permet d'obtenir une précision de $\pm 5 \, N \, mm^{-2}$
- Les résultats obtenus sont comparables d'une expérimentation à l'autre du fait de l'utilisation de géométries semblables.

Description (fig.1)

Les parties essentielles de l'assemblage, E1 et E2, constituant "l'éprouvette", sont assemblées dans un cadre composé de 2 longerons L et de 2 traverses T. L'éprouvette est réalisée à partir du produit faisant l'objet de l'étude, les longerons sont en même matériau ou dans un matériau similaire.
Les soudures de montage de T sur E1-E2 et L rendent l'assemblage rigide ; la soudure d'essai de E1 avec E2 mettra le système en charge sous l'effet de son retrait transversal.

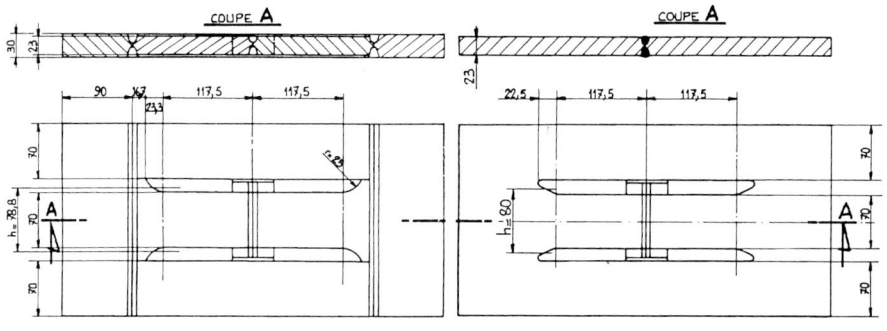

FIG.2 - ASSEMBLAGE PLAN, COMPOSITE.(ASSEMBLAGE B)   FIG.3 - ASSEMBLAGE PLAN, MONOBLOC (ASSEMBLAGE A)

La figure 2 présente un exemple de réalisation d'assemblage composite, réalisé avec des matériaux de même coefficient de dilatation.
Il est possible de réaliser un assemblage monobloc ; la figure 3 présente un exemple de réalisation, les figures 4 et 5 montrent l'allure des assemblages B et C et l'implantation de quelques bases de mesure.

ASSEMBLAGE B

FIG. 4 _ VUE D'ENSEMBLE   FIG. 5 _ BASES DE MESURES

Réalisation - Séquence des opérations et des mesures

Dans le cas de l'assemblage soudé, les soudures de montage des traverses T sur les longerons L et les éprouvettes E1-E2 sont réalisées de telle sorte que la préparation de la soudure d'essai S se trouve dans les tolérances qui auront été définies en fonction des objectifs de l'essai.

Les bases de mesures nécessaires à l'extensométrie mécanique sont implantées aux emplacements prévus à cet effet (par exemple en E sur les faces A et B de la partie éprouvette et des longerons).

Une première mesure donne la longueur initiale Lo de chaque base.

La mesure de la longueur L_1 de la base après réalisation de la soudure d'essai permet de connaître l'allongement relatif L_0-L_1 dû aux contraintes de retrait selon le sens transversal. On en déduit σ, contrainte moyenne sur la face concernée, selon le sens transversal.

La comparaison des valeurs de σ des faces A et B permet, par leur demi-somme et par leur demi-différence, d'en déterminer :
- n = contrainte moyenne en traction après soudage
- f = contrainte moyenne en flexion après soudage.
On procède à la mesure de la longueur L_2 de la base après traitement thermique. Il n'est pas aisé d'en déduire la contrainte moyenne après traitement, ainsi que cela est prévisible à l'examen du bilan présenté sur un graphique représentatif. Une détermination précise nécessite de passer au stade suivant.

Il est procédé à une coupe transversale de l'éprouvette en dehors des bases de mesure ; la mesure des bases donne alors la valeur Lu.

La différence L_2-Lu permet d'obtenir la valeur des contraintes moyennes après traitement thermique.

On peut en déduire un taux de relaxation qui permet de juger de l'éfficacité du traitement.
De même la mesure de l'écartement des bords de la coupe et la mesure de la flèche permettent de confirmer le résultat de l'extensométrie. Nous avons alors:

$$n = E \frac{R}{[(l_1 + l_2) + 2k'u + 2k''v + k'''(l_3 + l_4)] - B}$$

$$n = E.R \frac{1}{\Sigma l - B}$$

avec $\quad k' = \dfrac{j}{h} \qquad k'' = \dfrac{j}{g} \qquad k''' = \dfrac{j.e}{2me'''}$

n = contrainte moyenne en traction dans l'éprouvette,
E = module d'élasticité,
e''' = épaisseur des longerons,
R = valeur du retour élastique (ouverture de part et d'autre de la coupe)
Σl = somme des longueurs équivalentes des différentes parties de l'assemblage, ramenées à la section de "l'éprouvette".
B = longueur de la base de mesure.

Cette formule, évidemment, est applicable aussi pour confirmer la valeur de la contrainte moyenne dans l'éprouvette résultant du retrait transversal R de la soudure d'essai, en considérant que l'on reste dans le domaine élastique.

Résultats

Nous examinerons 3 cas où la partie essentielle de l'assemblage, "l'éprouvette", est constituée du même matériau (Z5 CN 18-10) :
- l'assemblage A est monobloc (fig.3)
- l'assemblage B (fig.2) est composite, constitué de 2 longerons et de 2 traverses en acier austénitique d'épaisseur plus forte (e éprouvette = 23 mm, e''' longerons = 30 mm)
- l'assemblage C est identique à l'assemblage B, mais les traverses ont leur largeur portée de 90 à 140 mm pour augmenter leur rigidité.

Nous utiliserons les résultats de l'extensométrie sur les longerons qui restent, eux, dans le domaine élastique ; nous les convertirons en contraintes moyennes dans l'éprouvette.
Nous pouvons alors en déduire un rapport n_u/n_1 de la contrainte moyenne en traction après traitement thermique à la contrainte avant traitement (tableau 1). Cette valeur brute est évidemment dépendante de la rigidité du système, nous lui apportons un terme correctif dans les conditions suivantes:

Repère	Après soudage (**) n_1 (daN.mm^{-2})	Après coupe (**) n_u (daN.mm^{-2})	n_u/n_1 (%)	L' (mm)	L'$_1$/L'$_u$	Tx (%)
A	13,5		45	956	1,29	58
		6,1		740		
B	14,3		35	927	1,78	61
		5,0		522		
C	15,5 (*)		24,5	928	2,21	54
		3,8		420		

(*) Valeur par excès, hors correction de déformation plastique.
(**) Valeurs déduites de l'extensométrie sur les longerons.
n : contrainte moyenne sens transversal, en partie courante d'éprouvette.

TABLEAU 1 _ DÉTERMINATION D'UN TAUX DE RELAXATION.

· Σ 1 ne tient pas compte de la déformation des traverses. Cette déformation n'est pas aisée à déterminer avec précision par le calcul. Par contre, la comparaison des valeurs de n_{epr} obtenues par extensométrie dont la détermination n'est pas influencée par les traverses, et des valeurs de n obtenues à partir de R permet d'apporter un terme correctif :

$$L' = \left[R - (\Sigma 1 - B)\frac{n_{epr}}{E}\right]\frac{E}{n_{epr}}$$

- nous posons $L'' = \Sigma 1 + L'$
et nous définissons un taux de relaxation:

$$T_x = \frac{nu}{n_1} \times \frac{L''_1}{L''_u}$$

dont les valeurs sont présentées dans le tableau 1.

Nous constatons alors un regroupement des résultats acceptable et qui est en bonne concordance avec les prévisions que l'on peut faire à partir des caractéristiques à chaud du matériaux.

ASSEMBLAGE FLEXION

Les taux de relaxation déterminés à partir des caractéristiques des matériaux (Reo,2 à 20°C et à la température du traitement, et essais de relaxation) ou des assemblages plans définis ci-dessus ne reflètent pas l'aptitude qu'une structure peut avoir à ne pas se déformer.

Afin de présenter des sujets de réflexion relatifs à la nécessité ou non de traiter, et, si le traitement s'avère intéressant, de permettre une bonne approche de son choix, et de juger de la valeur d'un procédé de fabrication ou de traitement, nous avons réalisé des assemblages dont le principe est présenté fig. 6.

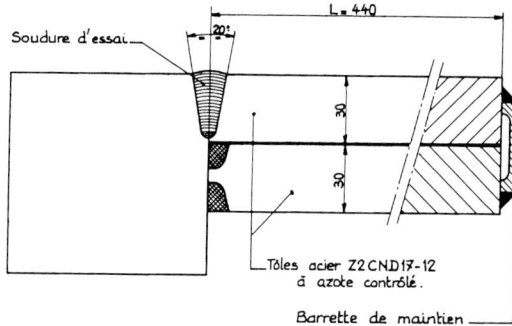

État	Z5CN 18.10 e=16 . L=350	Z2 CND 17.12 à azote contrôlé e=30 . L= 440 .
BRUT	f = 9,3 mm	f = 9,2 mm
Trait. VIBRATION	-	9,2
550°C _ 3 h .	7,2	8,2
650°C _ 10 h	3,1	6,2

FIG. 6 - ASSEMBLAGE FLEXION   TABLEAU 2 _ DÉFORMATION APRÈS COUPE

La barrette étant soudée, les déformations angulaires dues au retrait transversal de la soudure d'essai seront bloquées à l'extrémité des tôles, il en résulte un chargement de l'assemblage. Si l'on coupe la barrette, l'espace entre les tôles s'ouvrira (retour élastique) d'une valeur résultant du chargement initial (état brut) et de l'effet des traitements de relaxation essayés.

Le tableau 2 rassemble les résultats relatifs à deux assemblages, différents par l'épaisseur et la nuance des tôles (Z5CN18-10, e = 16 et Z2CND17.12 à azote controlé, e = 30).

Nous constatons que le traitement de relaxation par vibration n'apporte aucun gain selon ce processus d'essai, et que les traitements thermiques essayés laissent une possibilité de déformation relativement importante. Il apparaît que l'amélioration sensible de ces résultats doit être recherchée vers les techniques de réalisation (séquences de soudage, séquences de fabrication, procédés de soudage,...)

CONCLUSION

- Assemblage plan : Il est possible de chiffrer l'efficacité d'un traitement de relaxation à partir d'un assemblage de réalisation aisée et d'un matériel de mesure simple (extensométrie mécanique par exemple). Dans les exemples examinés, les taux de contraintes résiduelles se regroupent dans l'intervalle 54 % - 61 % pour le traitement à 550° C - 3 h de l'acier inoxydable austénitique Z5CN 18-10.

- Assemblage flexion : Une expérimentation simple démontre que pour un tel traitement, qui relaxe de 40 % ou plus, une structure réalisée sans précaution particulière, soudage d'un seul côté de la tôle dans le cas présenté, possède encore une aptitude évidente à se déformer au cours d'usinage par exemple. L'augmentation de la température du traitement améliore la situation. Nous n'avons pas constaté de relaxation résultant du traitement par vibration.

REFERENCES

[1] - K.Satoh, Y. Ueda, S. Matsui, M. Mastume, T.Terasaki, K. Fukuda, M. Tsuji - (1977) - Travaux japonais sur la sévérité de bridage en relation avec la fissuration des soudures - Documents IIS - 536 - 77

[2] - Pr. Koichi Masubichi (1975) : Rapport sur l'état des connaissances de l'analyse numérique des contraintes, des déformations et autres effets du soudage - Document IIS - 490 - 75

IV.10

Electromagnetic Monitoring of Residual Stress Relaxation During Heat Treatment

I. M. Zhdanov, V. V. Batyuk, A. A. Khriplivy, R. K. Gachik, K. B. Pastukhov, G. F. Kolot and A. V. Pulyayev

Kiev Polytechnical Institute, Kiev, USSR

ABSTRACT

Magnetoelastic technique of non-destructive inspection of residual stresses in welded structures for optimization and estimation of the heat treatment efficiency is suggested. The main requirements to the elements of electromagnetic transducers (sensors), providing an increase in the accuracy of residual stress measurement, are substaintiated. The devices for measuring stresses in the structures with the record of results on paper tape have been developed and the results of the inspection of residual stresses in the articles subjected to heat treatment are presented.

KEY WORDS

Electromagnetic transducers, residual stresses, magnetoelastic effect, non-destructive testing, heat treatment.

INTRODUCTION

One of the main problems, which have to be solved when developing efficient conditions for heat treatment of large-size parts, is to provide a minimum level of temporary and residual stresses. This problem is especially urgent in the case of heat treatment of welded structures of complicated geometrical shapes to size stability of which high requirements are imposed.

The solving of this problem is associated both with the study of the matters of initiation and relaxation of residual stresses and with the development of the means of measurement of them under actual conditions. In this case special attention is acquired by those techniques of residual stress measurement which provide the preservation of a structure integrity or lead minimum fracture at the level of permissible defects in the structure material.

One of such promising techniques of investigations and inspection of residual stresses is the magnetoelastic non-destructive technique. This technique is based on the dependence of the magnetic permeability of ferromagnetic materials on the material stressed state and is notable for a relative simplicity of the measuring means and the possibility to obtain quickly the results of measurement. At present the magnetoelastic technique is used not only in laboratory and research practice but also directly in the manufacturing processes of the fabrication of welded structures /2/.

To develop the system of technological inspection and introduce it to the manufacturing process of welding and heat treatment it is necessary to meet two main conditions:
- to develop the means of non-destructive inspection;
- to determine the standard values of residual stresses.

The development of the inspection means was made on the basis of the magnetic technique which has necessary sensitivity and makes the automation of the process possible.

ELECTROMAGNETIC TRANSDUCERS

The experience of the application of the developed measuring devices for non-destructive inspection of mechanical stresses in the structures made of ferromagnetic materials has shown that the accuracy of measurements using the magnetoelastic technique depends considerably on the value and constancy of a non-magnetic gap between the article and the electromagnetic transducer. It is possible to eliminate this drawback by making the instruments insensitive to the gap or allowing to fix a certain gap accurately.

In the magnetoelastic technique of stress measurement an a.c. induction coil (or the system of coils) is the exiter of eddy currents in the transducer. Depending on electric switching circuits, choke and transformer transducers are differentiated. In Fig. 1 ИНИ-I instrument consisting of the electromagnetic transducer and signal conversion unit, is presented.

ДМИ -I electromagnetic transducer is made according to Mekhontsev Yu.Ya. circuit /3/ and represents a differential transformer type transducer (Fig. 2) the primary winding of which is a magnetizing winding and a secondary one is a measuring winding. Both windings are placed on the U type cores located at an angle of 90°C one to another. A magnetic bridge is formed on the surface of the metal to be tested, in the zone of the electromagnetic transducer, the sections, enclosed between the points of conjugation of the cores with the surfaces to be inspected, being the bridge arms.

Electromotive force initiated in the measuring winding will depend on the disbalance and, consequently, on the value of magnetic anisotropy corresponding to the value of elastic stresses. The received signal in the form of an electromotive force to the electron meter input.

Fig. 1. ИНИ-I portable stress meter:
1 - electromagnetic transducers;
2 - meter.

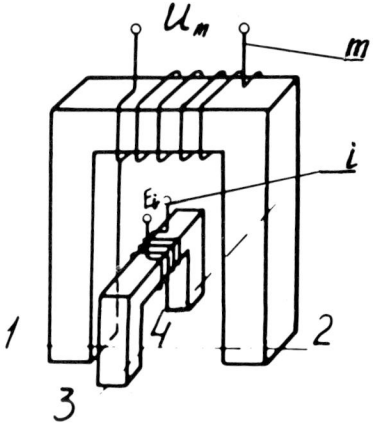

Fig. 2. Diagram of stress measurement with ДМА-I transducer.

The analysis of electromagnetic processes in the choke circuit of the electromagnetic transducers switching indicates a considerable effect of non-magnetic gap on the accuracy of measurement that requires whether a careful preparation before measuring or a check of the gap value. The principle of gap fixing is used in the developed H-I mechanical stress meter. Twopolar transducer of the instrument is switched into a resonant circuit in which at certain frequency the dependence of voltage on the gap at first increases in a monotone way and then reduces in a monotone way. A required gap value at which the transducer voltage is maximum is provided for each steel grade by selecting the circuit capacitance and signal frequency.

"Transducer-Capacitance" resonant circuit is fed with a polyharmonic signal, the 9th harmonic of which is close to the circuit resonant frequency. According to the value of the 1st harmonic a certain gap is set and according to the 9th harmonic the mechanical stresses in an article are measured. Minor changes in the 1st harmonic reading caused by mechanical stresses result in a minor error in the 9th harmonic reasing (Fig.3).

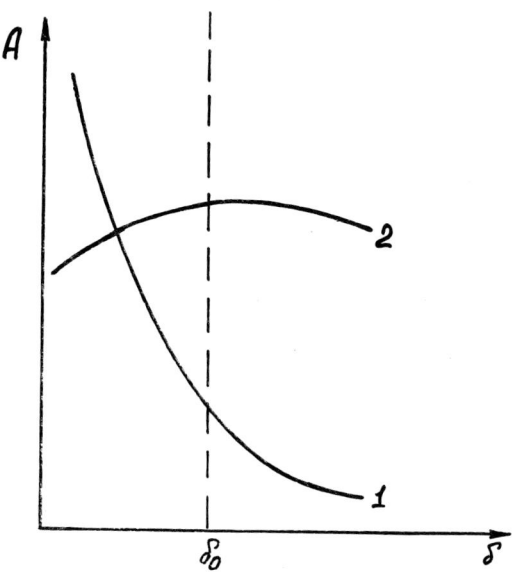

Fig. 3. Dependence of the 1st harmonic (1) amplitude increment and the 9th harmonic (2) amplitude increments on the size of non-magnetic gap.

For separating useful information and for tuning out from the direct components of the increment in amplitude or the voltage phase of the two-pole transducers not associated with the deformation of ferromagnetic materials (magnetic anisotropy, material structural conditions) use is made of special compensation circuits (Fig. 4). If T1 transducer is placed and T2 transducer is placed on an unloaded specimen made of the same material, in the case of a change in non-magnetic gap the transducers voltages will change as shown in Fig. 5. According to the hodograph curves (Fig. 5) the dependencies of Ui on the gap size for compensation voltages different in amplitude

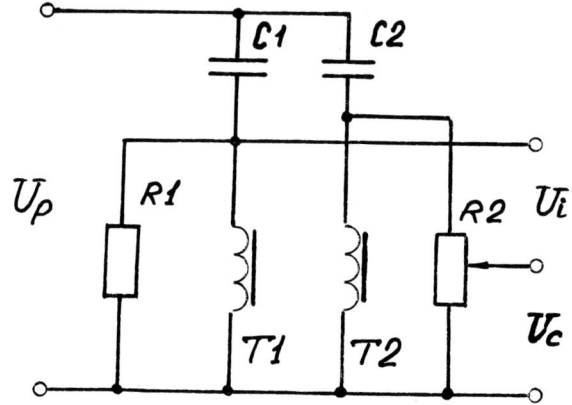

Fig. 4. Compensation circuit: $C_1 \simeq C_2$; $R_1 \simeq R_2$, T1 & T2 - similar two-pole transducers; Ui - information signal; Uc - compensation voltage.

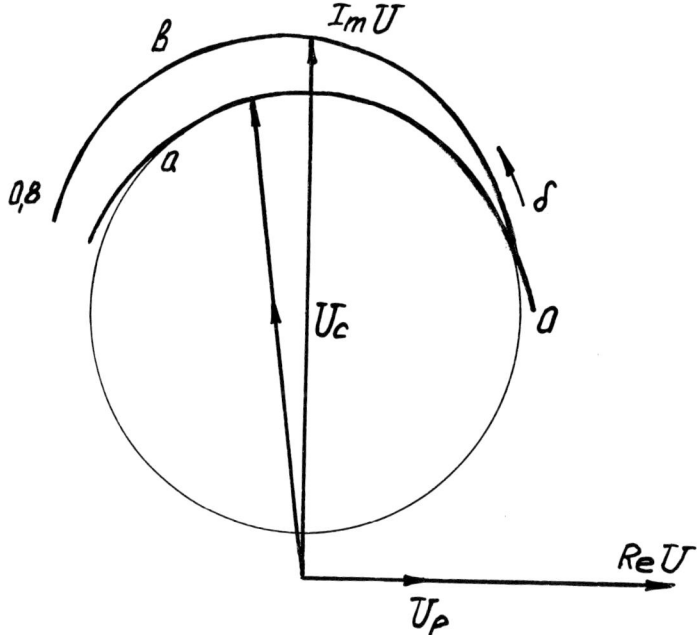

Fig. 5. Hodograph curves of stress vector:
a - with T1 transducer;
b - with T2 transducers.

are plotted (Fig. 6). Dependency 5 is obtained for the compensation voltage shown in Fig. 4. If T1 transducer is used as a measuring one and T2 transducer as a compensation one, measurements may be made in a rather wide range of the gap change. Such a principle is implemented in ИНГ-2.4 stress meter.

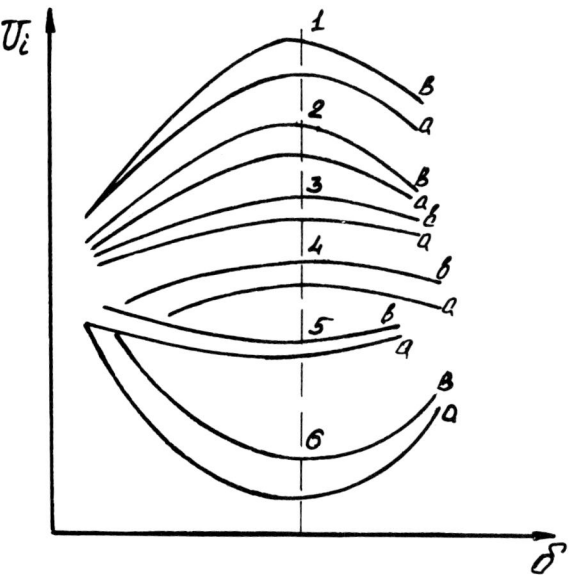

Fig. 6. Dependence of information signal on gap at different values:
a - stresses are equal to zero;
b - stresses are equal to $0.4\sigma_u$.

This instrument implements digital representation of data thus reducing a measurement error and allowing to use digital calculation means to process the results of measurement.

The experiments have shown that a change in the gap value within the limit of 0.03-0.05 mm introduces an error in the results of voltage measurement equal to about 10 MPa. Due to this a controllable setting of gap equal to 0.03-0.05 mm is the most reasonable for two-polar transducers from metrological point of view.

In order to determine an individual effect of residual stresses on magnetic permeability and to exclude the effect of nonuniform structure a technique of stress measurement using a differential circuit with ДМАД- 25 /2/ electromagnetic transducer has been developed (Fig. 7).

When measuring, the electromagnetic transducer with H-type core is oriented and moved so that the transducer poles conjugate with the like structural zones of the standard and the article to be tested.

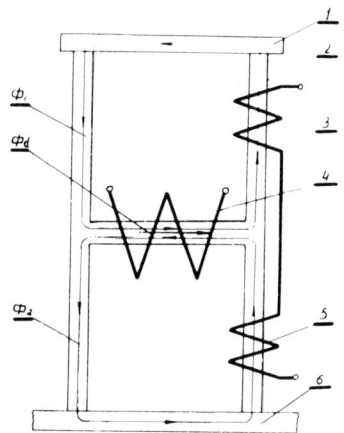

Fig. 7. Electromagnetic circuit of ДМАД-25 transducer, Φ_1, Φ_2 - magnetic fluxes in circuits; Φ_d - magnetic fluxes of disbalance: 1 - standard; 2-magnetic circuit; 3, 5-magnetizing coils; 4-indication coil; 6 article.

A template is used as a standard. The templet is cut out mechanically of the weldment which has undergone a corresponding cycle of manufacturing operations, the treatment conditions being observed thoroughly. As a result of cutting the template is relieved of the strasses acting in it, preserves a typical location serves of the structural zones. Owing to the differential circuit they estimate the change in the magnetic permeability which is conditioned by mechanical stresses in the article to be tested

RESIDUAL STRESS TEST STAND

The developed electromagnetic transducers are rather informative means of non-destructive measurement of residual stresses and they were used for developing the means and systems of stress inspection in cylindrical articles after welding and heat treatment.

СКОН-10 stand is one of such means, it is designed for inspecting stresses in welded shells (Fig. 8). It includes the rack for moving the magnetoelastic transducers, the set for torning an article being tested and the control panel.

Fig. 8. CKOH-10 set general view:
1 - rack; 2 - panel; 3 - article.

When CKOH-10 stand is used for testing, a measured information is recorded in the form of graphs on graph paper this making it possible to put it down in an article certificate and to use it many times by direct perception of man in order to analyse and estimate stresses in their relation to the manufacturing process parameters.

NON-DESTRUCTIVE INSPECTION OF STRESSES DURING HEAT TREATMENT

The effect of induction tempering conditions on metal stressed state in a welded joint may be visually observed in circular stresses in circumferential welds of cylindrical shells. The axial residual stresses are of low value and threfore the effect of tempering on the change in their value has not been estimated.

The measurements of residual circular stresses were made using CKOH-10 residual stress test stand for weld butt joints of 35XHM type steel. The batches of 3 articles per each heat treatment conditions have been investigated. The heat treatment conditions were: heating temperature T = 600, 650, 700°C; heating duration was 8.16 and 24 minutes.

The results of the investigation of tempering temperature effect on residual stresses are given in Fig. 9. Fig. 10 presents the effect of the duration of the holding on the value of the residual stresses during tempering.

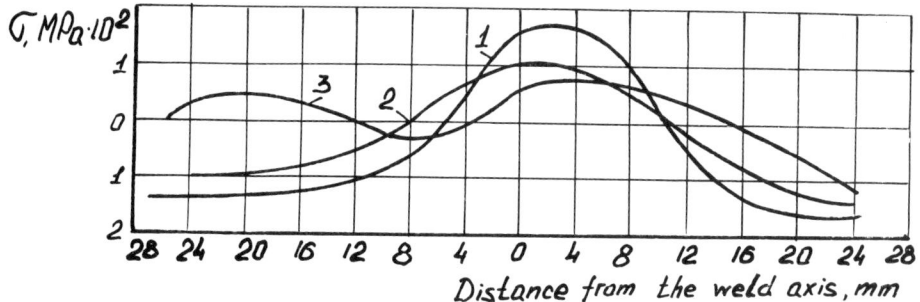

Fig. 9. Distribution of circular stresses depending on temperature during heat treatment:
1-tempering: T=600°C, holding for 8 min;
2-tempering: T=650°C, holding for 8 min;
3-tempering: T=700°C, holding for 8 min.

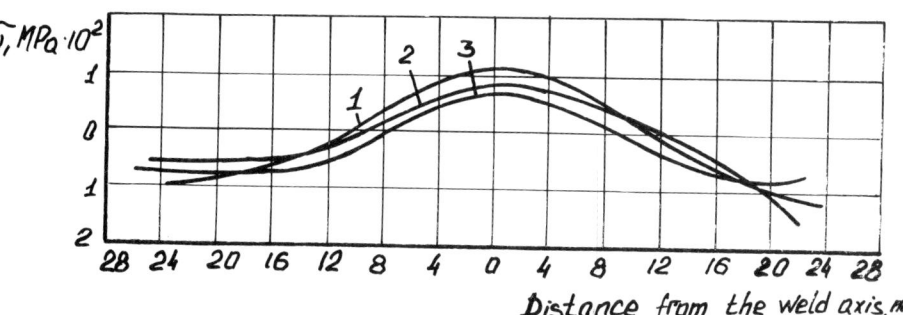

Fig. 10. Effect of holding duration during heating on residual stress distribution during heat treatment:
1-tempering: T=650°C, holding for 8min;
2-tempering: T=650°C, holding for 16min;
3-tempering: T=650°C, holding for 24min.

Depending on the degree of residual stress reduction the tempering at T = 700°C is the most favourable but the deterioration of mechanical properties caused by the effect of the heating high temperature makes its application impossible. The tempering at T = 650°C in comparison with the tempering at T=600°C is more suitable as to the stress value and to the pattern of stress distribution (maximum) (values are at further distance from weld).

The heat duration, as the results of measurements show (Fig.9), does not practically influence the level and pattern of residual stress distribution. Therefore the duration of holding when heating should be chosen proceeding from the conditions of efficient application of the equipment and minimum labour consumption when doing it.

The conducted investigations concerned with the study of the effect of tempering conditions on the decrease of residual stresses have made it possible to determine that the tempering at heating temperature T = 650°C for 8 minutes is optimal for these articles as to its effect both on the metal properties and the degree of residual stress decrease. These tempering conditions are recommended for heat treatment of the shell circumferential welds after welding.

The performed test of the distribution of residual stresses in circumferential welds of cylindrical shells with CKOH-10 set using the magnetoelastic technique has shown that after welding stresses reach their maximum values in tension (σ = 250 MPa) at a distance of 18...20 mm from the weld axis. The weld and heat-affected zone undergo compression, the stresses reach 150...180 MPa along the weld axis (Fig. 11, curve 1).

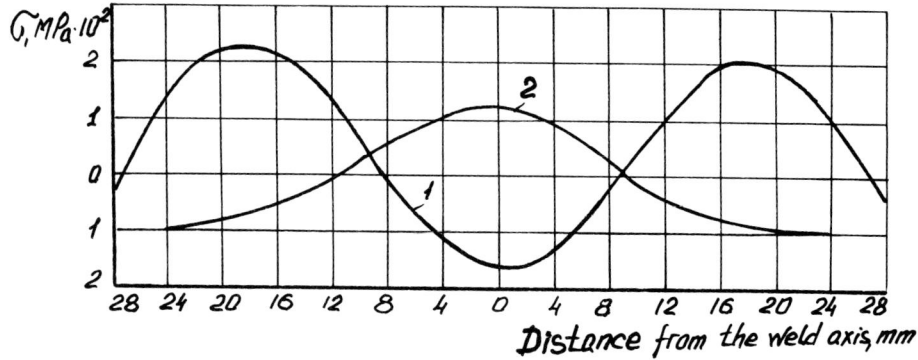

Fig. 11. Distribution of circular stresses after welding and heat treatment:
1 - after welding;
2 - after heat treatment.

Induction tempering redistributes residual stresses considerably (Fig. 11, curve 2). The stresses conditioned by structural transformations reduce, the stresses caused by shrinkage and plastic deformation during welding change. As a whole the induction tempering redistributes residual stresses after welding favourably resulting in the reduction of residual stresses in their value.

The given data show that the introduction of in-process inspection of residual stresses is reasonable first of all for circular stresses which are of a considerable value in welded joints of thin-walled shells and the degree of reduction of which by subsequent tempering is the most important proceeding from the conditions of strength.

Because of the fact that the residual stresses, caused by the application of local heating during such manufacturing operations as welding and induction tempering, are subjected to measurement using non-destructive techniques with great difficulty, till quite recently in the fabrication of welded structures there were practically no sufficiently simple and intermative techniques of residual stress inspection as an independent production operation which are suitable for shop conditions.

In the given study the specialization of measuring equipment, the application of a differential method of measurement using a structurally nonuniform standard, the development of a special sensor, the automation of measurement procedure, the implementation of measurement results recording have been the basis for the development of a manufacturing technique for residual stress inspection.

The realization of the above tasks has provided integrated solving of in-process inspection of residual stresses in welded structures and made the pre-condition for the application of the new inspection technique in the manufacturing processes of the fabrication of articles as a means providing statistical inspection of welding manufacturing processes and heat treatment. The data of the value and distribution of residual stresses after welding and heat treatment obtained with the help of CKOH-10 stand make it possible to solve more objectively the matters of articles acceptance for service or their final rejection to increase substantially the reliability of articles and reduce the amount of rejected articles.

NOVELTY

The developed technique and equipment have made it possible for the first time in the practice of diagnostics to implement the non-destructive inspection of residual stresses in the course of the fabrication and heat treatment providing recorded representation of measurement results. On the one hand the obtained results of the non-destructive inspection of residual stresses can be regarded as a variety of acceptance inspection providing the assigned level of article quality and on the other hand - as a variety of statistical inspection intended for current of heat treatment parameters.

REFERENCES

Majorov F.V. (1939). Magnetostiction technique and instruments for pressure measurement. Moscow, Zhukovsky Central Aviation State Institute.
Mekhontsev Yu.Ya. (1958). Elastic stress meter. Radio, 5.
Zhdanov I.M., and V.V.Batyuk (1978). Non-destructive technological inspection of residual stresses in the course of the fabrication of welded articles. Proceedings of the All-Union Conference of Higher Educational Establishments, 73-78.

Postweld Heat Treatment of C-Mn and Microalloyed Steels: An Evaluation on the Basis of C.T.O.D. and Wide Plate Tests

W. Provost*, A. Dhooge** and A. Vinckier***

*W.T.C.M. Ghent, Belgium
**Research Center of the Belgian Welding Institute, Ghent, Belgium
***Laboratory for Strength of Materials and Welding Technology, State University Ghent, Belgium

ABSTRACT

The effects of a post weld heat treatment (P.W.H.T.) on the toughness properties of weldments in C-Mn and microalloyed steels are discussed. Three pressure vessel steels (a CMn steel ASTM A 537 Cl.1 and two microalloyed steels TT St E 460) and one Nb-alloyed structural steel have been tested in the as welded and PWHT condition.
The data reveal the variation, due to PWHT in base metal, HAZ and weld metal properties such as strength, impact and C.T.O.D. toughness, hardness and microstructure. An assessment of the acceptable defect levels for both the as welded and PWHT condition has been performed on the basis of C.T.O.D. and wide plate test results. The test results indicate that, provided stress corrosion is not a matter of concern, PWHT of the TT St E 460 steels should be omitted because the PWHT results in a significant drop in HAZ and weld metal toughness. Acceptable toughness levels are obtained in the as welded condition for the C-Mn and Nb-alloyed structural steel tested, indicating that a PWHT is not necessarily always required to obtain adequate toughness properties.

KEYWORDS

Post weld heat treatment, microalloyed steels, structural steels, wide plate tests.

INTRODUCTION

The Belgian Regulations (1) and (2) for transport and storage of liquefied CO_2 at temperatures of about $-40°C$ require that all welded CO_2-containers have to be stress relieved, except when the following conditions are simultaneously met: $D/t \geq 0,2$ R_m, $R_m \leq 490$ N/mm^2 and $t \leq 15$ mm (R_m = minimum specified tensile strength).

The general regulations also state that all vessels have to be given an appropriate heat treatment in order to obtain their most favourable properties. It is thus intrinsically accepted that a stress relief heat treatment is always beneficial for commonly used steels.

In order to reduce their weight, modern CO_2-transport tanks are built in low alloyed high strength steels with a minimum specified tensile strength exceeding 490 N/mm^2, while the large dimensions of recently build CO_2 stationnary storage tanks result in plate thicknesses exceeding 15 mm. As in both cases one of the above mentioned conditions for omitting stress relieving is not fulfilled, practically all CO_2-vessels have to be post weld heat treated.

Recent research work (3-7), clearly indicates that stress relieving Nb and V-alloyed steels can give rise to an increase in transition temperatures. This is however not reflected in the fabrication codes, where no distinction is made between C-Mn steels and micro-alloyed steels.

Two accidents with CO_2 transport tanks constructed in high strength micro-alloyed steels have recently been reported (8) whereby in at least one case the low Charpy energy levels could be attributed to stress relieving. However, the influence of a PWHT on these tanks was not completely elucidated.

In this paper, the results of a closer investigation on the effects of a PWHT on the properties of weldments in low temperature steels are reported. The data record the variation in base metal, HAZ and weld metal properties such as strength, Charpy V and CTOD toughness, hardness and microstructure due to PWHT. Assessment of acceptable defect levels, both in as welded and stress relieved condition, is done on the basis of CTOD and wide plate test results.

More recently, the same testing techniques have been applied on welded ISO Fe510 D Mod. normalized structural steel plates in thicknesses of 40 and 70 mm in order to evaluate the weldability, fracture toughness and the need for a PWHT in case of offshore constructions with a design temperature of -10°C.

MATERIALS

a) Steels for Low Temperature Applications

Three steels for low temperature service have been tested. Materials A and B are two T St E 460 steels according to DIN 17102 (Fe E 460 KT according to Euronorm 113-72), 10 mm thick and manufactured by different steel mills. Steel C (T St E 355 according to DIN 17102 or Fe 355 KT according to Euronorm 113-72) is a 22 mm ASTM A 537 Cl 1 steel. The plate materials were in the normalized condition. Their chemical composition is given in Table 1.

b) Structural Steel for Offshore Applications

Test material D is a ISO Fe 510 D Mod. normalized structural steel, characterized by its low carbon content, Cu-alloyed and microalloyed with maximum 0,015% Nb. The steel was tested in two plate thicknesses, i.e. in continuous cast 40 mm and in ingot cast 70 mm. The chemical composition is given in Table 1.

WELDING AND POST WELD HEAT TREATMENT

All welding on plates A, B and C was performed by experienced manufacturers of transport and storage tanks, according to their qualified welding procedures. Materials A and B were submerged arc welded with an identical procedure with a heat input of 20 kJ/cm (S2Mo-wire, F8-A4-EAL flux). Steel C was submerged arc welded with a heat input of 14 kJ/cm (SD3-wire, F7-A4-EGC-flux). Part of the test plates were stress relieved in an electrically heated furnace at 570°C for 30 minutes (Steel A and C) and 60 minutes (Steel B) respectively.

The 40 and 70 mm thick plates in steel D were submerged arc welded in the 2G position with two different heat inputs: 30 and 40 kJ/cm (SD3-wire, F7-A4-EGC flux). The weld preparation is shown in Figure 1. The 40 mm thick plates were tested in the as welded condition while the 70 mm plates were tested in both the as welded and PWHT condition (600°C - 2hrs).

MECHANICAL TEST PROGRAMME

The following mechanical tests were performed:

- Weld procedure qualification tests for each material
- Charpy impact tests at various temperatures in the base material, HAZ and weld metal center (WMC) for steels A, B and C
- CTOD tests at -40°C on steels A, B, C and at -10°C on steel D on specimens with the fatigue crack positioned in the base metal, HAZ and WMC.
- Hardness traverses on each material.

All these tests were carried out both in the as-welded and in the stress relieved condition for steels A, B and C. Pellini dropweight tests have been carried out on steel A in the as-welded and stress relieved condition.

For steel A and C these tests were supplemented with the following wide plate tests :

- Wide plate tests on specimens provided with a 14mm Wells notch, applied before welding and tested in the as-welded condition.
- Wide plate tests on large scale panels, provided with a through-thickness notch located either in the HAZ or WMC, tested in the as-welded and in the stress relieved condition.

For steel D, seven wide plate tests were performed on large scale panels provided with either through thickness or surface notches in the HAZ. These tests were done only in the as welded condition.

TEST RESULTS

Qualification Tests

The results of all qualification tests are summarized in Tables 2a and 2b.

The yield strength of steel A is found to be significantly reduced by stress relieving: the minimum required 460 N/mm^2 is hardly obtained after PWHT. In contrast, steel B not only shows substantially higher strength values in the as-welded condition, but a slight increase in yield strength is observed after stress relieving. The strength properties of the CMn steel C and steel D are not affected by the PWHT.

No cracks were detected after side, face and root bend testing.

The -40°C impact test results pointed to a deleterious effect of stress relieving on both HAZ and weld metal toughness for the low-alloyed steels A and B. However, a favourable effect of the PWHT on HAZ and weld metal toughness could be observed for steels C and D. It must thereby be remarked that for materials A and B, the required 31J at -40°C was not always obtained in the stress relieved condition.

Impact Testing on Steels A, B and C

In order to determine the Charpy-V transition temperature, a set of three specimens was tested at various temperatures, ranging from -20 to -80°C. Their results are shown in figures 2a, b and c.

Excellent plate toughness down to -60°C is obtained for material A in the as received condition. After PWHT, the values obtained are even higher for all test temperatures. Material B shows much lower impact values in the as-received condition; the base metal impact properties appear to be fairly insensitive to the applied PWHT within the temperature range considered. The results obtained on steel C (figure 2c) indicate a rather high absorbed energy level in the as received condition, down to temperatures of -80°C. PWHT results in an increase in toughness of the base material.

An excellent WMC impact toughness is obtained in the as-welded condition on the welds in materials A and B, which were made with an identical welding procedure. An important increase in transition temperature is, however, found after PWHT, resulting in very poor absorbed energy levels. In contrast, the excellent WMC impact values obtained in the as-welded condition in steel C shift to higher values by stress relieving.

For material A, the lowest impact values in the as welded condition are found in the HAZ, but the 35 J/cm^2 (28 J) transition temperature is still below -40°C. After stress relieving, a pronounced drop in absorbed energy has been observed, resulting in an increase in transition temperature to about -20°C. Whereas the HAZ impact properties of material B are slightly higher in the as welded condition, they are hardly affected by the PWHT, except at very low temperatures (-60°C). It can be seen from figure 2b that an absorbed energy level of over 40 J/cm^2 is obtained at -50°C in both test conditions. The HAZ impact values of steel C decrease steeply with temperature below -40°C. It is apparent that a PWHT increases the HAZ-impact level for this material.

In any case, whereas the 35 J/cm^2 (28 J) transition temperature occasionally exceeds -40°C for materials A and B in the stress relieved condition, all impact values of material C exceed 50 J/cm^2 at -40°C in both test conditions.

CTOD Tests on Steels A, B and C

For each weldment, 18 CTOD specimens (B x   were tested at -40°C. Half of the specimens were stress relieved.

The data obtained are summarised in Table 3a. Steel A shows excellent base metal CTOD values in both test conditions. Substantially lower base material CTOD values were obtained for steel B, whereby one specimen, tested in the as welded condition, exhibited a totally brittle fracture at a CTOD value of only 0,1 mm. In general, a PWHT was found to have only a marginal effect on the base metal CTOD value of both steels, althought there seems to be a trend of increasing CTOD values after PWHT. The highest base material CTOD values are obtained for steel C; a PWHT has led to no appreciable variation in base metal CTOD properties of this steel.

The WMC-CTOD values of steels A and B are significantly influenced by PWHT. Although an identical welding procedure has been applied for these two materials it was apparent that steel A shows higher WMC-CTOD values in the as welded condition, although the WMC-CTOD values obtained for steel B are still satisfactory in the as welded condition. Applying a PWHT leads to an appreciable drop in CTOD values : unacceptably low CTOD values of 0,04 mm are found after stress relieving for both materials. In contrast, PWHT seems to be very favourable to the WMC-CTOD values of steel C, whereby excellent CTOD values are obtained in the PWHT condition.

Steel A shows rather low HAZ-CTOD values in the as welded condition, whereas low to very low HAZ-CTOD values are obtained for steel B. PWHT seems to have an adverse effect on the HAZ-CTOD values, whereby consistantly very low CTOD-values are found in the HAZ of steel B after PWHT. A high level of scatter is observed on the results of the HAZ-CTOD tests of steel C, tested in the as welded condition. After PWHT excellent values were obtained for all specimens.

CTOD Tests on Steel D

Both full thickness B x 2B and surface notched B x B CTOD specimens were extracted from the 40 mm as welded thick plates. Only B x 2B specimens were taken from the 70 mm plate which was tested in the as welded and PWHT condition. The specimens were notched either at the weld metal center or in the HAZ and tested at -10°C. The test results are summarized in Table 3b.

The base metal CTOD fracture toughness values at -10°C were found to be excellent. The minimum value recorded in the normalized condition was 1,36 mm. After PWHT, the obtained CTOD values were even better and all above 2 mm. Also the HAZ showed a high CTOD fracture toughness with a lowest value of 0,40 mm in the as welded condition. There was found to exist considerable scatter due to the inherent material heterogenity within the specific HAZ region sampled. The fracture toughness of the weld metal was found to be inferior to that of the HAZ. Despite the limited number of 70 mm PWHT specimens tested it is clear that stress relieving is slightly beneficial for the HAZ/FL fracture toughness, although the degree of improvement is somewhat variable and unpredictable.

Wells Wide Plate Tests

The susceptibility of materials A and C to strain ageing embrittlement was evaluated using Wells wide plate tensile specimens. Two test plates, provided with a symmetrical notch with a length of 14 mm each, were welded according to the above mentioned test procedure and tested at -40°C. The notch length has been selected such that a temperature of about 300°C is obtained during welding at the notchtip, resulting in the highest degree of embrittlement in that region.

Previous research work (9, 10 and 11) already indicated that a PWHT restores the ductility of strain aged material. Consequently, no Wells wide plate tests have been performed in the stress relieved condition. The results of these wide plate tests are summarized in table 4a.

The Wells wide plate steel A specimen could not be fractured as the maximum displacement capacity of the testing rig was reached after an overall deformation of 4% was obtained on the specimen gauge length (450 mm). This illustrates that the overall behaviour was unaffected by strain ageing. Stable crack growth by ductile tearing could be observed at both sides of the notch with a length of about 45 mm. Noteworthy is the high yield stress of the specimen exceeding the required minimum yield strength of the base material, despite the presence of such a severe notch (it must be remarked that all stresses are calculated on the gross section of the specimens). A local deformation at the notchtip of 13% could be derived from the moiré-pattern. This value is illustrative of the very high toughness of the material even at a temperature of -40°C.

The Wells wide plate specimen of steel C broke at an overall elongation of 1,34%. Again, stable crack extension by ductile tearing was observed at the notchtip, whereby very high local deformations were measured, exceeding 15%. The minimum specified yield strength of material C was also exceeded for this specimen. These results illustrate the low susceptibility of material C to strain ageing embrittlement. It is hereby to be remarked that the amount of strain ageing at the notchtip of a Wells wide plate specimen normally increases markedly with increasing plate thickness, resulting in lower overall strain values for material C.

HAZ and Weld Metal Center Notched Wide Plate Tests on Steels A and C

In addition to the impact and CTOD tests, tests at - 40°C on wide plate specimens provided with a 10 mm long through-thickness notch in the HAZ and in the WMC have been carried out to evaluate the influence of a PWHT on the toughness of the welded joints in steels A and C. Load-elongation and GCOD-elongation diagrams were recorded during the test whereby the overall elongation was measured on a gauge length of 450 mm, while the GCOD was measured at the cracktip over a gauge length of 8 mm straddling one of the notchtips. The local deformation at the notchtip was also measured using moiré-grids containing 10 lines/mm. The results are summarized in table 4a.

The test on the WMC notched wide plate specimen in steel A, tested in the as welded condition, was interrupted when an overall elongation of 4,5% was reached. The moiré pattern demonstrates (figure 3a) a gross section yielding deformation pattern whereby plastic deformation also occurs outside the notched section. Stable crack extension by ductile tearing was clearly visible at the notchtip. Noteworthy are the very high maximum stress and GCOD-values. The WMC-notched specimen tested in the stress relieved condition failed at an overall elongation of only 0,61% whereby no stable crack extension by ductile tearing could be observed at the notchtip. Much lower GCOD and local deformations at the notchtip were observed (figure 3b), compared to the as welded specimen. The fracture surface of the stress relieved specimen showed a fully brittle fracture appearance.

The two HAZ-notched wide plate specimens (steel A) showed a pop-in at lower stress and strain values than the actual fracture stress and strains. Again, the as welded specimens showed much higher toughness than the stress relieved ones.

All tests on WMC- and HAZ-notched wide plate specimens in steel C were interrupted after reaching an overall elongation of 4,2%. Because of the weld metal overmatching, the corresponding GCOD values are fairly low, indicating that most plastic deformation occurred in the unnotched area. Stable crack extension by ductile tearing was observed for all specimens. No difference in strength properties is noticeable between the as welded and the PWHT condition.

Wide Plate Tests on Steel D

Seven wide plates, four on 40 mm thick and three on 70 mm thick submerged arc welded plates containing artificial flaws (machined or fatigued) have been tensile tested at a temperature of -10°C. All weldments were tested in the as welded condition.

The main test results are summarized in Tables 4b and 4c. These Tables give the yield stress, maximum or fracture stress, overall strain (gauge length 350 mm and 800 mm respectively) and the clip gauge surface displacement GCOD at maximum load or at fracture.

Dropweight Tests on Steel A

The influence of a PWHT on the susceptibility of steel A to dynamic crack extension has been evaluated by means of dropweight tests. The NDT temperature was found to be below -60°C for both test conditions.

Hardness Measurements

Hardness traverses with a load of 500g were made on a macrosection extracted from each weld, both in the as welded and in the stress relieved condition for steels A, B and C. The hardness indentations sampled the weld deposit, the HAZ and the parent plate material, unaffected by welding.

The weld metal and HAZ hardness levels of steel A is found be increase slightly by a PWHT (from 270 HV to 300 HV), while a slight decrease of the plate hardness level is observed.

A PWHT has little influence on the HAZ and weld metal hardness levels of steel B, where hardness levels of only 225 to 250 HV were obtained in both test conditions. A significant decrease in hardness level occurs in the parent plate material where values of below 220 HV have been recorded in the stress relieved condition, compared to 240-260 HV in the as welded condition.

A noticeable effect of a PWHT has been observed in the HAZ hardness of steel C. The peak hardness level of 280 HV has been reduced to below 220 HV by PWHT. A slight reduction in weld metal and plate hardness has also been recorded.

An extensive number of hardness surveys (HV 10) have been performed on each individual weld combination of steel D. The maxium HV 10 values (240 HV) have been observed in the HAZ. Applying a PWHT resulted in a slight hardness decrease in both the HAZ (213 HV) and weld metal (205 HV).

Microstructural examination

The microstructure of the weldments were examined in the as-welded and in the stress relieved condition.

The coarse grained HAZ microstructures of steels A, B and C are shown in figure 4. These structures can be identified as mainly bainitic, consisting almost entirely of ferrite with aligned M-A-C. Preferential nucleation of ferrite along prior austenite grain boundaries is not observed and no significant amount of other phases is present. Compared with steels A and B, the coarse-grained HAZ of steel C shows a smaller grain size.

Little if no changes in microstructure could be identified after a PWHT. Only very marginal changes seems to have occurred, whereby the individual laths are slightly less well defined after PWHT. Additional precipitation on lath boundaries, although they most likely may have occurred, could not be observed on the optical micrographs.
The coarse grained HAZ microstructure of the structural steel D in the as welded condition is shown in Figure 5. It consists of upper bainite (ferrite with aligned M-A-C).

DISCUSSION

Application of the CTOD Design Curve

For a given CTOD value, allowable defect sizes can be evaluated on the basis of the CTOD design curve, provided that the design stress level and residual stress level are defined (12).

For steels A, B and C allowable defects sizes were calculated in both the as welded and stress relieved condition, whereby the following residual stress levels were adopted:
- as welded $\sigma_R = \sigma_{ys}/2$ and $\sigma_R = \sigma_{ys}$
- stress relieved $\sigma_R = 0$ and $\sigma_R = \sigma_{ys}/2$

Two design stress levels were selected, respectively 0,5 and 0,33 R_m.

The results of these calculations, applied to steel B, are summarized in Table 5. Allowable defect lengths corresponding to both the minimum and mean CTOD values have been calculated for the parent plate, the weld deposit and the HAZ. For simplification, we will stick further to the defect lengths corresponding to the minimum CTOD values. The results clearly indicate that acceptable defect levels based on the CTOD design curve approach strongly depend on the design stress and even more on the adopted residual stress level. In the extreme conditions that the residual stress level equals the yield stress of the material in the as welded condition, and that all residual stresses disappear by PWHT, higher acceptable defect levels are obtained in the stress relieved condition. This seems however not very realistic, not only because the residual stress level does not equal zero after stress relieving, but also as at the first proof test new residual stresses will also be generated in zones of high secundary stresses. When a more realistic residual stress level of half the yield stress is adopted in the PWHT condition, the highest acceptable defect sizes are obtained in the as welded condition.

The CTOD design curve concept can also be applied to the wide plate test results. It suffices to replace σ_1/σ_{ys} by $\varepsilon_0/\varepsilon_{ys}$ where ε_0 is the overall elongation at fracture of the wide plate specimen and ε_{ys} is the yield strain of the material at the notch location. The calculated allowable defect sizes can then be compared to the 10mm defect length of the wide plate specimens. The ratio of the actual notch length ($2a_c$) to the allowable defect length ($2a_{max}$) predicted by the CTOD design curve allows us to have a rough estimate of the accuracy of the latter.

The results, given in Table 6, clearly show that the design curve is conservative for all the wide plate tests and in some cases even overconservative. Taken into account that the test on the WMC notched wide plate specimen is the as welded condition had to be interrupted, it can also be concluded that the CTOD design curve is even more conservative for welds tested in the as welded condition. An evaluation of the necessity of a PWHT on the basis of the CTOD design curve is thus highly unreliable.

For steel D, the results of the wide plate tests have been compared with the prediction of the CTOD design curve. Values of $2a_{max}$ have been calculated using the minimum CTOD values. The analysis has been performed assuming no residual stress. The test results demonstrate again the very large degree of convervatism of the CTOD design curve approach. Apart from wide plate 2, which gave rise to a net section yielding deformation mode, the safety ratios vary between 3,8 and 33,7. These high safety factors are largely due to the yield strength overmatching of the weld metal and to the occurrence of the gross section yielding in all cases.

General Discussion

The results of the present study shows that the effect of a stress relief heat treatment on the toughness of the welded joints is highly material dependent.

Steel A

All fracture toughness tests, including impact, CTOD and wide plate tests point to the same conclusions. A reduction in HAZ toughness and an even more pronounced decrease in weld metal toughness, resulting from PWHT are demonstrated by all these tests. They show an acceptable toughness level of the welded joints in the as welded condition, while no significant strain ageing was observed.

Unacceptably low toughness values are obtained in the PWHT condition, both in the HAZ and WMC; mean impact values are below 35 J/cm^2, the mean CTOD value is below 0,2mm and the overall elongation of the wide plate specimen is less than 1%.

The decrease in HAZ and WMC toughness is accompanied by a slight increase in hardness. Metallographic examinations did not reveal any influence of the PWHT on microstructure. Changes which could affect hardness and toughness, such as precipitation and growth of carbides cannot be resolved by optical microscopy and would require transmission electron microscopy.

Steel B

The influence of PWHT on the toughness of steel B is similar to that on steel A. However, steel B shows lower initial base material toughness, whereby rather poor HAZ-CTOD values are obtained in the as welded condition. These low CTOD results are in contrast with the fairly good impact values; unfortunately, this could not be confirmed by wide plate testing. Moreover, no hardening by PWHT of the weld metal was observed in steel B while softening of the HAZ can be noticed.

Steel C

A significant improvement of the toughness of steel C by a PWHT was demonstrated by impact and CTOD testing, both in the HAZ and in the WMC. Simultaneously, a decrease in hardness in all zones of the weld was observed. However, impact and CTOD testing demonstrated an acceptable toughness level in the as welded condition. This was fully confirmed by wide plate testing, where all specimens showed a completely ductile fracture behaviour with overall elongations exceeding 4%. Therefore, it is reasonable to state that although a PWHT was found to be beneficial, it is completely superfluous for the tested plate thickness. These results are in line with those previously obtained on an other C Mn steel (11), where it was found that a PWHT can be omitted for plate thicknesses up to at least 30mm.

Steel D

Most tests on steel D have been performed on as received or as welded material. The small scale test results indicate that the steel possesses an excellent weldability and notch toughness, and no particular problems are to be expected during construction assuming that normal fabrication and welding procedures are applied. Large scale tests (CTOD and wide plates) have proven the excellent HAZ fracture toughness in the as welded condition. Indeed, large through thickness and surfach notches, introduced in the HAZ still gave rise to an overall plastification of the notch free zones prior to fracture. The results clearly indicate that the effect of residual stresses on HAZ and weld metal properties is negligible. Hence it might be justified to assume a zero residual stress level when assessing flaws in as welded structures with the CTOD design curve.

CONCLUSIONS

In the present study, the influence of PWHT on the toughness of welded joints in 3 low temperature steel grades and in one structural steel has been investigated. From the results obtained, the following conclusions can be drawn :

1. PWHT results in a significant drop in HAZ and WMC toughness for the two St E 460 steels A and B (thickness 10 mm). Consequently, stress relieving of this steel grade should be omitted.

2. All fracture toughness tests, including wide plate testing, show evidence of an acceptable toughness level of the welds in steel A in the as welded condition.

3. In contrast with the satisfactory Charpy V toughness, rather low CTOD values were observed in the HAZ of steel B in the as welded condition.

4. A significant improvement in Charpy V and CTOD toughness by PWHT is demonstrated for steel C (thickness 22 mm). It should however be noted that acceptable toughness levels were also obtained in the as welded condition. This was also reflected by the wide plate test results, giving evidence that PWHT is superfluous, at least up to the tested plate thickness.

5. Steel D shows excellent fracture toughness in the as welded condition. PWHT can be omitted for all thicknesses up to 80 mm.

ACKNOWLEDGEMENTS

The present research program was jointly funded by IWONL (Instituut ter Aanmoediging van het Wetenschappelijk Onderzoek in Nijverheid en Landbouw) and WTCM (Wetenschappelijk en Technisch Centrum van de Metaalverwerkende Nijverheid). The authors are indebted to Apragaz, Fabrique de Fer de Charleroi, ESAB, G & G International, Statoil (Norway) and Van Hool for their support and assistance. Also the help and cooperation of the staff of the Laboratory Soete of Ghent University in carrying out the experimental work is gratefully acknowledged.

REFERENCES

1. ADR - Belgian Amendments and Complements - Belgisch Staatsblad 1974-1980.

2. Algemeen Reglement op de Arbeidsbescherming - Belgisch Staatsblad.

3. J. DEGENKOLBE and D. UWER.
   Einfluss der Schweissbedingungen auf die Kerbschlagzähigkeit in der Wärmeeinflusszone von Schweissverbindungen hochfester Baustähle. Schweissen und Schneiden, Jahrgang 26/1974/Heft 11.

4. K. WELLINGER and A. KUUR.
   Untersuchungen über die Versprödungsneigung von Kesselbau- und Druckbehälterstählen. VDI-Z118 (1976) Nr. 17/18.

5. D. AURICH et al.
   Untersuchungen zum Problem der Spannungsarmglühgrenze. VGB-Conference "Research in Power Plant Technology 1980", Essen, May 1980.

6. G. BOTT, H. BANKSTAHL and E. SCRUBITZKI.
   Verhalten mikrolegierter Ferrikornbaustähle im Hinblich auf Vanadinversprödung beim Schweissen und Spannungsarmglühen. Schweissen und Schneiden, Jahrgang 32 (1980), Heft 3.

7. L. DORN and G. NIEBUHN.
   Untersuchungen zum Einflüss unterschiedlichen Energiezufuhr und Wärmenachbehandlung auf die Schweissgefügeeigenschaften von W St E 47 mittels Mikroscherprüfung. Schweissen und Schneiden, Jahrgang 32 (1980), Heft 6.

8. D. AURICH et al.
   Failure of tank waggon for liquified Carbon Dioxide. Mechanical and materials considerations. Proceedings of the Conference on transport and storage of LPG and LNG, Bruges, May '84, p. 283-294.

9. W. PROVOST
   Effects of a stress relief heat treatment on the toughness of welded joints of some high quality pressure vessel steels. WTCM-CRIF MT 112-1976. Welding Research Abroad, Aug.-Sept. 1976.

10. W. PROVOST
    Effects of a stress relief heat treatment on the toughness of pressure vessel quality steels welded with high heat input processes. WTCM-CRIF MT 134-1980. Int. Pres. Ves. & Piping, 9 (2), March 1981.

11. W. PROVOST.
    Effects of a stress relief heat treatment on the toughness of pressure vessel quality steels - Influence of the plate thickness. WTCM-CRIF MT 141-1981. Int. Journ. of Pres. Ves. & Piping, 10 (2), 1982.

12. PD 6493:1980 Guidance on some methods for the derivation of acceptable levels for defects in fusion welded joints. The British Standards Institution.

TABLE 1 : BASE MATERIAL CHEMICAL COMPOSITION.

	C %	Si %	Mn %	P %	S %	Cr %	Cu %	Ni %	Mo %	Al %	V %	Nb %	Ti %
Steel A	0,09	0,30	1,46	0,008	0,009	0,040	0,030	0,510	<0,010	0,021	0,140	<0,002	0,001
Steel B	0,14	0,40	1,35	0,009	0,003	0,109	0,259	0,528	0,038	0,035	0,064	0,021	-
Steel C	0,10	0,30	1,18	0,008	0,009	0,08	0,23	0,20	0,03	0,039	0,01	0,002	0,002
Steel D	0,09	0,44	1,25	0,010	0,002	0,07	0,28	0,117	0,030	0,032	0,004	0,012	0,001

TABLE 2.a : QUALIFICATION TEST RESULTS ON STEELS A, B AND C.

Steel	Cond.	Tensile tests				Impact values (-40°C)		
		Base material		Transverse		BM	HAZ	WMC
		σ_{ys_2} (N/mm^2)	σ_r (N/mm^2)	σ_y (N/mm^2)	σ_r (N/mm^2)	J	J	J
A	AW	495	615	503	623	153	46	99
	SR	463	594	472	610	213	31	23
B	AW	530	707	545	711	68	61	111
	SR	577	709	569	717	65	31	51
C	AW	354	522	350	493	103	74	103
	SR	364	511	356	499	88	94	177

TABLE 2.b : QUALIFICATION TEST RESULTS ON STEEL D.

Thickness (mm)	Condition	Heat input (kJ/cm)	Tensile tests						Impact values (-40°C)		
			Base material at R.T.		Transverse at R.T.		All weld metal at -10°C		BM	HAZ	WMC
			σ_{ys_2} (N/mm^2)	σ_r (N/mm^2)	σ_y (N/mm^2)	σ_r (N/mm^2)	σ_{ys} (N/mm^2)	σ_r (N/mm^2)	J	J	J
40	AW	30	360	484	355	503	526	631	238	57	51
		40			347	499	481	618		81	45
70	AW	30	346	490	348	499	464	587	233	85	46
		40			326	495	-	-		97	51
	SR	30	343	488	340	477	473	557	224	54	132
		40			338	493	-	-		107	114

IIWS-M

TABLE 3.a : CTOD TEST RESULTS ON STEELS A, B AND C (TESTING TEMPERATURE : -40°C)

Notch location	Condition	STEEL A		STEEL B		STEEL C	
		δt_{min} (mm)	δt_{mean} (mm)	δt_{min} (mm)	δt_{mean} (mm)	δt_{min} (mm)	δt_{mean} (mm)
BM	AW	0,46	0,53	0,10	0,21	1,19	1,17
	SR	0,58	0,61	0,28	0,28	1,12	1,14
WMC	AW	0,32	0,31	0,25	0,22	0,22	0,22
	SR	0,03	0,04	0,04	0,04	1,17	1,15
HAZ	AW	0,15	0,20	0,17	0,10	0,16	0,38
	SR	0,07	0,18	0,06	0,04	1,16	1,14

TABLE 3.b : CTOD-TEST RESULTS ON STEEL D (Testing Temperature: -10°C)

Plate Thickness (mm)	Heat input kJ/cm	Condition	Notch Location	δt_{min} (mm)	δt_{mean} (mm)
40	4	AW	WM	0,17	0,29
40	3	AW	HAZ	0,48	0,89
40	3	AW	WM	0,24	0,24
40 *	3	AW	WM	0,40	0,71
40 *	3	AW	HAZ	0,63	1,01
70	-	AW	BM	1,36	1,82
70	-	SR	BM	2,14	2,18
70	3	AW	HAZ	0,40	1,26
70	3	SR	HAZ	1,64	1,95

(*) B x B specimens

TABLE 4.a : SUMMARY OF WIDE PLATE TENSILE TEST RESULTS AT -40°C ON STEELS A AND C.

Steel	Condition	Notch type and location	Gross Yield. stress σ_{ys} (N/mm^2)	Gross stress σ_r (N/mm^2)	Overall strain ε_o (%)	Local strain ε_{loc} (%)	GCOD (mm)	Fracture initiat.	Fracture propagation
A	AW	Well's	494	538	>4,00	>13	-	ductile	-
	AW	WMC	479	646	>4,5	>13	>3,60	ductile	-
	SR	WMC	414	594	0,61	3	1,00	brittle	brittle
	AW	HAZ	465	573	2,56 (1,91*)	12	2,28 (1,40*)	ductile	brittle
	SR	HAZ	480	518	0,89 (0,48*)	6	3,00 (0,88*)	brittle	brittle
C	AW	Well's	379	400	1,34	15	-	ductile	brittle
	AW	WMC	398	>512	>4,22	>6	>1,32	ductile	-
	SR	WMC	402	>507	>4,22	>8	>1,87	ductile	-
	AW	HAZ	417	>517	>4,22	>6	>1,32	ductile	-
	SR	HAZ	397	>514	>4,22	>8	>1,73	ductile	-

* Pop-in

TABLE 4.b : SUMMARY OF WIDE PLATE TEST RESULTS AT -10°C ON STEEL D
(Plate thickness: 40 mm - As Welded)

Specimen Nr.	Heat Input kJ/cm	Notch Type and Location	Gross Yield Stress (N/mm^2)	Gross Stress (N/mm^2)	Overall strain (%)	Surface Displacement (GCOD) (mm)	Notch length (Length x depth)	Remarks
1	30	Parallel to weld TT-HAZ	345	419	2,54	1,99	40 x 40,3	Rough cleavage fracture which ran in HAZ. Small shear lips of 3 mm
2	30	Parallel to weld SN-HAZ	351	355*	0,51*	1,97	135 x 22,0	Rough cleavage fracture Small shear lips of 4 mm wide
3	30	Parallel to weld SN-HAZ	353	447* 448	3,52* 4,30	9,98	349,8x21,0	Ductile tearing of 3 mm Cleavage fracture with shear lips from 0 to 4 mm wide
4	40	Parallel to weld TT-WMC	>351	>461	>4.71	>5.73	40 x 40,5	Unbroken. Small cracks developed at the severely blunted notch tip

TT = Sawn through thickness notch
SN = Fatigued Surface Notch
* = Load drop

TABLE 4.c : SUMMARY OF WIDE PLATE TEST RESULTS AT -10°C ON STEEL D
(Plate thickness: 70 mm - As Welded)

Specimen Nr.	Heat input kJ/cm	Notch Type and Location	Gross Yield Stress (N/mm^2)	Gross Stress (N/mm^2)	Overall strain (%)	Surface Displacement (GCOD) (mm)	Notch length (Length x depth) (mm^2)	Remarks
5	30	Parallel to weld TT-HAZ	369	>430	>3,03	>2,05	40 x 70	Unbroken. Notch tips were severely blunted.
6	30	Parallel to weld SN-HAZ	376	424* >445	2,47* >3,13	1,00 >1,46	797 x 5	Unbroken. Small brittle cracks of 600 mm length and 3 mm in depth was observed
7	30	Parallel to weld SN-HAZ	378	>419	>2,40	>2,40	280 x 12	Unbroken. Notch tips were severely blunted

TT = Sawn through thickness notch
SN = Fatigued Surface Notch
* = Load drop

TABLE 5 : CTOD DESIGN CURVE DEFECT ASSESSMENT ON STEEL B.

Notch loc.	Cond.	σ_{ys} (N/mm^2)	σ_d (N/mm^2)	σ_R	σ_1 (N/mm^2)	$\frac{\sigma_1}{\sigma_{ys}}$	$\delta c_{(min)}$ (mm)	$\delta c_{(mean)}$ (mm)	$2\bar{a}_{max}$ min (mm)	mean (mm)
BM	AW	530	280,0	$\sigma_{ys}/2$	545,0	1,03	0,10	0,21	15,9	33,3
				σ_{ys}	810,0	1,53			9,7	20,3
			186,6	$\sigma_{ys}/2$	451,6	0,85			20,6	43,3
				σ_{ys}	716,6	1,35			11,2	23,6
	SR	577	280,0	0	280,0	0,49	0,25	0,28	118,3	132,5
				$\sigma_{ys}/2$	568,5	0,99			38,4	43,0
			186,6	0	186,6	0,32			277,4	310,7
				$\sigma_{ys}/2$	475,1	0,82			49,8	55,8
WMC	AW	605	280,0	$\sigma_{ys}/2$	582,5	1,01	0,20	0,22	28,5	31,4
				σ_{ys}	885,0	1,46			17,9	19,7
			186,6	$\sigma_{ys}/2$	489,1	0,81			38,7	42,6
				σ_{ys}	791,6	1,31			20,4	22,5
	SR	658	280,0	0	280,0	0,41	0,04	0,04	21,5	21,5
				$\sigma_{ys}/2$	609,0	0,93			5,9	5,9
			186,6	0	186,6	0,28			50,8	50,8
				$\sigma_{ys}/2$	515,6	0,78			7,5	7,5
HAZ	AW	530	280,0	$\sigma_{ys}/2$	545,0	1,03	0,06	0,17	9,5	27,0
				σ_{ys}	810,0	1,53			5,8	16,5
			186,6	$\sigma_{ys}/2$	451,6	0,85			12,4	35,0
				σ_{ys}	716,6	1,35			6,7	12,0
	SR	577	280,0	0	280,0	0,49	0,03	0,04	14,2	18,9
				$\sigma_{ys}/2$	568,5	0,99			4,6	6,1
			186,6	0	186,6	0,32			33,3	44,1
				$\sigma_{ys}/2$	475,1	0,82			6,0	8,0

TABLE 6 : CTOD DESIGN CURVE ASSESSMENT OF WIDE PLATE TEST RESULTS FOR STEEL A.

Notch location	Condition	$2a_c$ (mm)	ϵ_o (%)	$\delta_c min$ (mm)	$2\bar{a}_{max}$ (mm)	$2a_c/2\bar{a}_{max}$
HAZ	AW	10	1,91	0,15	2,58	3,87
HAZ	SR	10	0,48	0,07	5,26	1,90
WMC	AW	10	>4,50	0,27	<1,94	>5,15
WMC	SR	10	0,61	0,03	1,79	5,57

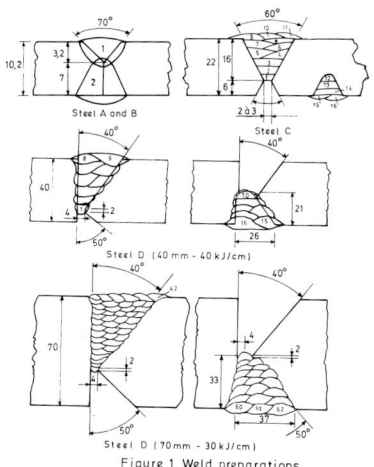

Figure 1 Weld preparations.

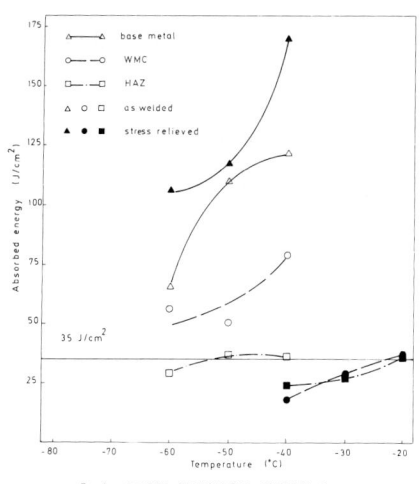

Fig 2a CHARPY TRANSITIONS, MATERIAL A

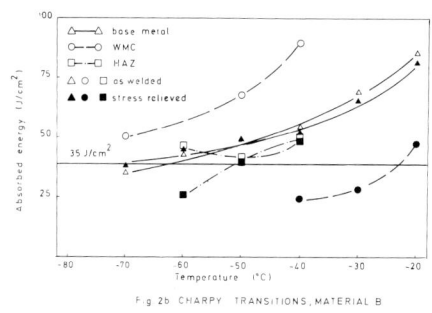

Fig 2b CHARPY TRANSITIONS, MATERIAL B

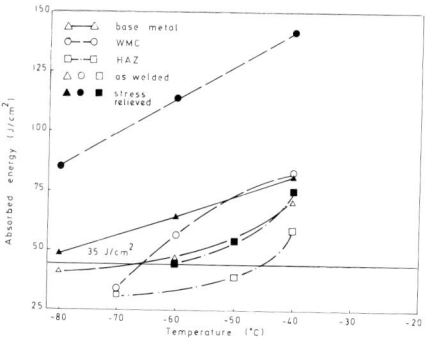

Fig 2c CHARPY TRANSITIONS, MATERIAL C

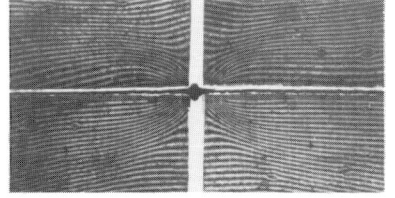 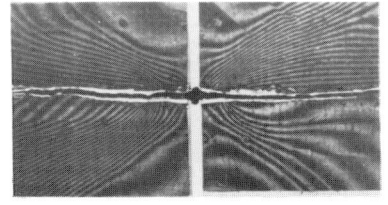

a. AS WELDED   b. STRESS RELIEVED

FIGURE 3 : MOIRE PATTERNS AT THE NOTCH TIP OF
HAZ NOTCHED WIDE PLATE SPECIMENS (STEEL A)

STEEL A   STEEL B   STEEL C

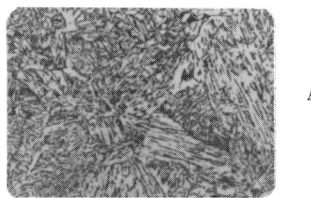

STEEL D

FIGURE 4 : MICROSTRUCTURE OF COARSE GRAINED HAZ
x 250

IV.12

Is High Tempering Always Needed for Low-carbon and Low-alloyed Steel Structures

A. E. Asnis and G. A. Ivashchenko

E. O. Paton Electric Welding Institute of the Ukrainian SSR Academy of Sciences, Kiev, USSR

In designing and fabrication of welded structures there often arises a question of a need to provide for the measures to decrease residual stresses. In these cases high tempering is used. The latter is the efficient means to reduce the stress level in the entire structure, thus advantageously distinguishing it from the other methods, where the changes in the stress level are of a nonuniform nature.

The tempering of welded structures is the energy-consuming and rather expensive operation /1, 7/ it requires the longer production cycle, furnaces or even special buildings. The necessity of tempering these or those groups of welded structures is generally one of the most complicated problems.

Quite often, at the sufficient capacity of heat treating shops, the welded structures, which do not require tempering, are subjected to it. Numerous investigations /2-5, 8, 9, etc./ show that in a number of cases at the stage of design and technology development it is possible, if necessary, to avoid tempering in the fabrication of welded structures due to the proper design and technological solutions and as well as steel selection.

The main requirements to metal for welded structures are the sufficient toughness and ductility.

These requirements are met practically by low-carbon and low-alloy steels delivered for welded structures according to the existing standards. In static tests under the positive and negative temperature conditions the failure of defect-free welded joints of the said steels is accompanied, as a rule, by the significant plastic strains. Of great importance is ageing of steels in the regions of thermal strain concentrations /6/ which results in brittle fractures.

The necessary heat treatment of welded joints in the fabrication of metal structures resulted from the fact that the sufficiently cold-resistant steels were not available in the metallurgical industry at the beginning of the development of welding engineering. Low-carbon steels of various degrees of deoxidation were mainly used for welded structures, high tempering required to relieve welding residual stresses and to improve serviceability of the structures having been justified to a certain extent.

Lately, the cold-resistant weldable steels have been used for welded structures /5/. The application of such steels allows to eliminate in many cases the expensive operation, i.e. high tempering.

The possibility to avoid the heat treatment can be confirmed by the great experience gained in construction and operation of welded railway and motor bridges, converters, blast furnaces, cars and locomotives, etc., made of weldable steels. It should be noted that designers at one time raised a question about the heat treatment during the development of these structures. Only due to the absence of furnaces of such enormous dimensions these structures had to be constructed without relieving the welding residual stresses. More than 30 years supervision proves the reliable performance of these structures.

It is appropriate to emphasize that failures of bridge structures (Belgium, Germany), tanks and ships which occurred as far back as in thirties, a few months later after their construction, were associated mainly with the poor quality of steels which could not meet the current requirements of standards on steels for the welded structures.

To verify the abovesaid, the tests were performed on the special specimens of various grades of steels. Susceptibility to brittle fractures at lower temperatures and under impact loading was studied by using the procedure and the specimens developed at the E.O.Paton Electric Welding Institute of the Ukrainian SSR Academy of Sciences /5/. The cracked part of a section between two notches was taken as a criterion. Fig. 1 shows the results of the tests performed at different temperatures.

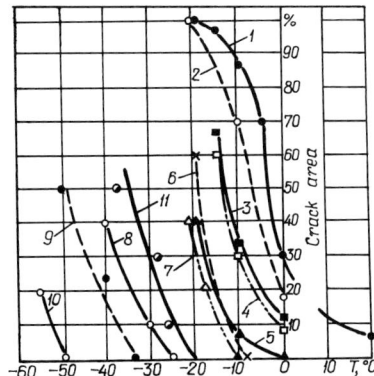

Fig. 1. The character of changes in cold resistance of a welded joint depending on

the steel grade:1,3, 4-СтЗкп,
СтЗпс, СтЗсп , respectively;
2 - 15ГФД ; 5 - 14Гпс ; 6 -
10Г2Б ; 7 - 09Г2 ; 8 -15ХСНД;
9 -18Г2АФМ ; 10 - 15Г2БМ ;
11 - СтЗсп after high tempering.

It can be seen that cracks initiate at 0°C in the ВСтЗсп steel specimens and at -40 and -50°C in the 18Г2АФМ and 15Г2БМ type steel specimens, respectively. The specimens of rimmed steel with higher nitrogen content failed completely between the two pairs of holes at the -10°C temperature.

Heat treatment is frequently recommended to increase the toughness of the welded joint HAZ metal /7, 8/.

To study the effect of high tempering on the critical brittleness temperature and mechanical properties the 600X400X20 mm specimens were butt welded under the conditions which provided the cooling rate W_{cool} = 3.5°C/s at 800...500°C temperatures. With such cooling rate the required impact toughness of the HAZ metal above 30J/cm^2 can not be provided at -40°C for steel ВСТ3сп and at -70°C for steels 09Г2С , 10ХСНД and 15Г2АФ.

A number of welded joints were subjected to high tempering under the following conditions: loading into a furnace at 300°C, heating up to 620-650°C at the 50-60°C/h rate, soaking for 2 min per 1 mm of thickness, furnace cooling down to 300°C at the 50-60°C/h rate and, further on, air cooling.

The 0.8 mm dia. specimens for coarse grain effect micromechanical testing of the HAZ metal and also the 10X10X55 mm specimens for V-notch impact bend KCV testing at different temperatures and the COD specimens (S_c)were cut out of the welded joints. Notches were made within the coarse grain regions.

The test results are given in Fig. 2. As one can see, the decrease in yield and ultimate strength and the increase in ductility are observed within the HAZ of the welded joints of steel grades ВСтЗсп , 09Г2С and 10ХСНД after high tempering.

The tests showed that the high tempering did not practically exert the positive effect on impact toughness of the HAZ metal of the ВСтЗсп and 09Г2С steel welded joints, slightly raised the absolute values of impact toughness of the 10ХСНД steel joint HAZ metal, however, at the -40°C temperature the impact toughness value remained at the previous level. The critical brittleness temperature of the 15Г2АФ steel joint HAZ metal shifted to the side of the positive temperatures.

These results were verified by the COD (S_c) specimen tests, Fig. 2.

Some authors recommend heat treatment for relieving the weld-

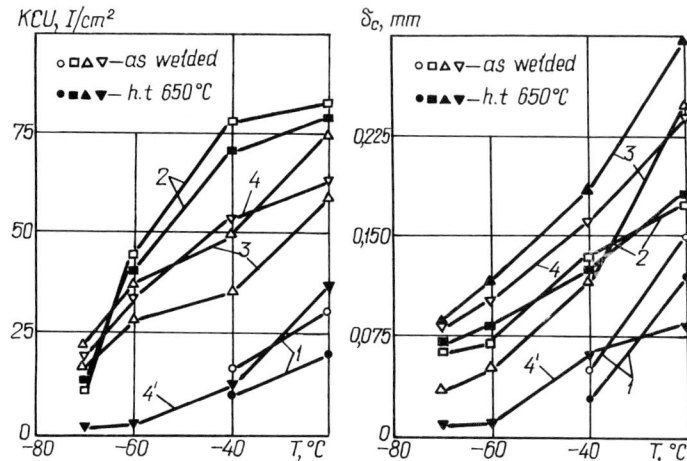

Fig. 2. Dependence of impact toughness (KCV) and COD (δ_c) on the test temperature:
I-ВСт3сп; 2-09Г2С; 3-I0ХСНД; 4-I5Г2АФ.

ing stresses as the means to improve the performance of welded structures subjected to cyclic loading. Sometimes it can be useful when the endurance of a structure is only 25-30% of that of the base metal due to the irritational design with sharp stress raisers. And even in this case the application of heat treatment to increase the structure life which is only 5-15% can hardly be justified. It should be noted that the fatigue crack propagation rate increases after high tempering.

Therefore, the heat treatment can be applied for stress relieving to improve the performance of the welded structure only in those cases when the effect of welding residual stresses is negative.

One of those cases is the performance of structures in corrosive environment when there exists the risk of stress corrosion occurrence.

The other cause for the welded structure heat treatment requirement consists in the fact that residual stresses present in the structure can deteriorate the accuracy required in its mechanical treatment and in the process of service. The 3,5 years supervision of the machine-tool welded frames showed no changes in geometrical dimensions. Apparently, deformations under the conditions of subsequent service are hardly probable. The investigations performed allowed to conclude that tempering of machine-tool frames can be avoided. Heat treatment should be carried out only in the case when high accuracy is required, and heat treatment can be performed at the 520... 550°C temperature. Such heat treatment provides the reliable performance of the welded joints and no changes in geometrical dimensions are observed.

SUMMARY

The investigations performed allow to conclude the following. The proper selection of steel and welding consumables provides the sufficiently high structural strength in as-welded condition and the high tempering operation can be avoided in many cases. High tempering causes slight embrittlement of the metal, thus raising the crack propagation rate at alternate loadings in the presence of a moderate, usually allowable stress raiser. The decrease in toughness is observed in the HAZ of a number of steels.

For the low-carbon and low-alloy steel structures which require high tempering to relieve the welding residual stresses (in particular, to keep the structure geometrical dimensions and corrosion resistance constant) tempering should be carried out at the lower temperatures, i.e. at 520...550°C instead of 650...670°C /11/.

REFERENCES

Faerman A.I.(1967). Analysis of expenditures for heat treatment of welded structures. Proceedings of Leningrad Polytechnical Institute, 283,41-49.
Asnis A.E., Ivashchenko G.A., and M.M.Malova (1967). On the possibility to avoid high tempering of building-block machine support frames. Avtomaticheskaya svarka, 1, 22-25.
Vinokurov V.A., Fishkis M.M., and V.V.Chernykh (1967). On the elimination of high tempering. Avtomaticheskaya svarka, 2, 26-30.
Nikolaev G.A. (1952). On the elimination of heat treatment of low-carbon steel weldments. Vestnik inzhenerov i tekhnikov, 5, 217-219.
Asnis A.E., and G.A.Ivashchenko (1985). Improvement in strength of welded structures. Naukova Dumka, 256.
Grigorenko V.S., and E.V.Kotenko (1968). The effect of residual stresses and strain ageing on brittle crack initiation resistance of steel. Avtomaticheskaya svarka, 2, 34-37.
Fick J.I.J., and J.H.Rogerson (1978). The effect of stress relief heat treatment on the toughness of C-Mn submerged arc weld metals. "Weld. and Metal Fabr.", 46, 2, 85-89.
Robinson J.L. (1978). Factors controling root run toughness in multipass manual metal arc welds in C-Mn steels. "Weld.Res. Inst.", 8, 6, 425-509.
Salkin R.V. (1979). Effects metallurgiguss et mecaniques d'un traitement thermique de relaxation des tensions residuelles. "Rev. Soudure", 35, 3, 145-150.

IV.13

Considerations for the Post Weld Heat Treatment of Pressure Parts

I. G. Hamilton and A. R. G. Abbott

Babcock Power, Technical Group, Renfrew, U.K.

ABSTRACT

The inadequacies of current code requirements for P.W.H.T. are briefly considered and the need for P.W.H.T. established. Technical consequences of the operation are listed. Recent U.K. activity to improve codes for P.W.H.T. is described, specifically with regard to dissimilar material joints, austenitic steels and local heat treatment. The importance of temperature measurement and its improvement by the introduction of microprocessor control are discussed. Two practical examples of P.W.H.T. of complex units where finite element analysis was used to prescribe the heating rates conclude the paper.

KEYWORDS

P.W.H.T., code requirements, technical effects, dissimilar materials, austenitic steels, local heat treatment, temperature measurement, microprocessor control, finite element analysis.

INTRODUCTION

Post weld heat treatment (P.W.H.T.) is an operation carried out during the manufacture of a component to improve its integrity. As such it results in additional costs, not only in executing the heat treatment but in the effects on non-destructive examination, materials testing and component delivery times. Moreover it is not without its technical problems. In general the component manufacturer has to follow particular national code(s) for design, materials, fabrication, inspection and testing, specified by the clients. Most of these codes are explicit only in certain aspects of requirements for post weld heat treatment and there are variations in these requirements for like components, not only between international codes, but between national codes.

As an example, it is sufficient to cite the question of temperature, where the requirements range from nil specification of temperature, to specifying only a minimum, to required ranges, which can vary significantly from code to code. Attempting to understand these different approaches is difficult because the rationale of any code making group is seldom available. Since this may involve commercial as well as technical considerations, it is hardly surprising that the differences exist. However, it underlines the importance of developing the technical understanding of the operation such that the code making bodies will have better data on which to base their requirements. The fact that so many welded structures are not P.W.H.T. raises the question concerning the need for this operation. In a recent review of pressure vessel failures by Smith (1986), it is interesting to note that in 23 out of 42 cases, individually reported, either lack of P.W.H.T. or improper P.W.H.T. is mentioned. It is not suggested that this was the prime reason for failure since the failures were multi-casual, but the frequency of reporting this factor appears highly significant. Additionally, a number of the failures occurred at material thicknesses which in many codes would not have required P.W.H.T. This underlines that thickness alone is not a satisfactory criterion for omitting P.W.H.T. without paying due attention to material properties, pressure vessel design details and operating environment.

TECHNICAL CONSIDERATIONS

P.W.H.T. is complex and may involve the following phenomena:-

(1) The relaxation of residual stresses. (2) Distortion. (3) The tempering of the weldment, ie. the weld metal, H.A.Z. (visible and non visible) and the parent material. (4) The reduction in hardness of the H.A.Z. (5) The elimination of hydrogen from the weldment. (6) The stabilisation of the micro-structure for use in the creep range by heating to temperatures significantly higher than service temperature. (7) A progressive variation in mechanical properties of the joint, tensility, ductility and fracture toughness. (8) Thermal stress cracking. (9) Reheat cracking.
(10) Embrittlement of weld metal, parent material or H.A.Z. during slow cooling by the segregation of residual elements to the grain boundaries.
(11) High temperature strain embrittlement.

RECENT U.K. CODE ACTIVITIES

With the recognition that the codes were incomplete in their requirements, during the last few years activity has attempted to clarify a number of factors:-

(1) General Recommendations for P.W.H.T. of Dissimilar Ferritic Steel Joints (BS5500 : 1985, Appendix H). Most codes tend to deal solely with welds between similar materials and give little help with dissimilar materials, commonly used in the manufacture of complex pressure parts. In the U.K. the approach attempts to reconcile the following basis conditions:-

(1) P.W.H.T. should be compatible with the parent materials being welded.

(2) P.W.H.T. should be compatible with the relative importance of the pressure parts being welded.

Materials have been classified into five groups, each of which has constant P.W.H.T. temperature ranges. P.W.H.T. within groups is obviously permissible and between different groups, where T1'- T2' is not greater than $10^{o}C$. T1' is the lower temperature of the material requiring the higher temperature and T2' is the upper temperature of the material requiring the lower temperature. Where T1' - T2' is greater than $10^{o}C$, the P.W.H.T. is subject to agreement between the purchaser and manufacturer.

Where the weld is between pressure parts of equal importance, P.W.H.T. should be in the higher temperature range. Where the importance differs, P.W.H.T. should be as for the more important part.

P.W.H.T. between different material groups in the higher temperature range should aim to keep the temperature as near to the minimum as practicable and vice versa.

(2)   P.W.H.T. of Austenitic Steels.   In general, national and international codes tend to specify a full solution treatment for P.W.H.T. of austenitic steels. Practically this is a difficult operation and there is understandably a reluctance to apply P.W.H.T. Guide Notes on the safe use of stainless steel in Chemical Process Plant (1978) state "Stress relieving heat treatment which can be used to prevent stress corrosion cracking of carbon steel, should not normally be considered to be a satisfactory way of preventing stress corrosion in austenitic steel."   The notes indicate that the heat treatment operation may induce residual stresses sufficient to promote stress corrosion cracking, as well as other undesirable features such as embrittlement or carbide precipitation.

An analysis of the effects of P.W.H.T. at various temperatures has been made by the Water Tube Boilermakers Association, in collaboration with the Central Electricity Generating Board,(1983),Table 1.

From these submissions it is obvious that P.W.H.T. of austenitic steels is not a matter which is adequately covered in existing codes.

(3)   Problems of local P.W.H.T.   The problems of local P.W.H.T. are generally understood and heated band widths are specified.   Problems arise with interfering structural discontinuities, such as nozzles and flanges within the band width.   A particular instance of this is where pipes are butt welded to branches on a pressure shell.   In BS1113:1985, where the branch is of length <5 $\sqrt{rt_1}$ (r is the internal branch radius and t_1 is the thickness of the branch at the butt weld) elaborate precautions are required to avoid undesirable thermal stresses.   This has had the general effect of increasing the lengths of branches welded to pressure vessels.

FURTHER PRACTICAL CONSIDERATIONS

It must be appreciated that although the codes imply a uniform P.W.H.T. this may not be the case in the manufacture of large pressure parts.   Different portions of the pressure part may receive different times at final P.W.H.T. temperature, depending on manufacturing sequence and the fabricator's philosophy of intermediate P.W.H.T.   Moreover, P.W.H.T. at a unique

TABLE 1 Effect of various heat treatments on austenitic steel weldments

	Heat Treatment	Stress Relaxation	Effect on Mechanical Properties	Microstructural Effects
A	Solution Anneal	Maximum	Possible reduction in tensility.	Maximum grain growth. Maximum sensitisation for solution treatments. Transforms delta ferrite to austenite.
B	Solution Normalise	Good		Some grain growth. Less sensitisation than A. Transforms delta ferrite to austenite.
C	Solution Accelerated Coolint	Welding Stresses removed	Properties maximised.	Some grain growth. Minimum sensitisation. Transforms delta ferrite to austenite.
D	$800^\circ C - 850^\circ C$	~90%	Creep rupture life may be reduced, although ductility better than as welded. Impact properties poor.	No grain growth. Sensitisation high. Delta ferrite transforms to intermetallics (chi and sigma) + $M_{23}C_6$.
E	$700^\circ C - 750^\circ C$	~75%	Impact properties better than D.	No grain growth. Sensitisation less than D. Delta ferrite transformation as for D, but more slowly.
F	$550^\circ C - 600^\circ C$	A stabilisation treatment before machining to reduce peak stresses. Little or no effect on mechanical properties or microstructure.		

temperature is impracticable, due to the temperature control achievable in
large heat treatment furnaces and the error variations in temperature
measurement. Since temperature and time affect the mechanical properties
of the welded joints, it may be necessary to establish that the design
requirements are met (i) at the lowest temperature(s) for the shortest
time and (ii) the highest temperature(s) for the longest time.

The I.I.W. recommended that for C-Mn Al treated steels the maximum P.W.H.T.
temperature should be reduced to 580°C, since in these conditions the
residual stresses were significantly relaxed and the deleterious effects
of P.W.H.T. minimised. Much of the data on which this recommendation was
based derived from laboratory uniaxial stress relaxation tests. In the
U.K. the recommendation was not adopted, primarily because it was felt that
the effect of P.W.H.T. on mechanical properties, at the lower end of the
temperature range, had not been sufficiently investigated and a lingering
suspicion of the validity of stress relaxation tests to simulate production
welds. A recent report by Leggat (1985) showed on submerged arc welded
panels (a) significant residual stresses after 2 hours at 550°C, ~ 200N/mm2
(b) measured residual stresses up to 50 N/mm2 greater than the upper band
of stress relaxation data for corresponding P.W.H.T. temperatures and hold
times. In view of the importance of residual stress, especially with
regard to the application of fracture mechanics, it is suggested that this
situation should be clarified.

TEMPERATURE MEASUREMENT

Thermocouples (Chromel Alumel) can either be the mineral insulated metal
sheathed or fibre coated types. The latter type originally used an
asbestos coating but this has now been changed to glass fibre for
temperatures up to 750°C and ceramic fibre up to 1250°C.

In most codes, the requirement for attachment of thermocouples is that
they should be in effective contact with the vessel. Wide temperature
variations have been experienced with the metal sheathed thermocouples
even using the slotted nut method of attachment and this has been
attributed to direct exposure of the thermocouple to the heating medium.
The proven method for fibre coated thermocouples is capacitance discharge
welding of the two wires directly onto the component with a preferred
spacing of 6mm.

With respect to future developments, the use of infra-red pyrometers is
being studied since improved accuracy and response times, particularly
in the P.W.H.T. temperature range, have been noted in manually operated
instruments for hot and warm forming applications. A dual pyrometer
system is envisaged to compensate for the radiation effects of the
heating source and furnace atmosphere.

TEMPERATURE CONTROL SYSTEMS

Temperature control systems for P.W.H.T. of pressure parts are invariably
based on a single control thermocouple which can be either atmosphere or
metal attached. There is no doubt that a system based on a number of
metal attached thermocouples is difficult to achieve but the economic
advantage of an automatic and predictable cycle together with the improvement in temperature tolerances at all stages in the cycle are worth
pursuing.

From experience in specialised applications, the system would need to operate on both metal temperatures and heat source. The latter can be represented by the atmosphere temperatures provided the response to heat source fluctuations is fast enough. The principle of the system is that the set point of the atmosphere controller responds to the metal temperature controller thus increasing heat input at lower temperatures but minimising the possibility of overshoot at the start of soak. Overshoot control can be further refined by the Proportional Integral Derivative (P.I.D.) mode in the controller. A summary of results obtained in simple geometries is shown in the attached table and shows the improvement in iniformity which can be achieved for the same cycle time.

TABLE 2   Comparison of Temperature Uniformity

COMPONENT	Soak Temp.	Average Heating Time to Temp. Rate	TEMPERATURE UNIFORMITY			
			BEFORE		AFTER	
			Heating	Soak	Heating	Soak
FLAT PLATE	900°C	130°C/Hr	70°C	15°C	50°C	10°C
CYLINDER	"	130°C/Hr	120°C	20/25°C	80°C	15°C
DRUM HEAD	"	180°C/Hr	75°C	20/25°C	50°C	15°C
BEAM SECTION	"	220°C/Hr	75°C	10/15°C	40°C	5°C
FLAT PLATE	600°C	100°C/Hr	40°C	10/15°C	20°C	5°C
CYLINDER	"	100°C/Hr	80°C	25/30°C	40°C	15°C
BEAM SECTION	"	100°C/Hr	40°C	10°C	30°C	5°C

EXAMPLES OF P.W.H.T. ON COMPLEX UNITS

A coolant seal assembly for a nuclear boiler required P.W.H.T. with thickness variations of 3mm (thermal sleeve) to 210mm (main forging) involved. Extremely close dimensional tolerances were required and finite element analysis was carried out to establish maximum allowable temperature differentials, some of which were extremely demanding e.g. $\Delta T = 48°C$. A third complicating factor was the need to avoid oxidation due to the limited access for cleaning after P.W.H.T. The furnace used had the same operating principle as a coil annealing furnace, i.e. the atmosphere (nitrogen/5% hydrogen) being heated indirectly in an annulus by electrical heating elements and then directed over the component. Control was based on a 24 point comparator which enabled each thermocouple to be assessed sequentially against the master control thermocouple which itself could be changed if required. For this application, an upper thermal sleeve was selected as the master thermocouple. The results are summarised in Fig. 1.

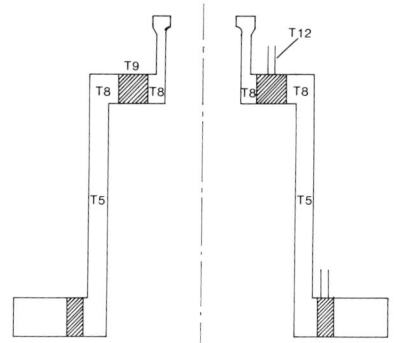

SUMMARY OF RESULTS

EVENT	SPECIFICATION	ACTUAL
Heating Rate RT-500°C	15°C/Hr.	13°C/Hr.
Heating Rate 500°C - 690°C	10°C/Hr.	21°C/Hr.
Soak Temperature	690°C ± 15°C	676°C - 693°C
Cooling Rate 690°C - 500°C	10°C/Hr.	21°C/Hr.
Cooling Rate 500°C - 260°C	15°C/Hr.	11°C/Hr.
∇T (T12-T9)	37°C at 200°C	5°C
∇T (T12-T9)	31°C at 600°C	10°C
∇T (T8-T5)	23°C at 200°C	12°C
∇T (T8-T5)	19°C at 600°C	4°C

FIG. I

Another example which involved large variations in size was the P.W.H.T. of a fully machined tubeplate 2235mm dia. x 400mm thick across which a D shaped baffle, section 25mm thick, 900mm high had been welded. Tight dimensional tolerances had been set for both the baffle plate assembly and the tubesheet and obviously a scale free surface after P.W.H.T. was necessary. Since no suitable atmosphere controlled furnace was available, the component was placed in a gas tight box with an argon purge and located in an electric furnace. Temperature control was achieved by relating the metal attached thermocouple on the component to the furnace zone control thermocouples which were located on the outside of the gas box. Finite element analysis was used for both temperature distribution and thermal stresses using the Von Mises equivalent elastic stress. The results showed that for the heating arrangement, means of reducing the heat absorbed by the baffle was essential and further analysis showed that both insulation and a reflective shield would be necessary. The maximum allowable temperature differentials for the three main component parts were also calculated. The results are shown in Fig. 2.

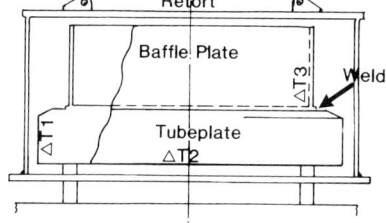

TIME INTO TREATMENT (HRS)	TEMP. OF ITEM (°C)			
	RETORT	TUBEPLATE	WELD	BAFFLE PLATE
10	200	42	58	69
28	400	300	315	330
44	600	550	568	580
56	690	660	650	670
60	600	628	630	630
72	400	470	474	490

FIG. 2

CONCLUDING REMARKS

Whilst the existing codes give general information there is no doubt that they are technically inadequate to specify explicitly the P.W.H.T. requirements for complex fabrications in alloy steels. Moreover, the existing information is often spread throughout a number of codes and the problem would be greatly clarified if there were national codes devoted solely to the subject. At this stage, with many of the technical problems still unresolved, it is too much to expect an acceptable international code.

REFERENCES

Contribution on post weld heat treatment of austenitic steels (1983) Water Tube Boilermakers Association, London.
International Study Group on Hydrocarbon Oxidation (1978). Guide notes on the safe use of stainless steel in chemical process plant. The Institution of Chemical Engineers, U.K.
Leggat, R.H. (1985) Relaxation of residual stresses during post weld heat treatment of submerged arc welds in a C-Mn-Nb-Al steel. The Welding Institute, Members Report No. 288/1985.
Smith, T.A. (1986) Pressure vessel failure experience (to be published).

Argon-ARC Treatment Application in Welded Structure Fabrication

A. E. Asnis and G. A. Ivashchenko

E. O. Paton Electric Welding Institute of the Ukrainian SSR Academy of Sciences, Kiev, USSR

An efficient method of improving the strength of welded joints subjected to alternating loads in service is known to be the treatment of weld-to-base metal transition areas, i.e. mechanical cleaning, cold-work hardening, grounding with an abrasive wheel, etc. Sometimes, the high-temperature tempering at 620-650°C is recommended. This process can produce both positive and negative effects /1, 2/.

An efficient method of improving the strength at dynamic loading is argon-arc treatment of the weld-to-base metal transition area. The testing results and the experience of argon-arc treatment application in welded structures subjected to cyclic loads show that the endurance limit was 1.4-1.8 times increased. A particular method of improving the cyclic strength should be specified taking into account the fact that besides cyclic loads the welded structures can also be subjected to impact ones both at positive and negative temperatures.

The E.O.Paton Electric Welding Institute of the Ukrainian SSR Academy of Sciences studied the effect of the high-temperature tempering and welded joint argon-arc treatment on properties at repeated impacts.

The tested specimens were box girders (Fig. 1) of low-carbon ВСт3сп steel and low-alloyed 14Г2АФ type steel, δ =10mm. The transverse weld in the girder lower chord was made under conditions, at which the cooling rate at 800...500°C temperature was w = 2.5°C/s. At such cooling rates it is practically impossible to achieve the required toughness of the HAZ metal and, consequently, the impact strength of the welded joint at negative temperatures.

Before testing the structure element was subjected to high-temperature tempering at 650°C temperature and in the other element the weld-to-base metal transition areas were treated by the argon arc. The results are presented in Fig. 2. It is

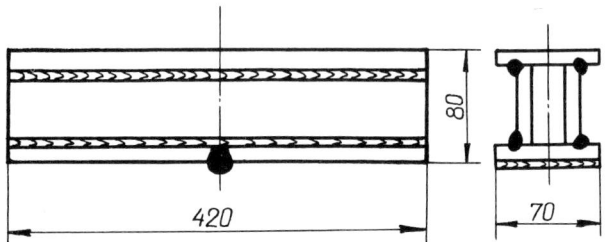

Fig. 1. Girder specimen for single impact testing.

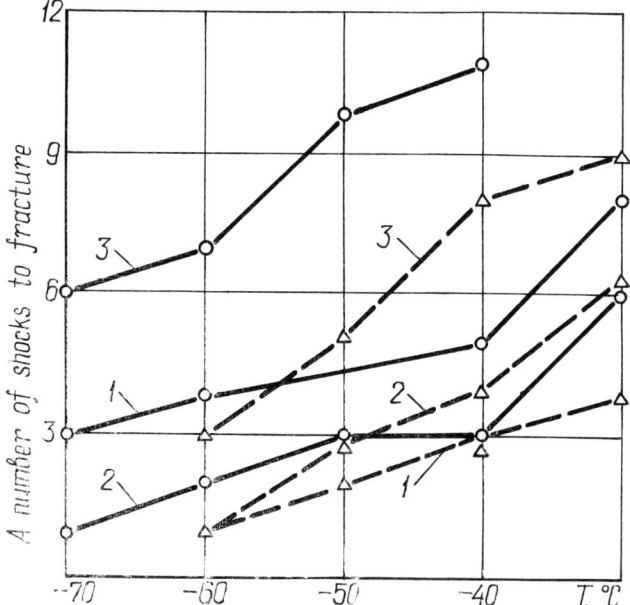

Fig. 2. Treatment effect on welded joint impact strength: 1-without treatment; 2-high-temperature tempering; 3-argon-arc treatment; --- - ВСт3сп; ——— - 14Г2АФ.

seen that argon-arc treatment considerably improves the welded joint impact toughness.

The high-temperature tempering produces different effects on the welded joint strength at impact loads. The specimens, namely ВСт3сп steel girders, after tempering have somewhat higher impact strength, than in the initial as-welded condition. The 14Г2АФ steel specimens have lower impact strength after tempering.

The obtained results permit to affirm, that when specifying the high-temperature tempering to lower the residual welding stresses, not only the degree of stress level lowering should be determined, but also the possible changes of welded joint service properties depending upon alloying should be taken into account.

By applying the surface treatment methods, which allow to control the structural parameters of the metal outer layers and to simultaneously change the geometry of the weld to base metal transition zone, the welded structure serviceability both under cyclic and impact loads can be considerably improved.

REFERENCES

Asnis A.E., and G.A.Ivashchenko (1985). Improvement of welded joint strength. Naukova Dumka, 256.

Asnis A.E., and G.A.Ivashchenko (1973). On thermal treatment of welded structures, subjected to alternating loads. Tekhnologia, organizatsia i mekhanizatsia svarochnogo proizvodstva, 2, 10-17.

Recent Studies on Reheat Cracking of Cr-Mo Steels

K. Tamaki and J. Suzuki

Department of Mechanical and Materials Engineering, Mie University, Kamihama-cho, Tsu, Mie 514, Japan

ABSTRACT

The combined effects of Cr, Mo and P on the reheat cracking sensitivity were shown in Cr-Mo contents diagram using the critical restraint stress. The tolerable P contents were determined on five Cr-Mo steels.

KEYWORDS

Reheat cracking sensitivity; combined effect; critical restraint stress; cracking-sensitive field; critical P content.

INTRODUCTION

A new idea for assessing the reheat cracking sensitivity of several Cr-Mo steels is proposed in this paper from the view point of the combined effects of Cr, Mo, P, V and Ti. A tolerable P content, below which the reheat cracking will hardly occur, is also proposed for some typical Cr-Mo steels.

EXPERIMENTS

Five groups, eighty-four specimens were used; they cover a wide chemical composition range of 0 to 5%Cr, 0.25 to 1.5%Mo, 0.004 to 0.11%P, 0.06%V and 0.07%Ti. The reheat cracking sensitivity of each steel specimen was evaluated by the magnitude of the critical restraint stress for producing cracking, $\sigma_{AW-crit}$. This was obtained by the modified implant test[1].

DIAGRAM OF CRITICAL RESTRAINT STRESS

The contour lines of $\sigma_{AW-crit}$ values are drawn in Cr-Mo contents diagram as shown in Fig.1[2,3,4] for four steel groups; (a)Cr-Mo-0.02%P steels, (b) Cr-Mo-0.01%P steels, (c)Cr-Mo-0.02%P-0.06%V steels and (d)Cr-Mo-0.02%P-0.07% Ti steels.

In the case of Cr-Mo-0.02%P steels(Fig.1(a)), the cracking sensitivity is highest in the field of about 1%Cr and Mo$\geq$0.5% ($\sigma_{AW-crit}\leq$400N/mm^2, shaded area).

The cracking-sensitive field of Fig.1(a) is moved toward the right hand side with decreasing the P content from 0.02 to 0.01% as shown in Fig.1(b).

By the addition of 0.06%V, the cracking-sensitive field of the basic Cr-Mo steels (Fig.1(a)) is remarkably expanded to the left hand side as shown in Fig.1(c).

By the addition of 0.07%Ti, the taper of contour lines of the basic Cr-Mo steels is increased in the field of Cr$\leq$1% and Mo$\leq$1% as shown in Fig.1(d).

CRITICAL P CONTENT

The critical restraint stress of 1%Cr-0.5%Mo steel decreases steeply with increasing P content when it exceeds a critical value of 0.008%, as shown in

Fig.2. This critical P content, Pcrit, is regarded as the threshold value for preventing the reheat cracking by means of decreasing P content. Pcrit values of five Cr-Mo steels are shown in Table 1[4]. 1%Cr-0.5%Mo and 2%Cr-1%Mo steels exhibit very small Pcrit values.

REFERENCES
[1]Tamaki,K.and J.Suzuki, Trans.Japan Welding Soc.,14,117-122(1983)
[2]Tamaki,K.and J.Suzuki, Trans.Japan Welding Soc.,14,123-127(1983)
[3]Tamaki,K.,J.Suzuki and M.Tajiri, Trans.Japan Welding Soc.,15,17-24(1984)
[4]Tamaki,K.and J.Suzuki, Trans.Japan Welding Soc.,16,117-124(1985)

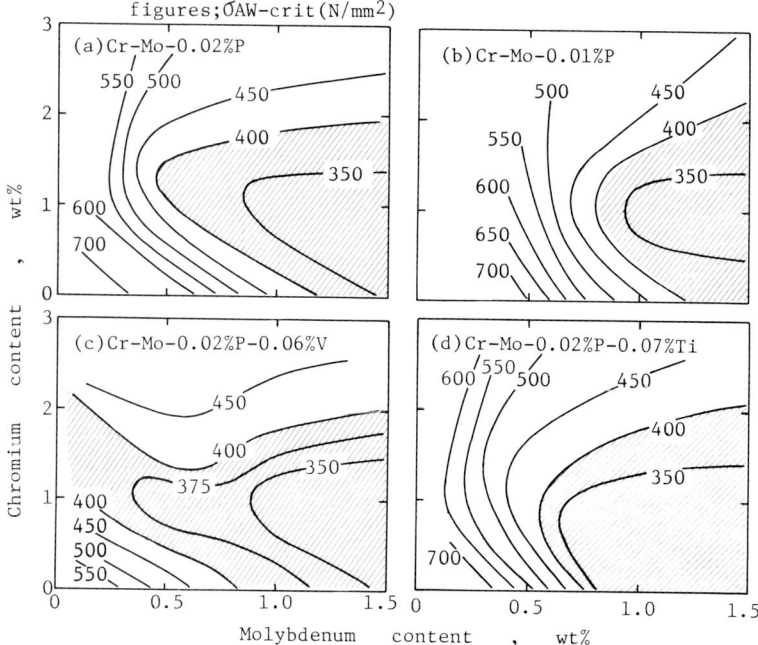

Fig.1 Contour lines of critical restraint stress, σAW-crit of four groups of Cr-Mo steels shown in the Cr-Mo contents diagram

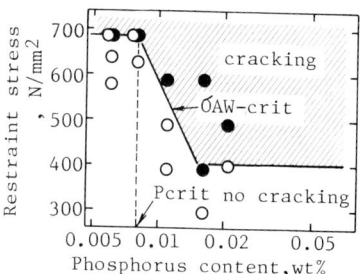

Fig.2 Influence of phosphorus on the critical restraint stress, σAW-crit of 1%Cr-0.5%Mo steel

Table 1 Critical phosphorus content, Pcrit of Cr-Mo steels

Steel type	Pcrit
0 %Cr-0.5%Mo	0.05
0.5%Cr-0.5%Mo	0.01
1 %Cr-0.5%Mo	0.008
1.3%Cr-0.5%Mo	0.014
2 %Cr- 1 %Mo	0.005

Traitement Thermique de Joints Soudes de la Conduite Principale de Circulation D'Une Centrale Nuccleaire

I. I. Varbenov*, P. I. Raykov* and D. D. Dimitrov**

*KZU, Bulgarie
**S. O. Energétique, Bulgarie

RESUME

Ce poster montre les résultats du traitement thermique /TTh/ de joints soudés de la conduite principale de circulation /CPC/ et les équipements y adjacents: le corps du réacteur et le générateur de vapeur d'un bloc nuccléaire d'une puissance de 1000 MW. Le préchauffage est téalisé par induction avec des appareils de fréquence moyenne-1000 et 2000 HZ; puissance 120,200 et 400 kW. Dans les conditions spécifiques de construction et de montage du bloc nuccléaire sont résolus les problèmes de répartition des appareils de TTh, leur alimentation en électricité et en eau de refroidissement, sinsi que les produits utilisés, la préparation du joint soudé pour le TTh, les régimes de TTh, l'enregistrement des températures.

MOTS-CLE

Traitement thermique; soudage; induction; réacteur nuccléaire.

INTRODUCTION

La conduite principale de circulation /CPC/ est réalisée d'une tuyauterie Dy 850 qui connecte le réacteur nuccléaire, le générateur de vapeur et les pompes. Lors de la construction ont été soudés 40 joints. Un intérêt particulier pour le TTh représentent les soudures de tubes sur réacteur et tube sur générateur de vapeur, étant donné les grandes différences de géométrie et d'épaisseur du métal de base des deux cotés du joint. De plus, la zone du soudage atteint une température de 650°C, tandis que la température á l'intérieur du réacteur ne doit pas dépasser 230°C et 400 á 500°C á l'extérieur; ce qui impose un refroidissement en même temps quele préchauffage. Un autre probleme - de tels équipements uniques sont traités thermiquement pour la premiére fois en Bulgarie et nous n'avons pas pu élaborer un modele expérimental sur lequel déterminer les nombreux parametres et facteurs influant le régime thermique. No us avons utilisé six appareils de Brown-Bovery, situés de cote -4 á +36. L'alimentation en électricité est assurée par un sous-répartiteur de 2 MW provenant des blocs nucléaires déjá en exploitation. L'eau de refroidissement venait d'une station de pompage et d'un réservoir de 300 m^3. Les températures sont enregistrées par 6 thermo - couples, dont un est accouplé sur le computer muni de 8 programmes pour les régimes. Sur chaque joint soudé sont réalisés 4 types de traitement thermique: préchauffage á 150-250°C durant 5 á 7 jours et nuits, revenue á 150-250°C pendant 12h, revenue intermédiaire á 650°C pendant 3h et revenue finale á 650°C pendant 7h.

<u>Conclusion</u>: Un traitement thermique compliqué de joints soudés a été réalisé en conditions tres séveres et exigences élevées.

Development of Alloying Systems of Structural Steels Having a High Resistance Against Overheating, Eliminating the Post Electroslag Welding Normalization

S. V. Egorova, Yu. A. Sterenbogen, A. V. Yurchishin, E. N. Solina and N. G. Zotova

E. O. Paton Electric Welding Institute of the Ukrainian SSR Academy of Sciences, Kiev, USSR

The majority of low-alloy steels, used in industry, possess the high sensitivity to overheating in welding. It is especially high in electroslag welding (ESW) resulting in the fact that this the most economic and efficient method of welding the thick metal, providing the 100% weld quality, is used either with a subsequent normalization or, in spite of an expediency, is not generally used.

During recent years the E.O.Paton Electric Welding Institute is carrying out works on the development of new cold resistant low-alloy steels with a high resistance against overheating in ESW. Thus, the new low-alloy heat resistant steel 10x2Г M (10x2ГHM) containing ≤ 0.10% C, 0.17...0.37% Si, 1.9...2.3% Mn, 0.4...0.6% Mo, 0.3...0.6% Ni was developed and introduced into industry. This steel possesses high heat resistance up to 560°C, resistance against high-temperature hydrogen fracture, cold resistance down to -60...70°C, including that in HAZ at ESW without post normalization. The high cold resistance of 10x2ГM (10x2ГHM) steel joints, made by ESW without normalization is testified by low temperature fracture tests of experimental vessels by using the criteria of fracture mechanics.

As a result of a large number of experiments in which more than 60 steel compositions were studied, the systems of alloying the structural general-purpose steels with a high resistance against the overheating in ESW were developed. In addition,the important regularity of effect of manganese content in steel on the HAZ metal impact strength was used. It was stated that it is sharply increased at manganese content above 1.8% (Fig. 1).

The ferrite impact strength is increased in steels containing about 2% of this element, the dispersity of second phase precipitation being considerably suppressed by a formation of the Widmanstätten structure and procutectoid ferrite at the grain boundary in HAZ. At an additional microalloying of similar steels with aluminium and cerium the grain growth in HAZ is retarded, the shape and distribution of sulphide inclusions being improved. The following steels with approximately 2 % Mn possess the high resistance against overheating in ESW: 09Г2СЮЧ, 09ХГ2СЮЧ (~1% Cr), 09Г2СМЮЧ (up to 0.7% Mo), 09Г2СФЮЧ (0.03% V), 09Г2СРЮЧ (up to 0.003% B).

At optimum alloying even a simple silicon-manganese steel 09Г2СЮЧ can withstand overheating at the most powerful processes of welding, including electroslag one. Moreover, the coarse grain in the overheating zone is not drastically hazardous. Metal with such grain can be cold resistant enough first of all in that case if: 1. the ferrite base, composing the larger part of structure in the low-alloy steel, possesses a high impact strength; 2. the second phase is sufficiently low-dispersed. The ferrite-bainite structure without or with small areas of pearlite, meets thos requirement best of all; 3. the second phase precipitations are coagulated, and the second kind stresses are relaxed by a high tempering.

The fracture toughness of the overheating zone metal in ESW of 09Г2СЮЧ steel after a single high tempering at 650°C is given in Fig. 2. Thousands of tons of special-purpose welded structures serviced at pressure and at temperatures down to -60...-70°C were manufactured of this steel by using the ESW without a post normalization.

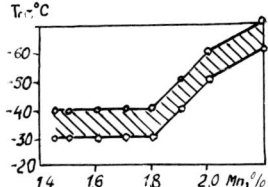

Fig. 1. Effect of Mn content in silicon manganese steel with 0.05...0.12% C on the T_{Cr} value of overheating zone metal.

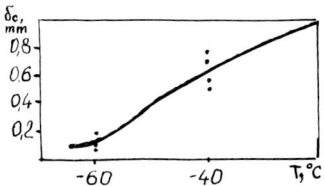

Fig. 2. Fracture toughness of overheating zone metal in ESW without normalizing of 09Г2СЮЧ steel.

Producing of Cast Bimetallic Die Billets by Using the Electroslag Heating with Accompanied Annealing and Selection of Heat Treatment Conditions

V. A. Nosatov, Yu. A. Sterenbogen, T. Kh. Ovchinnikova, O. G. Kuz'menko and A. V. Denisenko

E. O. Paton Electric Welding Institute of the Ukrainian SSR Academy of Sciences, Kiev, USSR

The methods of producing the bimetallic horizontal castings providing the successive pouring of two molten metals under the slag always attracted the attention of metallurgists by their simplicity and economic efficiency. However, till now it was not possible to produce the quality bimetallic billets by using these methods mainly due to the fact that during the metal shrinkage the voids are formed in the casting and it is difficult to obtain the reliable fusion of layers. It is possible to provide the reliable melting of layers and to avoid the void formation in the casting if to create the conditions for a directed metal solidification. For this purpose it is necessary to keep the slag temperature at the interface exceeding that of the metal melting and to provide the heat removal onto the bottom plate from the poured metal.

The E.O.Paton Electric Welding Institute has suggested the method that permits to produce the quality bimetallic castings by a successive pouring and metal solidification by using the electroslag heating.

The principle of the method is as follows. Slag and then a first portion of metal are poured into a mould or another forming device of preset dimensions. One electrode or a group of non-consumable electrodes supplied from a.c. power source, are immersed into slag with a help of a specialized unit and the process is maintained in such a way that the slag temperature throughout the interface was higher than the metal melting temperature. The slag, being overheated on the metal surface, prevents the solidification of the poured portion from above and benefits the directed solidification of metal from the bottom plate side. Moreover, if the heat from the metal was mainly removed onto the bottom plate the metallic pool would be shallow and the metal shrinkage would be uniformly distributed all over the surface. Due to the electroslag process at the poured metal surface it is possible to keep the metal within the required temperature range for any desired period. In pouring the next portion a reliable melting between the layers and a constant ratio of their thickness are provided. The metal pouring by portions can be performed many times depending upon the required number of layers.

If the pouring is performed into the refractory brick lined mould on the cast bottom plate then the delayed cooling of metal occurs under the hot slag layer with a rate which is close to that of cooling at annealing and this is known to favourably affect the distribution of stresses in the casting.

The method, developed on the basis of the electroslag heating, is successfully used for producing the bimetallic die billets of die and carbon steels. It is shown that to improve the mechanical properties of cast die steel it is rational to perform the homogenizing annealing at 1150°C temperature for 10-15 hours or the normalization with a subsequent tempering before the final heat treatment.

The austenite grains grown during the homogenization are refined after quenching to 8-9 size that meets the requirements for die steels. The temperatures of quenching and tempering are set depending upon the steel grade. After the final heat treatment the cast die steel by mechanical properties at operating temperatures (400-600°C) is not inferior than the same forged steel and by wear resistance it is superior than the forged one. The testing of large-sized cast bimetallic dies on 5-16 t hammers showed that by the resistance they 1.5-2 times exceed the forged ones.

The Influence of the Thermal Treatment on the Welding Stress Relieving and the Improving of Properties of Welding Joints of Large-sized Pressure Vessels

A. G. Lamzin, N. M. Kabanov and P. M. Korol'kov

VNII Montazhspetsstroj, Moscow, USSR

ABSTRACT

The influence of notfurnace volumetric (complete) heat treatment of large-sized spherical vessels in erection zone on the welding stress relieving and properties of welding joints is analysed. It is shown that heat treatment decreases the level of welding stresses and improves the properties of welding joints.

INTRODUCTION

Welding spherical vessels made of steel 09Г2С having a capacity of 600 m³ and operating under a pressure of up to 1.8 MPa are 10.5 m diameter, have a wall thickness of up to 36 mm and mass up to 100 t (Fig. 1).

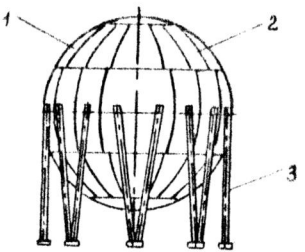

Fig. 1. Spherical vessel (600 m³ capacity): 1-shell, 2-welds, 3-supports.

In the erection zone these vessels are assembled and welded from separate blocks and are subjected to volumetric heat treatment in order to decrease the welding stress and to improve the properties of welded joints. The welding joints are made usually by automatic submerged-arc welding and also by manual arc welding with heavy-coated electrodes. High-temperature tempering is made under heating up to 580-650°C at the rate of up to 30°C/hr, holding for 2 hours and subsequent cooling at the rate of up to 30°C/hr till 300°C, the total time of the heat treatment cycle is 38 hours /1/. Such heat treatment is accomplished for welding spherical vessels with the wall thickness more than 30 mm and in the case of corrosion medium the wall thickness may be whatever you like.

The investigations aimed at defining the influence of the heat treatment on the stress relieving and properties of welded joints were undertaken on the test pieces with thickness of 30 mm.

Notfurnace volumetric heat treatment of spherical vessels is conducted by gas-flame heating from the inside using the equipment and procedures worked out by VNIIPTCHimnefteapparatura (affiliated to the Ministry of Chemical Engineering), together with VNIImontazhspetsstroj (affiliated to the Ministry of Erection and Special Construction).

EXPERIMENTAL

30 mm thick plates of 09Г2С steel were made by automatic submerged arc welding and by manual arc welding. Some of them remained in the original state, while others were subjected to heating in an electric furnace according to the heat treatment regime of spherical vessels described above. To imitate the production conditions these test pieces were welded in the rigid frame made of angle metal (dimension 50X50 mm). The final dimensions of the test pieces were equal to 400X360X30 mm. The plates prepared to welding had a double vee butt. The welding joints under automatic submerged arc welding of AH-348A type of arc flux were accomplished in 4 layers by coiled electrode of C_B-I0HЮ type of 4 mm diameter with the feed speed 103 m/hr. The speed of welding was 20 m/hr, while the arc current was 650-700 A and the arc voltage was 40-42 V. The welding joints under manual arc welding by electrodes УОНИ -13/55 type were accomplished in 12 layers while the arc current was 160 A and the arc voltage was 22 V. Residual stresses were studied using the electrotensometry method /2/.

Relative (residual) deformations were recorded in points, where rosette-type strain gauges had been placed. For this, the test pieces to be studied were cut into elements (destructive method) each containing one gauge /2/. To measure the welding stresses in surface layers of the test pieces relative deformations were found. Resistance strain gauges were glued onto the test pieces to form rectangular rosettes made up of the three resistors (Fig. 2). The pattern of the rosette-type strain gauge installation (see Fig. 2) was worked out according to the tentative analysis of stress distribution and their

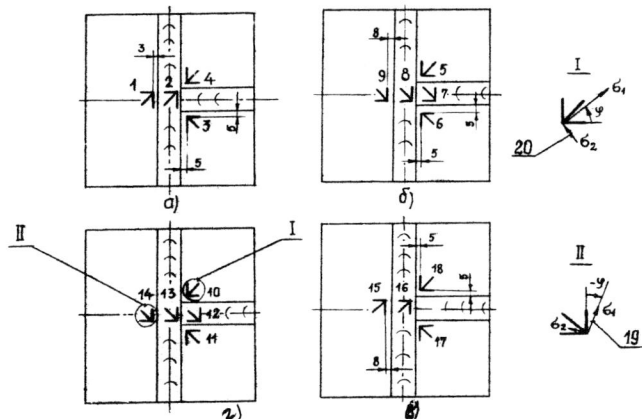

Fig. 2. Location of resistance strain gauges on test pieces: a - test piece N 1; b - test piece N 2; ᵬ - test piece N 3; г - test piece N 4; 1-18 - rosette-type strain gauges; 19-residual tension stresses; 20 - residual compression stresses; G_1 and G_2 - main residual stresses; ψ - the angle determining the direction of the main residual stresses.

likely concentration in welded joints. Four welded joints were used for investigations:
- test piece 1 - produced by automatic welding in the original state;
- test piece 2 - produced by manual arc welding in the original state;
- test piece 3 - produced by automatic welding after heat treatment;
- test piece 4 - produced by manual arc welding after heat treatment.

The main residual stresses σ_1 and σ_2 and their direction (Fig. 2, ψ angle) and also the equivalent ones σ_{eq} were found using the well-known formulas /3, 4/ (the results of the calculations are summarizes in the Table).

The investigation comes to the following results:
- joints made by automatic submerged arc welding have the greatest stresses in the original state (60-70% of the yield point of the basic metal), the character of welding stresses is stretching (see Table);
- as a result of heat treatment, residual stresses in test pieces made by automatic welding go down by 80% and in case of manual arc welding - by 60%.

Mechanical tests showed that ultimate strength went down by about 10% (from 520 to 460 MPa) as a result of heat treatment. This high-temperature tempering also reduces the hardness of the weld and basic metal approximately of about 10-15% (correspondently from 176 to 154 BH and from 155 to 141 BH), bending angles of all types are equal to 180°. Charpy impact tests were carried out at temperatures of +30°C to -70°C and showed that under heat treatment the characteristic transition temperature of weld metal from the viscous state into the brittle state went down from -15°C to -20°C.

SUMMARY

Notfurnace volumetric heat treatment of spherical vessels reduces the level of residual welding stresses of 4-5 times (from 200-245 MPa to 40-60 MPa), also improves cold resistance of welding joints, decreasing simultaneously the ultimate strength to go down by about 10% and hardness of weld metal and basic metal by about 10-15%.

TABLE. Main Residual and Equivalent Stresses

N of test piece	Type of welding; heat treatment	N of rosette-type gauge (Fig.2)	Welding stresses, MPa		
			σ_1	σ_2	σ_{eq}
1.	Automatic welding; without heat treatment	1	246	169	218
		2	142	103	127
		3	91	56	79
		4	156	86	135
2.	Manual arc welding; without heat treatment	5	78	49	68
		6	101	43	87
		7	100	-13	107
		8	115	70	100
		9	163	99	143
3.	Automatic welding; heat treatment	15	17	6	15
		16	32	-23	48
		17	38	17	33
		18	40	9	36
4.	Manual arc welding; heat treatment	10	50	-18	61
		11	32	8	29
		12	48	13	43
		13	85	62	75
		14	25	2	24

REFERENCES

Bierger, I.A. Ostatochnye napryazheniya (Residual stresses). In Russian (1963). GNTI, 231.
Korol'kov, P.M. et al. (1986). Ob'emnaya termoobrabotka sfericheskikh rezervuarov. (Complete heat treatment of spherical vessels), In Russian. Montazhnye i spetsialnye raboty v stroitel'stve, 3, 13-15.

Makarov R.A.(1985). Tensometriya v mashinostroenii (Tensometry in engineering), In Russian. **Mashinostroenie**, 286 p.
Ponomarev S.D. et al (1956). Raschety na prochnost v mashinostroenii (Strength analysis in engineering), In Russian. **Mashgiz**, 883.

Problems in Welding and Heat Treatment of Thick-walled Martensitic Steel Steam Lines

B. Kovačević

T P K — Zagreb, Žitnjak bb, Yugoslavia

1. INTRODUCTION

Problems in martensitic steel welding processes are very complex. The welding process requires a very high technological discipline as beginning from preheating, root welding, hot root quality control, temperature maintenance, groove filling up to the heat treatment which must follow immediately after welding according to strictly established regime. The heat treatment course should be strictly respected, especially in case of thicknesses greater than 50 mm. The work deals with the problems of heat treatment during welding of the $\emptyset$ 427 x 53 steam line made of X 20 Cr Mo V 121 material (according to DIN 17175) and with the way of attestation and control of the correctness of the heat treatment regime maintenance. The work also deals with metallurgical and mechanical properties of the material in relation to the manifold repeating of heat treatment which sometimes should be repeated because of repairing the weld joint defects.

2. WELDING AND HEAT TREATMENT TECHNOLOGY

The basic rules that must be sticked to before welding such a quenched steel are:
- the welded joint must have such a groove preparation that a good access to the electrodes is made possible while the quantity of the melted metal must be as smaller as possible (see Fig. 1);
- the welding process must proceed with continuous preheating and keeping up of the temperature of intermediate layers from $300^\circ C$ to $400^\circ C$;
- in case of the manual electric arc welding there should be used coated electrodes and the welding to be performed with the current $I = 30$ d (d - electrode diameter);
- the welded joint must be made in as more as possible bonded layers (not by swinging) and the weld thickness should neither be greater than 4 or 5 mm nor wider than 10 or 14 mm;

- the welding process must be accompanied by a continuous visual inspection and X-ray inspection from time to time, so that the noted defects can instantaneously be repaired;
- immediately after welding the heat treatment must be carried out, i.e. low relieving at the temperature of about 150°C for 1 hour at least and then annealing at the temperature from 730 to 750°C for 2 to 4 hours (thermal cycle is given in Fig. 2);

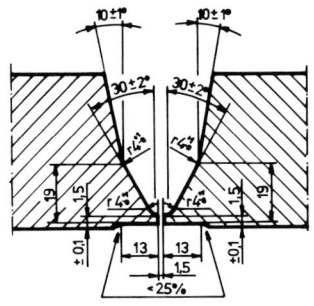
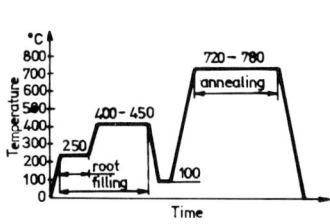

Fig. 1 Welding groove preparation

Fig. 2 Heat treatment cycle

3. INSTEAD OF CONCLUSIONS

When performing technological tests, welding and heat treatment procedure qualifications and especially on the object itself there have been noticed some details for which it can be said in principle that they are the basic conditions for a successful overcoming of welding problems in the mentioned materials.

The noticed problems could be systematized as follows:
- technological discipline
- maintenance of the heat treatment cycle without any corrections
- additional training of welders
- working conditions.

In addition, the whole heat treatment repeating results in the decrease of mechanical and metallurgical and other properties of the described steel, what is proved by experiments. After four sequential heat treatments of the welded joint, material properties have been worsened. This implies that the number of the welded joint repairs is also limited because repairs should be carried out under the same heat treatment regime, too.

REFERENCES

1. G. Kalwa, E. Schnabel:
Wärmebehandlung und Eigenschaften dickwandiger Bauteile aus warmfesten Röhren stählen. VGB KRAFTWERKSTECHNIK 58, 8/1978
2. K. Kussmaul, D. Blind:
Schweiss-und Bruchmechanikversuche an Proben und Bauteilen aus Werkstoff X 20 CrMoV 12 1 bei Wandicken bis 270 mm
VGB KRAFTWERKSTECHNIK 60 - 4/1980

Amelioration des Regles de Detensionnement des Joints Soudes

Programme de recherche coopératif français

USINOR (Unirec, Usinor-aciers, CFEM, Creusot-Loire-Industrie)
IRSID, Institute de Soudure, Cetim, Veritas, Framatome, Ecole Centrale de Paris, Ecole des Mines de Paris,
Ecole Polytechnique, CEA/SRMA

RESUME

Proposition d'un critère pour le détensionnement des joints soudés fondé sur la connaissance des conditions de rupture avant et après traitement de détente

MOTS CLES

Joint soudé, détensionnement, ténacité, approche locale.

INTRODUCTION

Les codes de construction prévoient la détente des assemblages soudés au-delà d'une certaine épaisseur, dans le but principal de supprimer les contraintes résiduelles. Ce traitement coûteux, qui n'améliore pas nécessairement la ténacité du joint soudé, peut être évité si les contraintes résiduelles sont relaxées par plastification de la structure. Par contre, si la structure rompt de manière fragile, cescontraintes modifient la charge de rupture. Pour décider s'il faut ou non détendre, il est donc important de savoir dans quel cas se produirait la rupture avant et après détente.

CRITERE DE DETENSIONNEMENT

La charge d'instabilité plastique P_L, fonction de la géométrie de la structure et du défaut et de la limite d'élasticité R_p de l'acier, décroit lorsque la température de service augmente. La charge de rupture fragile P_R, qui peut être calculée par la mécanique de la rupture en élasticité linéaire et en élasto-plasticité en fonction de la géométrie de la structure et du défaut,des contraintes résiduelles et de la ténacité K_{IC} (ou J_{IC}), croît avec la température. On peut définir une température de transition fragile-ductile T_D à l'intersection des courbes représentant les évolutions des charges P_L et P_R. Les contraintes résiduelles n'agissant plus au-dessus de la température T_D, on peut décider de ne détendre que si la température de service est comprise entre les températures T_D ainsi définies avant et après détensionnement. Ce critère s'applique au cas où la détente abaisse T_D ; dans le cas contraire, le détensionnement est néfaste. Ce type d'analyse se heurte à plusieurs difficultés : estimation des contraintes résiduelles, relaxation de celles-ci lors de la plastification de la structure, détermination de la variation de K_{IC} (J_{IC}) avec la température, prise en compte de la dispersion de la ténacité inhérente au domaine de transi-

tion, réalité de la notion de ténacité pour une zone hétérogène (Z.A.T).
L'utilisation de la méthode de l'approche locale permet de s'affranchir des
trois dernières difficultés. Cette méthode consiste à mesurer la contrainte
critique de clivage ainsi que sa distribution statistique à l'aide d'essais à
basse température sur petites éprouvettes entaillées. Il est possible de prévoir les conditions de rupture d'une pièce fissurée à l'aide de ces résultats
en appliquant cette méthode fondée sur la statistique de WEIBULL. En conditions de plasticité confinée et déformations planes, par exemple, la probabilité de rupture est donnée par :

$$P_R = 1 - \exp \left\{ \frac{R_p^{m-4} K_I^4 B C_m}{V_o \sigma_u^m} \right\}$$

où V_o est un volume élémentaire choisi arbitrairement, C_m un coefficient de
distribution des contraintes dans la zone plastifiée, B la longueur du front
de fissure, σ_u la contrainte critique de clivage, et m l'exposant de WEIBULL
caractérisant l'hétérogénéité. L'expérience montre que la contrainte de clivage est indépendante de la température, de sorte que seule la baisse de R_p
lorsque la température augmente contrôle la variation de K_{IC} (J_{IC}).

RESULTATS

Les figures a et b représentent, pour deux joints soudés de limite d'élasticité différente, les variations en fonction de la température des charges provoquant la rupture fragile, déterminées à partir de l'approche locale, et des
charges provoquant l'instabilité plastique, avant et après détensionnement,
pour une fissure dans le métal fondu. Les valeurs expérimentales obtenues sur
plaques sont également portées. σ_u et m ne sont pas affectés par le détensionnement. Pour le joint a, la présence de contraintes résiduelles de compression
et l'augmentation de la limite d'élasticité apportée par le détensionnement
rendent celui-ci néfaste. Pour le joint b, la situation est inversée.

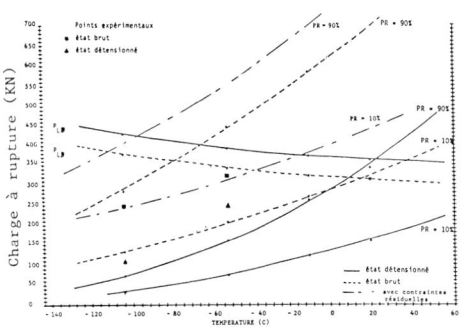
Fig. a. Joint SE 701 (Rp = 700 MPa)

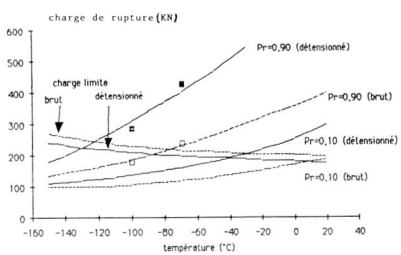
Fig. b. Joint E355 Ni (Rp = 355 MPa)

REMERCIEMENTS

Les participants remercient le Ministère de la Recherche et de la Technologie
pour l'aide apportée à ce programme.

REFERENCES

FONTAINE, A., MAAS, E. - Analyse de la rupture par clivage d'un métal fondu
à l'aide d'une approche locale. Rapport IRSID RE 1271, Mai 1986.

Efficiency of Application of Heat- and Vibro-treatment to Reduce Residual Stresses in Weldments

V. V. Batyuk, A. A. Khriplivy, S. K. Fomichev, S. N. Minakov and P. N. Gansky

Kiev Polytechnical Institute, Kiev, USSR

ABSTRACT

The report covers some specific features of heat treatment as the most wide-spread technological operation after welding - that of change energy input. The decrease of tensile residual stresses during heat treatment from 200-240 MPa to 60-120 MPa was shown by means of pipes with diameter 168-720 mm made of steel -20. Vibrotreatment is expedient from viewpoint of energy when compared with heat treatment.

KEY WORDS

Heat treatment, vibrotreatment, residual stresses, pipes, circumferential welds, energy input.

Process of welded joint formation can be represented as a number of sequential stages: energy input from the heat source - accompanying technological operations of energy input change - formation of structure - formation of stresses-strained state - formation of microdefects - technological operations of energy input after welding. Control of residual stresses may be introduced at any of above mentioned stages. Optimal from the viewpoint of energy consumption is the treatment which provides localization of energy input into the weld zone and does not need to return to initial stages of welded joint formation to have an effect on stressed-strained state, which is formed at the final stages.

Circumferential welds of pipes with diameter 168 to 720 mm made of steel -20, wall thickness 12 ti 22 mm were chosen as the subject of investigation. Magnetoelastic testing method was employed to assess the level of residual stresses.

The results of measurements showed that after welding the level of maximum tensile residual stresses in circumferential welds is as high as 200 to 240 MPa (Fig. 1,a) for the entire range of diameters under consideration. In post-welding treatment from the point of view of energy input localization of energy input zone is expedient in the weld zone. Mechanism of formation of residual stresses as a result of nonuniform heating of the workpiece predetermines the necessity of expansion of neat zone or heating of the whole workpiece. In this connection from the point of view of energy, heat treatment is unprofitable. As a result of heat treatment the level of residual stresses is considerably reduced to as low as 60 to 120 MPa (Fig. 1,b), and the value of reduction does not depend on the pipe diameter. But nonuniformity of local heating does not allow to remove residual stresses in circumferential welds. Besides, heating of a workpiece in the process of heat treatment is a return to the first stage of weld joint formation, after which all other stages are implemented. In this connection the promising methods of post welding treatment are those which provide for application of mechanical energy to affect mechanical stress, excluding intermediate stages of structure formation.

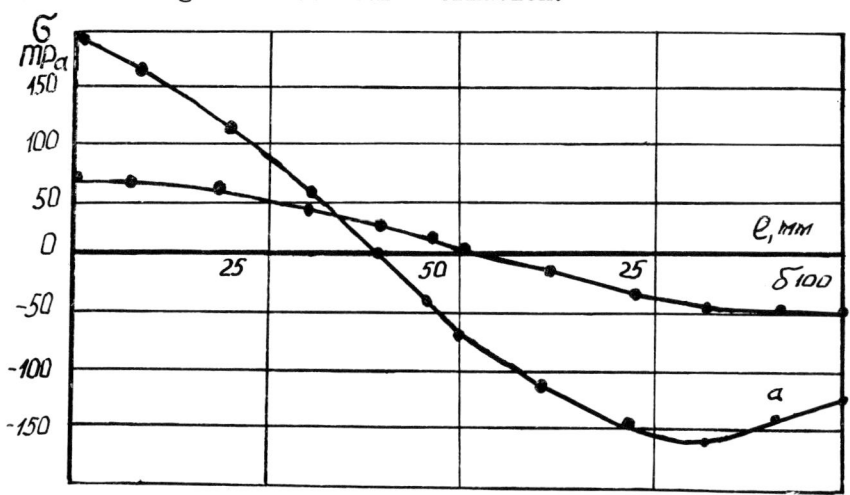

Fig. 1. Distribution of tangential residual stresses in a circumferential weld of a pipe 720 mm in a diameter, wall thickness 22 mm, after welding(a) and after high-temperature tempering (b).

Vibration treatment is one of such techniques. In vibration treatment localization of energy input in weld zone is possible. The major procedure condition in vibrotreatment is the necessity to provide resonance frequency in different zones of a workpiece. The most efficient feedback in this case is control of residual stresses. Control of residual stresses with magnetoelastic method in the process of vibrotreatment makes it possible to determine resonance frequency for a given zone as well as vibration time (Fig. 2). In case of alternating rigidity of a structure, for example, in places of welding on manholes and pipe unions, simultaneous control in several zones is needed.

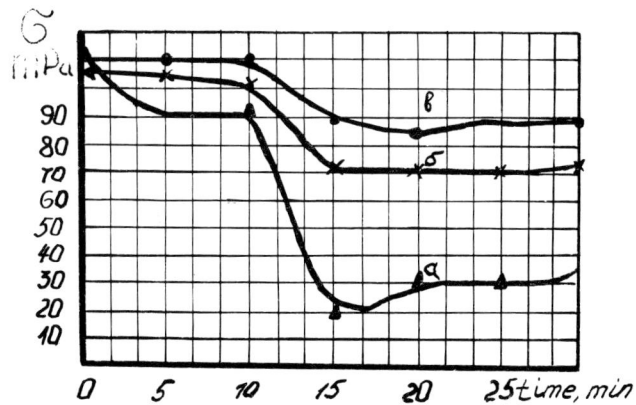

Fig. 2. Variation of tangential residual stresses in a circumferential weld 720 mm in diameter, wall thickness 22 mm, in the process of vibrotreatment at a distance from vibrator of 200 mm (a), 50 mm (b) and 100 mm (c).

It is recommended to use three-channel stress meter. Measurements showed that decrease of residual stresses to 140-150 MPa can be achieved through vibrotreatment.

This vibrotreatment is expedient from viewpoint of energy when compared with heat treatment. Development of equipment and techniques of vibrotreatment would enable to completely exclude heat treatment in cylindrical structures.

Crankshaft Overlaying Without Post Heat Treatment

A. M. Slivinsky, V. T. Kotyk, P. F. Petrov and V. I. Prokhorov

Kiev Polytechnical Institute, USSR

The reclamation of worn engine crankshafts by surfacing makes it possible to increase considerably the service life of these expensive and difficult-to-manufacture parts.

To obtain high hardness of the deposited metal use is made of high-carbon surfacing wires of 30X CA type, flux-cored wires /1/, ceramic fluxes, containing graphite and ferrochrome. After surfacing a part is subjected to heat treatment (tempering) /2/, high-frequency current treatment /3/ increasing hardness and wear resistance of the deposited layer. This technology requires expensive surfacing consumables and makes the process of crank-shaft reclamation considerably complicated.

A composition of the filler wire for surfacing has been developed at Kiev Polytechnical Institute. The wire is composed of CB08A metallic base with ferroalloy filler which makes up 12-15% of the wire weight. The wire with filler is very suitable in operation, it is strong, readily fed with usual one-roller feeding mechanisms. The absence of slag-forming and gas-forming components in the filler composition makes the wire practically insensitive to atmospheric effect. The manufacture of the wire is much simpler than that of flux-cored wire and does not need complicated equipment. The wire 2 mm in diameter was used for the surfacing of crankshafts. The surfacing of it was made under flux AH-348A under the following conditions: $U = 26$-28 V, $I_w = 160$-180 A, $V_w = 18$ m/h. The suggested conditions are sufficiently productive, do not cause considerable deformations and can be used for crankshaft journal crankpins 40-80 mm in diameter. The slag crust detaches itself spontaneously during surfacing. The surfacing process is performed along a spiral line with overlapping the bead by 30%. As a result of surfacing a smooth glittering metal layer without pores and cracks is obtained (Fig. 1).

To obtain high hardness of the deposited layer without heat treatment it is necessary to obtain the metal with martensite structure which hardens in air in retarded cooling. These requirements are met by the metal obtained by surfacing with wire with filler, metal contains 0.35-0.40% of carbon, of manganese 0.6-0.8%, 0.8-1.0% of silicon, 0.8-1.0% of molibdenum, 3.5-3.8% of chromium. The metal being cooled in air without heat treatment has the hardness equal to 52-54 HRC. The technology of surfacing with wire with filler, which has been developed, simplifies the reclamation of crankshafts and provides high hardness and wear resistance of deposited layer without heat treatment.

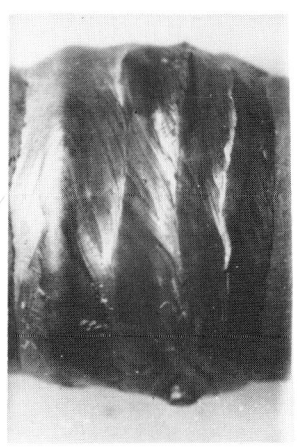

Fig. 1. The appearance of deposited layer.

Juzvenko Ju.A., A.P.Pakholyuk, G.A.Kirilyuk, and S.Yu.Krivchickov (1980). Surfacing of automobile engine crankshafts with self-shielding flux-cored wire. Automatic Welding, 2, 67-68.
Kuckuyevitsky V.A., and D.G.Guseynov (1964). Fatigue strength of automobile engine crankshafts reclaimed by electric arc surfacing. Automatic Welding, 3, 71-75.
Nalivckin V.A., T.P.Nuykina, V.V.Popandopulo. at al. Submerged surfacing of crankshafts of ГАЗ-51 and ЗИЛ-120 automobile engine crankshafts. Automatic Welding, 2, 70-74.

Thermal Deformation Process Control in Welding of Thin Sheet Pieces with Application of Heat Sink

I. M. Zhdanov, V. V. Lysak, B. V. Medko, V. V. Kononchuk and V. N. Nifantov

Kiev Polytechnical Institute, Kiev, USSR

Application of heat sinks in welding of structures is a form of conrtollable heat treatment the purpose of which is to reduce the level of residual stresses.

Investigations carried out at Kiev Polytechnical Institute showed that in order to provide effective heat removal from the weld zone heat sink devices must be placed with in the range of conventional welding are heat spot. Reduction of instantaneous volume of plastically deformed metal determined by isothermic line for metal transition into plastic state, causes significant reduction of tensile stresses /1/.

At the same time the width of plastic deformation zone also decreases. The value of heat flow removed with help of specialized equipment is determined by the following expression:

depending on thermal diffusivity (a) and thermal condictivity ( ) of metal to be welded and material of equipment their temperatures $T_0 = 1/2$ (T_1-T_2) and interaction time

Placement of heat sinks within the range of conventional heat spot has an effect on the character of arc burning. Mechanical contraction of the arc with heat sink devices increases the coefficient of heat source concentration and efficiency of thermal effect on metal. Combination of arc contraction processes and intensive heat removal with welding process can be possible only in case of application equipment.

The developed equipment performes intensive heat removal from the weld zone due to guaranteed provision of optimal step of contracting regions. Parameters of equipment can be calculated by means of statistical and probability techniques.

Experimental investigations proved optimality of thickness of heat removing components of equipment which does not exceed

1/4 of the weld pool length /2/.

Application of specialized equipment provides for the possibility of directed control of heat effect in metal, and by combining this effect with welding process enables to considerably increase accuracy of fabrication of far welded structures, to reduce residual stresses, and in welding aluminium alloys to decrease the width of weakened zone 3-4 times.

REFERENCES

Zhdanov I.M., B.V.Medko, V.V.Lysak, and Melnikov (1982). Shape and size effect of high-temperature plastic zone on the value of residual stresses in a weld. Avtomaticheskaya Svarka, 4, 41-44.
Zhdanov I.M., V.N.Nifantov, B.V.Medko, a d V.V.Lysak (1986). Intensification of argon-arc welding process of thin-sheet structures. From the book "Saving of material, energy and labour resourses in welding production", 55-56.

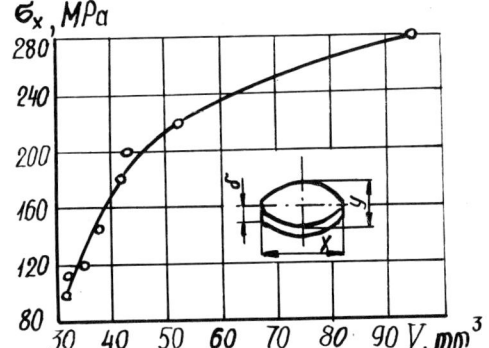

Fig. 1. Relationship between residual stresses and instantaneous volume of plastically deformed metal.

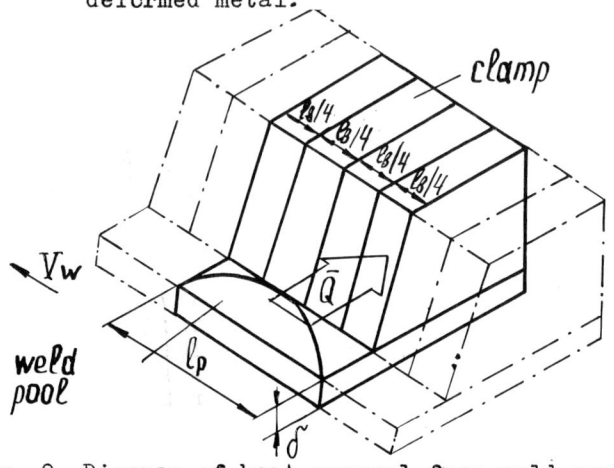

Fig. 2. Diagram of heat removal from weld zone.

Evaluation of Prefracture State During Welding of Large-sized Thick-wall Constructions

E. A. Kirillov, V. I. Panov, I. J. Ievlev, V. P. Pokatilov and V. V. Volkov

Production Amalgamation "Uralmash", Sverdlovsk, USSR

Production of welded large-sized thick-wall constructions with multipass welds is connected with the high degree of possibility of brittle fracture yet in the course of welding. The intermediate heat treatment is one of the means to ensure the technological strength. However, the scientifically proved recommendations good enough for engineering practice are not available.

In the present investigations the need of the intermediate heat treatment was defined by appearance of the prefracture state.

The following steels were used in tests:
- low-carbon steel with yield point   350 MPa;
- low-carbon low-alloy steel with yield point   550 MPa;
- low-carbon low-alloy steel with yield point   700 MPa.

Electrodes of the required strength were used for welding.

Study of the cases causing brittle fractures in constructions during welding, allowed the following joints to be taken as typical ones:
- welded joints with free deformation of work pieces in the process of welding;
- welded joints with compliance of metal adjoining to the groove (i.e. change of the groove shape is possible in the course of welding);
- welded joints with inchangeable contour.

Welding was done by five methods of multipass arc welding. The experiments were conducted on process samples and dummies. Some investigations were carried out directly in welding of thick-wall large-sized constructions. The conditions of cracks initiation were also explored on the massive constructions which had fatigue cracks after long service.

Also examined is the kinetics of development of the factors which influence formation of the prefracture stages such as:
- phase transformations in the heat affected zones;
- deformations and stresses as the groove is filled;
- peculiarities of heat propagation at various methods of multipass welding and their influence on the formation of stress-deformation fields;
- change of the stress-deformation fields under influence of welding of the subsequent welds.

The formation of initial cracks and their further growth were defined by an acoustic emission method. The obtained data were checked by other non-destructive methods and by cutting the check specimens.

Comparison of kinetics of opening the cracks to the critical limit in the samples under test showed that there was considerable difference in behaviour of cracking defect in the course of multipass welding between the samples providing the condition of plain-strain deformation in the vicinity of cracks and standard samples when testing them for out-of-center tension and bending, respectively.

When testing the welding joints with free deformation of workpieces being welded, the edges of the surface cracking defect in the root of weld undergo temporary alternating strain. There exist conditions for thermal strain ageing of the weld metal since the angular deformations occur more intensively at a temperature of about 400°C. In comparison with the other tested welded joints the incubation period of formation of initial cracks is the greatest one and the process of crack growth is the easiest one.

The formation of initial cracks in welding of compliant and flexible welded joints is of the same character and their formation occurs in the vicinity of stress concentrators while the process of their growth is different. The compliant welded joints in the process of welding undergo considerable residual deformation effects - nonequal covering in length of the upper portion of the groove, complex bending of the bottom portion. This influences the store of energy and the further growth of cracks is retarded. In rigid welded joints the further growth of cracks takes place quite easily and temporary deformations act on them.

The investigations performed allowed to establish the norms which afford to estimate the necessity of fulfilment of intermediate heat treatment depending on the level of strength of parent metal, the type of construction, its importance and the type of welded joints.

The important constructions were welded with the use of a series of instruments operating on the principle of acoustic emission which allowed to avoid the cases of brittle fractures in the process of welding.

Heat-treatment Effect on HAZ Metal Crack-resistance of Anticorrosion Cladded Low Alloy Steels

E. G. Starchenko, S. N. Navolokin, D. M. Shur, A. E. Runov and A. V. Zalinov

NPO TSNIITMASH, Moscow, USSR

The process of a high-temperature tempering of steels alloyed by active carbide-forming elements (V, Cr, Mo) is accompanied by their dispersion hardening. This phenomenon is specially emphasized in investigation of the possibility of initiating intergranular fracture in a coarse grain area of heat-affected zone (HAZ) under residual stresses /1, 2/.

The effect of tempering regimes and stress level on susceptibility to cracking of Mn-Ni-Mo-V, Cr-Ni-Mo-V, Cr-Mo-V and Mn-Mo-Ni structural steels was studied on compact tubular samples (Fig. 1)

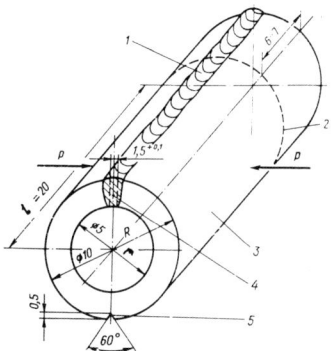

Fig. 1. Compact tubular sample: 1-weldjoint; 2-line of cutting a microsection metallographic specimen; 3-HAZ metal; 4-slot; 5-stress concentrator.

It was studied with model HAZ coarse grain structure characteristic of such a zone in the result of strip electrode cladding under heat-input range of 6-8 mJ/m. Modelling of cladding thermal cycles was carried out up to T_{max} 1350±2°C and postheat treatment in 550-680°C/0.5-24 hr regimes. Loading of samples before slot welding up (Fig. 1) gave a definite stress level in metal at a concentrator tip. In the process of post heat treatment relaxation of stresses defined by calculation method took place. Hereat it is shown experimentally that under heating up to 620-650°C and not less than 5 hr tempering residual stress maximal values decreased by 1.5-2 times.

Metallographic analysis of compact sample metal continuity at a concentrator tip allowed to define temperature time limits of investigated steels susceptibility to cracking. It is established that Mn-Ni-Mo-V and Cr-Mo-V steels are more susceptible to intergranular cracking, in particular, under tempering regimes of 580-620°C/1-5 hr and 620-650°C/1-3 hr, respectively. Hereat stresses total to 0.8-0.9 $\sigma_{0.2}^T$ ($\sigma_{0.2}^T$ - metal yield strength under test temperature). Susceptibility of such an area on Cr-Ni-Mo-V steel HAZ to cracking under reheat is considerably lower even at a low stress level of $\sigma_{0.2}^T$. It is established that in the investigated test temperature range and heat-treatment regimes steels with minimal harmful inclusion content are characterized by the largest susceptibility to possible cracking.

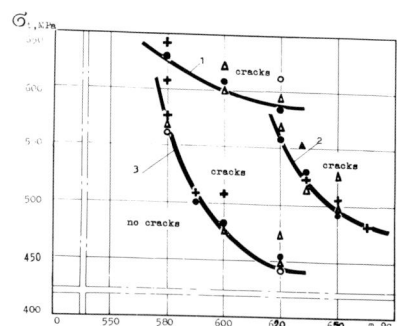

Fig. 2. Dependence of HAZ metal crack resistance of low-alloy steels on quenching temperature and stressed state: 1-Mn-Ni-Mo-V steel; 2-Cr-Mo-V steel; 3-Cr-Mo-Ni-V steel.

REFERENCES

Dhooge P., Dolby R.E., Seibille J., et al (1978). A review of work related to reheat cracking in nuclear reactor pressure vessels steels. *Inst.Jour.Pressure Vessels and Piping*, 6, 5, 329-409.

Zemzin V.N., I.P.Zhitnikov (1972). Conditions of crack formation in a near weld zone under heat treatment. *Automatic welding*, 2, 1-5.

The Effect of Heat Treatment on the Structure and Properties of Welded Joints Made of Heat-Resistant Cr-Mo-V Steels

R. Z. Shron

VTI, Chelyabinsk, USSR

Different aspects of the effect of heat treatment on the supporting capability of structures made of heat-resistant steels should be considered when making choice of heat treatment conditions for welded joints.

The metallurgists commonly require that postwelding tempering temperature be lower than the base metal tempering temperature. In application to the welded structures made of heat-resistant Cr-Mo-V steels which are intended for long-term service in the temperature range of 540-570°C this requirement is shown not to be valid. Brittle cracks initiation in the coarse grains area of HAZ and in the weld metal at the beginning of service may be reliably prevented only when the postwelding tempering temperature equals or is even higher than that assigned for the base metal. At the same time repeated tempering is not dangerous for the base metal, for the resulting softening is gradually decreasing as the test duration increases. The long-term strength considered on the time basis comparable to the estimated service life is actually not affected by repeated tempering. The reliable operation of welded structures exposed to high temperature stress relief has been confirmed by service experience.

To prevent cracking during heat treatment of welded joints made of heat resistant Cr-Mo-V steels one must define the heating conditions taking into account kinetics of stress relaxation and dispersion hardening processes. The preferred heating conditions will be those allowing to separate time-temperature ranges of these two processes.

Stress relaxation in the welded joints mentioned takes place mostly due to the deformation of the soft interlayers where creep-resistance is lowered. The less is the relative thickness of the interlayers, the lower becomes the relaxation rate. It is accounted for by the fact that the creep rate of the soft interlayers is restrained by a more rigid metal. At $\mathcal{X} =$

0.1 creep strain within the interlayers is negligible as compared to the specimens with thick interlayers ($\mathcal{X} > 0.5$). This must be taken into account when analyzing relaxation test results for weld specimens and estimating welded joints tendency to cracking during heat treatment.

In some applications austenization with subsequent high temperature stress relief is prescribed for welded joints made of heat-resistant Cr-Mo-V steels. Contrary to the commonly accepted opinion this heat treatment does not necessarily lead to the refinement of coarse grains in weld metal and to the formation of a homogeneous structure in welded joints. After normalization as well as after annealing the welded metal structure is comparable to the as-welded structure. This is accounted for by structure memory effects in the former case and by secondary liquation in the latter. Without considering all these factors when choosing the austenization conditions the desired service reliability of the welded joints would not be obtained.

AUTHOR INDEX

Abbott, A R G 313
Anik, S 91
Asnis, A E 307, 321

Batyuk, V V 277, 343
Blauel, J G 79
Blondeau, R 53
Bosansky, J 29
Bourges, Ph 53
Burget, W 79

Carter, W P 199
Chen, H 169
Chen, L 169
Chene, J J 37
Chur, D M 353
Cotton, H J 61
Cundev, S 131

Debiez, S 69
Denisenko, A V 331
Dhooge, A 289
Dikicioglu, A 91
Dimitrov, D D 327
Dong, X 169

Egorova, S V 329

Fomichev, S K 343

Gachik, R K 277
Gansky, P N 343
Gatovskij, K M 117
Groß, H G 117

Hamilton, I G 313
Heeschen, J 231

Hristov, S 161
Hrivnak, I 13, 189

Ievlev, I J 351
Ivashchenko, G A 307, 321

Jesensky, M 153

Kabanov, N M 333
Kalev, L 239
Kalna, K 177
Kasuya, T 61
Khriplivy, A A 277, 343
Khromchenko, F A 221
Kirillov, E A 351
Kolot, G F 277
Kononchuk, V V 349
Korol'kov, P M 333
Kotyk, V T 347
Kovacevic, B 339
Krustev, A 239
Kudryavtsev, Yu F 101
Kuz'menko, O G 331

Lamzin, A G 333
Lancos, J 189
Leclou, A 269
Leggatt, R H 247
Lysak, V V 349

Makhnenko, V I 257
Markov, S P 117
Maynier, Ph 53
Medko, B V 349
Mihailov, V 239
Mikheev, P P 101
Minakov, S N 343

Mraz, L 29
Müller, H H 125

Navolokin, S N 353
Nifantov, V N 349
Nosatov, V A 331

Okumura, M 61
Ovchinnikova, T Kh 331

Panov, V I 351
Pastukhov, K B 277
Paton, B E 141
Petrov, P 161
Petrov, P F 347
Pochinok, V E 257
Pokatilov, V P 351
Prokhorov, V I 347
Provost, W 289
Pulyayev, A V 277

Raykov, P I 327
Runov, A E 353

Sedek, P 145
Seyffarth, P 117
Shakeshaft, M E J 37
Si, Z 169
Sklyarevich, V E 141
Slivinsky, A M 347
Solina, E N 329

Starchenko, E G 353
Sterenbogen, Yu A 329, 331
Suzuki, J 325

Tamaki, K 325
Trufiakov, V I 101
Tülbentçi, K 91

Varbenov, I I 327
Vasileva, L 109
Vejvoda, S 189
Velikoivanenko, E A 257
Velikonja, M 207
Velkov, K 109
Vinckier, A 289
Vinokurov, V A 1
Volkov, V V 351

Wang, Y 169
White, C M 199
Wohlfart, H 231

Yurchishin, A V 329
Yurioka, N 61
Yushchenko, K A 13

Zalinov, A V 353
Zhdanov, I M 277, 349
Zotova, N G 329
Zvetanov, S 161

SUBJECT INDEX

Annealing 331
Argon-arc treatment 321
Austenitic steels 313

Boiler steels 91
Brittle failure 177

C-Mn steels 247
Carbon migration 37
Circumference welds 343
Code requirements 313
Coherent precipitates 29
Cold delayed cracking 109
Cr-Mo-V steel welded joints 221
Cr-Mo steels 37
Cracking-sensitive field 325
Crankshafts 347
Critical P content 325
Critical restraint stress 325
CTOD-testing 79

Dimensional stability 145
Dislocations 29

Electromagnetic transducers 277
Electron microscopy 29
Electrostag heating 331
Electrostag welding 329
Energy input 343
Explosive relaxation treatment 13
Explosive system 161

Fatigue resistance 101
Finite element analysis 117, 239, 313
Fracture mechanics 79
Fracture toughness 37

Grinding 13
Gyrotron 141

Hammer peening 13
Hard explosion 169
Hardness 61
Heat affected zone 61, 79
 toughness 91
Heat input conditions 221
Heat pretreatment 109
Heat sink 349
Heat treatment 239, 277, 343
Heterogenous predecomposition
 austenite 109
High strength steel 207
High-temperature tempering 353

Impact toughness 109
Induction heating 125, 131
Internal stress 189

Liquified gases 189
Local explosive treatment 161
Local heat treatment 313
 of welds 125
Local microstresses 109
Low alloy steel 61, 189, 307, 329
Low-carbon steels 307

Magnetoelastic effect 277
Martensite 61
Metal anelasticity 145
Microalloyed steel 289
Microprocessor control 313
Microrelaxation 145
Microwave radiation 141

Natural ageing 145
Nd-microalloyed steel 29
Nitrogen 239
Non-destructive testing 277
Notch and fracture toughness 153
Nuclear components 125

Optimization 1
 of postweald annealing 231
Overheating 329
Overloading 13

Peak stress reduction 231
Phase transformation 117
Pipeline construction 125
Pipes 343
 steam pipes 221, 339
Plastic deformation 169
Plasticity 29
Post weld heat treatment 37, 61, 207, 247, 289, 313
Precipitation hardening 61
Prefracture state 351
Preheating 117
Pressure test 189
Pressure vessels 177, 333

Rapid ageing 145
Reclamation 347
Reheat cracking sensitivity 325
Repair welding 189
Residual stress 13, 101, 117, 161, 169, 177, 231, 247, 257, 269, 277, 307, 343, 349
 measurement by x-rays 231
 measurement of volume in welded joints 153

Short-time post-heating 117
Shot peening 13
Soft explosion 169
Steel
 C-Mn steels 247
 austenitic steels 313
 boiler steels 91
 high strength steel 207
 low-alloy steel 61, 189, 307, 329

low-carbon steels 307
microalloyed steel 289
Nd-microalloyed steel 29
stainless 239
structural 289
Steam pipes 221, 339
Storage tank 189
Strain measurement 153
Strength tests 153
Stress corrosion cracking 239
Stress intensity 239
Stress relaxation
 tests 247
 treatment of weldments 13
Stress relief 79, 247
 annealing 91, 207
 explosion treatment 169
 heat treatment 29, 231
 vibratory 13, 153
Structural steels 289
Successive pressure cycles 189

Temperature measurement 313
Tempering 307
 high-temperature 353
Tensile and hardness properties 37
Thermal deformation 349
Transfer medium 161
Transformation stress 117
Transient thermal stress 13
TTT-diagram 117

Vessel fabrication 125
Vibratory stabilization 145
Vibrotreatment 343

Warm pre-stressing 177
Warm pressure test 177
Weld metal 29, 79, 247
Welded joint 29
Welding heat input 91
Welding procedure qualification test 79
Wide plate tests 289